Understanding Variable Stars

Variable stars are those that change brightness. Their variability may be due to geometric processes such as rotation, or eclipse by a companion star, or physical processes such as vibration, flares, or cataclysmic explosions. In each case, variable stars provide unique information about the properties of stars, and the processes that go on within them.

This book provides a concise overview of variable stars, including a historical perspective, an introduction to stars in general, the techniques for discovering and studying variable stars, and a description of the main types of variable stars. It ends with short reflections about the connection between the study of variable stars, and research, education, amateur astronomy, and public interest in astronomy. This book is intended for anyone with some background knowledge of astronomy, but is especially suitable for undergraduate students and experienced amateur astronomers who can contribute to our understanding of these important stars.

JOHN R. PERCY is a Professor of Astronomy and Astrophysics at the University of Toronto, based at the University of Toronto in Mississauga (UTM). His research interests include variable stars and stellar evolution, and he has published over 200 research papers in these fields. He is also active in science education (especially astronomy) at all levels, throughout the world. His education interests and experiences include: teaching development at the university level, development of astronomy curriculum for Ontario schools, development of resources for educators, pre-service and in-service teacher education, lifelong learning, public science literacy, the roles of science centres and planetariums, the role of skilled amateurs in research and education, high school and undergraduate student research projects, international astronomy education and development, and multicultural astronomy. He is Director of the undergraduate Science Education program, and the Early Teacher Program at UTM, and is cross-appointed to the Ontario Institute for Studies in Education.

He has served as President of the Royal Astronomical Society of Canada, the Royal Canadian Institute, the American Association of Variable Star Observers, the International Astronomical Union Commissions on Variable Stars, and on Astronomical Education, and of the Astronomical Society of the Pacific. He was recently the recipient of the Royal Canadian Institute's Sandford Fleming Medal for contributions to public awareness and appreciation of science and technology, the U of T School of Continuing Studies' Citation for Exceptional Commitment and Achievement in adult learning, and the Distinguished Educator Award of the Ontario Institute for Studies in Education. In 1999, he was elected a fellow of the American Association for the Advancement of Science. In 2003, he received the University of Toronto's Northrop Frye Award for exemplary linkage of teaching and research. His most recent book, co-edited with Jay M. Pasachoff, is *Teaching and Learning Astronomy* (Cambridge University Press, 2005).

Understanding Variable Stars

John R. Percy
University of Toronto, Toronto, Ontario, Canada

CAMBRIDGE UNIVERSITY PRESS
Cambridge, New York, Melbourne, Madrid, Cape Town,
Singapore, São Paulo, Delhi, Tokyo, Mexico City

Cambridge University Press
The Edinburgh Building, Cambridge CB2 8RU, UK

Published in the United States of America by Cambridge University Press, New York

www.cambridge.org
Information on this title: www.cambridge.org/9781107403703

First published 2007
First paperback edition 2011

A catalogue record for this publication is available from the British Library

ISBN 978-0-521-23253-1 Hardback
ISBN 978-1-107-40370-3 Paperback

Janet Akyüz Mattei (1943–2004) (Photo by Michael Mattei, courtesy of the AAVSO.)

This book is dedicated to the memory of my colleague and friend Dr Janet Akyüz Mattei (1943–2004). She was born in Bodrum, Turkey, and educated at Brandeis University (BA 1965), University of Virginia (MS 1972), and Ege University, Turkey (MS 1970, Ph.D. 1982). She served as Director of the American Association of Variable Star Observers for 30 years, from 1973 until her death. She led the AAVSO through a period of unprecedented growth, in the volume of data submitted by observers and requested by professional astronomers, and in the diversity and complexity of research projects supported. The AAVSO became internationally respected for its treasury of data and information, and for its international network of volunteer observers. She won a dozen major awards in countries around the world. She served in scientific and educational organizations and committees at every level, up to the International Astronomical Union. As a professional astronomer, she was an expert on cataclysmic variables and Mira stars, as well as on the general topic of amateur-professional collaboration – an area in which she made a profound contribution in many ways, and many places. But Janet was more than a scientist and administrator. She was a teacher, with an infectious enthusiasm for astronomy – and wildflowers. She was a diplomat and a leader, an exemplary human being, and a dear friend of every amateur or professional astronomer or educator she worked with. I had the pleasure of working with her on many projects, notably *Hands-On Astrophysics*. My interests in variable stars, and their role in science and education, have been indelibly affected by my 30 years of collaboration and friendship with Janet. I know that she has left a similar mark on hundreds of others, worldwide.

John R. Percy

Contents

Boxes

Figures

Tables

Preface

The roots of this book go back over forty years. As an undergraduate in 1960, I was exposed to the variable star research at the David Dunlap Observatory of the University of Toronto. Mentors such as Don Fernie, Jack Heard, and Helen Sawyer Hogg brought the field to life. As a graduate student, I sampled both theory (with Pierre Demarque) and observation (with Don Fernie). Then I was fortunate to obtain a faculty position at the University of Toronto's brand-new Erindale Campus in Mississauga, west of Toronto. I was concerned with teaching, supervising students, and building a new university campus.

My research continued, and my graduate teaching responsibility was a course on variable stars. This book evolved from that course. The 1970s were in many ways the 'golden age' of variable stars at the University of Toronto. A dozen graduate students undertook M.Sc. and/or Ph.D. theses on variable stars. The David Dunlap Observatory, being a 'local' observatory under our control, enabled both large-scale surveys, and long-term studies to be carried out – both of which are almost impossible at modern-day national observatories. The observatory was equipped with both a 1.88m spectroscopic telescope, and 0.6m and 0.5m photometric telescopes, and many of these thesis projects combined these techniques in a very effective way. I learned much from these graduate students, and owe much to my colleagues, including a succession of Directors of the David Dunlap Observatory - who were also Chairs of the Department of Astronomy and Astrophysics.

My urge to write a book on variable stars developed around 1980. I had become active in the American Association of Variable Star Observers. I had also begun to concentrate on the supervision of undergraduate research projects on the observation and analysis of variable stars. There was a need for a textbook suitable for high-level amateur astronomers, as well as for graduate and undergraduate students. There was an excellent book in German by Cuno Hoffmeister *et al.*, but, by the time it was translated into English by Storm Dunlop, it was becoming out of date.

At that point, I acquired two potential co-authors – Janet Mattei and Lee Anne Willson. Janet was a professional astronomer who was overseeing the remarkable growth of the AAVSO as it entered the space age. Lee Anne, like me, worked in a university setting, and was deeply involved with the AAVSO as well as with students. Lee Anne actually spent a year in Toronto, at the newly established Canadian Institute for Theoretical Astrophysics at the University of Toronto. As I looked back, twenty years later, I realized that we had *almost* finished an excellent book! But we had become distracted. I edited conference proceedings on *The Study of Variable Stars using Small Telescopes*, and co-edited *Variable Star Research: An International Perspective* with Janet and with Christiaan Sterken, as well as proceedings of astronomy education conferences. To some extent, the 'book bug' was out of my system.

But by the twenty-first century, there still remained a need for a book on variable stars, and here it is. Unfortunately (or fortunately), the field of variable stars has changed drastically since 1984. The spirit of the Mattei–Percy–Willson book remains, but little of the substance. At my own university, as with most others in North America, graduate students' interest in stars has been replaced by interest in extragalactic astronomy and cosmology, so it has been many years since I offered my graduate course on variable stars. My knowledge of the field, at the graduate level, is no longer comprehensive. But you cannot understand galaxies if you cannot understand stars! And variable stars, of course, are the tools by which we understand the age and distance scale of the universe, as well as the structure and evolution of the stars.

One of the problems in writing this book is to decide what to include and what to leave out. On the one hand, there are thousands of pages of details which could have been included. On the other hand, I recognize that there is a wealth of good print and on-line material on variable stars which is already available, including on the AAVSO's website. And I admit that this book reflects my own research interests – perhaps more than it should. I hope, however, that it establishes the 'big picture' of variable stars, and gives the reader background and context for further learning about this ever-exciting field.

I thank the many individuals who have contributed to this book, in various ways, in addition to the late Janet Mattei and Lee Anne Wilson who got things started.

Several of my colleagues have kindly read through specific chapters, and contributed useful comments: Johannes Andersen, Christine M. Clement, William Herbst, John B. Lester, Geraldine J. Peters, Aleks Scholz, Matthew Templeton,

David G. Turner, Marten van Kirkwijk, Elizabeth Waagen, and Thomas R. Williams. Special thanks to Andy Howell, Slavek Rucinski, and Michael Shara for reading and commenting on substantial sections where my expertise was limited. Any errors or deficiencies in this book, however, are strictly my own.

Other colleagues have contributed illustrations: Tim Bedding, Jocelyn Bell Burnell, Gary Billings, Christine M. Clement, Peter Cottrell and Ljiljana Skuljan, Alex Filippenko and Tom Matheson, Gilles Fontaine, Doug Gies, Richard O. Gray, Gerald Handler, Arne Henden, Gregory W. Henry, Bill Herbst, Andy Howell and the CFHT Supernova Legacy Project, Alexandra Kalasova and Petr Harmanec, Laszko Kiss, S. Lefebre and Tony Moffat, Michael Mattei, Stefan Mochnacki and Slavek Rucinski, Arto Oksanen, Optec Inc., Geraldine Peters, Mercedes Richards, Michael Shara, Aleks Scholz, and David Turner.

I thank them and all those others who have given me permission to use figures from their publications. This includes the editors of the journals from which many of the figures are taken. And I thank all those others who I might have forgotten.

The AAVSO Headquarters staff have been extremely helpful, not only in supplying illustrations for this book, but in so many other aspects of my research and education work. Arne Henden, Matthew Templeton, Elizabeth Waagen, and the late Janet Mattei – to whom this book is dedicated – are the most visible of the names. I also thank the AAVSO observers, without whom there would be no AAVSO, and without whom my research would have been much more difficult – or impossible. Thanks especially to Howard Landis who contributed, in so many ways, to the success of the AAVSO photoelectric photometry program.

I thank my research collaborators Peter Cottrell, Don Fernie, Petr Harmanec, Greg Henry, Bill Herbst, Christiaan Sterken, Endre Zsoldos and others, and my dozens of students – mostly undergraduates and outstanding senior high school students in the University of Toronto Mentorship Program. Their work is highlighted in several parts of this book, and has been an inspiration to me.

I thank the Natural Sciences and Engineering Research Council of Canada for supporting my research, and NSERC Canada and the Ontario Work-Study Program for supporting many of my undergraduate research students.

Three decades ago, Simon Mitton introduced me to Cambridge University Press, and I am proud to have them as publisher of "my book." Special thanks to editors Jeanette Alfoldi, Lindsay Barnes, Vince Higgs, Jacqueline Garget, Sally Thomas, and especially to copy editor Anne Rix for her careful work.

Joseph B. Wilson and especially Byron Desnoyers Winmill have also provided careful, editorial help, especially in mastering the intricacies of Cambridge style and formatting. I thank them, both for this and for their patience in dealing with my usually disorganized approach!

Finally, I thank my wife Maire and our daughter Carol for their inspiration, patience, and love. I have certainly benefitted from their examples as scholars, authors, teachers, and mentors. I hope that some of the joy of variable star astronomy, its organizations and people, has rubbed off on them.

1

History and development

Sometime, far back in prehistory, people first began watching the sky. They must certainly have noted the daily and yearly motions of the sun, the waxing and waning of the moon, and perhaps even the wanderings of the planets. They might even have noted changes in the stars, not changes of *position* but changes of *brightness*. There are hundreds of these *variable stars* among the naked-eye stars, and the changes of some of them are so striking that they could not escape the notice of a careful eye.

Written records of skywatching begin in the Near East around 2000 BC, and in the Far East around 1000 BC. Unwritten records may also exist: paintings, rock and bone carvings, alignments of giant stones. The study of these has been called *archaeoastronomy*. The interpretation of *any* ancient records is a challenge to both the astronomer and the historian, and many interesting books and articles have been written on the topic.

The Babylonians laid the foundations of Western astronomy through their mathematics and through their systematic observations of the sun, moon, planets, and stars (albeit for astrological purposes). Babylonian astronomy was absorbed into Greek culture, where it eventually led to 'models' of the visible universe. In some schools of philosophy, these models were intended as mathematical conveniences, designed only to represent or predict solar, lunar, or planetary motions. In other schools, these models took on wider significance: they were intended to represent physical reality. Aristotle (384–322 BC), for instance, wrote that the world was made of four elements: earth, water, air, and fire. Bodies beyond the sphere of the moon were made of a fifth element or 'quintessence', which was ingenerable and incorruptible, and underwent only one kind of change: uniform motion in a circle. Variable stars could not exist in Aristotle's universe.

Aristotle's works had tremendous impact on Western thought. They survived in translation even after the original versions (such as those in the famous library at Alexandria) were lost. Perhaps for this reason, there are few if any records of variable stars in Western literature. The 'chronicles' maintained by monasteries in mediaeval times record many astronomical and meteorological phenomena, but variable stars are not among them. There is only one European record of the 'new star' of 1006 AD, now classified as a supernova – the brightest one known – and that record is controversial.

In the Orient, observers were not inhibited by Aristotle. In fact, their cultures often placed great importance on omens, such as unexpected events in the sky. Chinese, Japanese, and Korean records are a fruitful source of information on supernovae, novae, comets, eclipses, and other such events, if one can interpret the records correctly. It appears that about 80 'new stars' were recorded up to about 1600, eight being supernovae and the rest being ordinary novae. Other cultures, such as aboriginal North Americans, may also have recorded these 'new stars', but they apparently did not do so in the precise, methodical fashion of the Orientals.

1.1 Tycho's and Kepler's stars

In the West, the sixteenth century brought a renaissance and revolution in scientific thought. Ironically, the seeds were contained in the writings of Aristotle: by laying the basis of the scientific method, he developed the weapons with which his philosophy was later attacked and overthrown. By happy coincidence, Nature provided two rare events – supernovae – which would help to usher in the modern age of astronomy. The first occurred in 1572. In November of that year, Tycho Brahe (1546–1601) recorded a 'new star' in Cassiopeia. He (and independently the English astronomer Thomas Digges) established its position and its fixed, stellar nature, and measured its changing brightness relative to planets and stars. At maximum brightness, it rivalled Venus! He published his results in a book entitled *On a New Star, Not Previously Seen Within the Memory of Any Age Since the Beginning of the World*. His observations were so careful and systematic that later astronomers (notably Walter Baade) could reconstruct the changing magnitude of Tycho's star to an accuracy of ± 0.2. Since the distance to the remnant of the supernova can now be determined, the absolute magnitude of the supernova at maximum brightness can be found. This provides a bench mark which can be used to determine the distance to other supernovae.

Just over three decades later, in October 1604, Johannes Kepler (1571–1630) recorded another 'new star.' He (and independently David Fabricius, pastor in Osteel in Ostfriesland) established its position and its fixed, stellar nature, and recorded its changing brightness. Kepler's star was somewhat fainter than Tycho's, but still at maximum brightness it rivalled Jupiter.

Only a few years later, the telescope was first applied to astronomy by Galileo Galilei (1564–1642), and since then astronomers' technical resources have increased by leaps and bounds. Unfortunately, no supernova in our galaxy has been seen since 1604.

1.2 The beginnings of modern astronomy

Variable stars (or variables for short) continued to be discovered and observed sporadically, occasionally through deliberate, systematic measures but more often by chance. In 1596, Fabricius noted that *o* Ceti (subsequently called *Mira*, the wonderful) was sometimes visible, sometimes not. Furthermore, the cycle of visibility repeated regularly every 11 months. The star was a periodic variable.

Towards the end of the eighteenth century, another burst of progress was made. Some of this must be attributed to William Herschel (1738–1822), the 'father' of modern stellar astronomy. His development of large reflecting telescopes, his efforts to gauge the distances of the stars from their apparent brightnesses, and his interest in the physical nature of stars and nebulae, all contributed to the development of the study of variable stars. He discovered the variability of two stars – 44i Bootis, and α Herculis. It is interesting to read what his son John Herschel (1792–1871) said about variable stars in the 1833 edition of his *Principles of Astronomy*:

> this is a branch of practical astronomy which has been too little followed up, and it is precisely that in which amateurs of the science, provided with only good eyes, or moderate instruments, might employ their time to excellent advantage. It holds out a sure promise of rich discovery, and is one in which astronomers in established observatories are almost of necessity precluded from taking a part by the nature of the observations required. Catalogues of the comparative brightness of the stars in each constellation have been constructed by Sir William Herschel, with the express object of facilitating these researches.

These words are still true today. We must remember, though, that many eminent scientists of a century or two ago were amateurs by our definition. William Herschel was a good example. So too were John Goodricke (1764–86) and Edward Pigott (1753(?)–1825), two well-to-do Yorkshire gentlemen who made a special study of variable stars. Pigott is credited with the discovery of the variability of η Aquilae in 1784, of R Coronae Borealis in 1795, and of R Scuti in 1796. Goodricke is credited with the discovery of the variability of δ Cephei and β Lyrae in 1784, and of the periodicity of β Persei (Algol) in 1782–1783 (though the latter was apparently discovered independently by a German farmer named Palitzch, who had an interest in astronomy). Together, Goodricke and Pigott also proposed that the variability of Algol might be caused by eclipses (but by a planet!). Goodricke, a deaf mute, died at the untimely age of 21, and Pigott, after the death of his friend, gradually gave up astronomy, but, during the brief period of their collaboration, 'the Yorkshire astronomers' laid the foundations for the study of variable stars as a branch of astronomy.

1.3 Systematic visual observations

Throughout the nineteenth century, variable stars continued to be observed. In 1786, Pigott had listed 12 definite variables and 39 suspected ones. By about 1850, F. W. A. Argelander (1799–1875) listed 18 definite variables and numerous suspected ones. Furthermore, many of the definite variables had good light curves (plots of brightness *versus* time) thanks to the techniques of measurement which William Herschel had devised and which Argelander had refined. Argelander's work marks another high point in the study of variable stars because, through his compilation of the *Bonner Durchmusterung* catalogue and charts, he made hundreds of thousands of visual measurements of the brightnesses of stars, providing magnitudes for comparison stars and discovering many new suspected variables in the process.

Not the least of Argelander's contributions was his well-known appeal to 'friends of astronomy'. It appears at the end of an article in Schumacher's *Astronomisches Jahrbuch* for 1844, in which Argelander discusses the history, importance, methods of study, and idiosyncracies of variable stars:

> Could we be aided in this matter by the cooperation of a goodly number of amateurs, we would perhaps in a few years be able to discover laws in these apparent irregularities, and then in a short time accomplish more than in all the 60 years which have passed since their discovery.
>
> Therefore do I lay these hitherto sorely neglected variables most pressingly on the heart of all lovers of the starry heavens. May you become so grateful for the pleasure which has so often rewarded your looking upward, which has constantly been offered you anew, that you will contribute your little mite towards the more exact knowledge of

> these stars! May you increase your enjoyment by combining the useful and the pleasant, while you perform an important part towards the increase of human knowledge, and help to investigate the eternal laws which announce in endless distance the almightly power and wisdom of the Creator! Let no one, who feels the desire and the strength to reach this goal, be deterred by the words of this paper. The observations may seem long and difficult on paper, but are in execution very simple, and may be so modified by each one's individuality as to become his own, and will become so bound up with his own experiences that, unconsciously as it were, they will soon be as essentials. As elsewhere, so the old saying holds here, 'Well begun is half done', and I am thoroughly convinced that whoever carries on these observations for a few weeks, will find so much interest therein that he will never cease. I have one request, which is this, that the observations shall be made known each year. Observations buried in a desk are no observations. Should they be entrusted to me for reduction, or even for publication, I will undertake it with joy and thanks, and will also answer all questions with care and with the greatest pleasure.*

Argelander's appeal did not go unheeded. The Variable Star Section of the British Astronomical Association was founded in 1890, and has been active ever since. In America, the study of variable stars was supported especially by E. C. Pickering (1846–1919) at the Harvard College Observatory. In 1882, Pickering had published *A Plan for Securing Observations of Variable Stars*, pointing out, among other things, that such work might especially appeal 'to women'. In 1911, the American Association of Variable Star Observers (AAVSO) was founded in Cambridge, Massachusetts. Originally, it was very closely associated with the Harvard College Observatory. Now, though some links remain, the AAVSO is an independent, international association, receiving hundreds of thousands of observations each year from members and associates around the world. The study of variable stars has indeed been a fruitful area of astronomy for women, in part because Pickering encouraged and hired women astronomers (albeit at low salaries). Henrietta Leavitt, Margaret Mayall, Cecilia Payne-Gaposhkin, and Helen Sawyer Hogg are among many examples.

* This translation, by Annie J. Cannon, appears in *Popular Astronomy* for 1912, and in *A Source Book in Astronomy*. Several types of variable stars were particularly well suited to visual observation by amateurs: unpredictable variables of the R CrB, U Gem and nova type, and large-amplitude regular and semi-regular variables such as the long-period or Mira variables.

1.4 The photographic revolution

Several developments occurred in the late nineteenth century which had a profound effect on the study of variable stars. One was the development of photography. This had begun many years earlier (John Herschel being one of the pioneers), and was first applied to stellar astronomy in 1850, but it was not until about 1880 that emulsions became sensitive enough, and telescope drive systems became accurate enough, for astrophotography to make its full impact. The photographic plate, which could now integrate starlight for up to several hours, could record stars which were far too faint to be seen visually in the telescope. Furthermore, thousands of stars could be recorded on a single plate, making discovery and measurement more efficient. Photography provided a permanent record of star brightness, capable of being checked or re-measured at any time. If a newly discovered variable had been recorded on earlier plates, its previous history could be investigated. Archival collections such as that of the Harvard College Observatory contain a gold mine of information on the past variability of stars – and of more exotic objects such as quasars.

Many observatories embarked on systematic photographic surveys of variable stars, notably Harvard, Heidelberg, Leiden, Sonneberg, and Sternberg. Each observatory accumulated its own collection of variables, with consequent problems of nomenclature. Selected regions of the sky were more closely surveyed: the northern and southern Milky Way, the region of the galactic bulge in Sagittarius, globular clusters, the Magellanic Clouds, and later more distant galaxies such as M31. The number of known variables increased rapidly to over 1000 in 1903, over 2000 in 1907, and over 4000 in 1920. Today, there are tens of thousands of confirmed variables.

1.5 Spectroscopy

The development of astronomical spectroscopy and spectral classification also led to a better understanding of variable stars. Unusual aspects of the spectra of variable stars were noted visually as early as 1850: the absorption bands in the spectra of cool, long-period variables, and the bright emission lines in the spectra of novae and of Be stars such as γ Cassiopeiae. The development of photographic spectroscopy by H. Draper in 1872 led to large-scale projects in spectral classification, notably by Pickering and his associates at the Harvard College Observatory. It became possible to classify variables according to temperature, and later according to luminosity; this enabled astronomers to learn how variable stars compared and fit in with normal stars. This in turn led to improvements in the classification schemes for variable stars. Spectroscopy also provided the raw material for deducing the chemical composition of the stars, and the differences in composition between different groups of stars. But it required modern astrophysics to do the interpretation correctly.

1.6 Classification and explanation

The earliest classification scheme dates back nearly two centuries: Pigott divided variables according to the nature of their light curve into novae, long-period variables, and short-period variables. A century later, Pickering devised a more detailed scheme: (Ia) normal novae: primarily nearby ones in our own galaxy; (Ib) novae in nebulae: now known to be primarily supernovae in distant galaxies; (IIa) long-period variables: cool, large-amplitude pulsating variables; (IIb) U Geminorum stars: dwarf novae; (IIc) R Coronae Borealis stars: stars which suddenly and unpredictably decline in brightness; (III) irregular variables: a motley collection; (IVa) short-period variables such as Cepheids and later including the cluster-type or RR Lyrae stars; (IVb) Beta Lyrae type eclipsing variables; and (V) Algol type eclipsing variables.

Pickering's classification scheme contains some hints as to the nature and cause of the variability. Classification and explanation go hand in hand (in principle), and in the late nineteenth and early twentieth centuries, progress was made in understanding the physical nature and the physical processes in variable stars, and in stars in general. This culminated in Arthur S. Eddington's (1882–1944) monumental book *The Internal Constitution of the Stars.*

Two centuries ago, Goodricke and Pigott had speculated that the variability of Algol and similar stars might be due to eclipses. A century later, the eclipse hypothesis was firmly established: in 1880, Pickering carried out a mathematical analysis of the orbit based on careful observations of the light curve, and by 1889 the orbital motion was directly observed by H.C. Vogel using astronomical spectroscopy and the Doppler effect. Nevertheless, some eclipsing variables defied explanation and, even today, Goodricke's β Lyrae is one of the most enigmatic objects in the sky. Furthermore, the eclipse hypothesis was often overextended: it was used to explain the Cepheids, long-period variables and even the R Coronae Borealis stars! Ironically, the importance of binarity in explaining the variability of such other types as novae has only recently been appreciated.

In a sense, Goodricke anticipated yet another type of variability. Unable to explain the behaviour of Beta Lyrae under the eclipse hypothesis, he suggested that 'the phaenomenon seemed to be occasioned by a rotation on the star's axis, under a supposition that there are several large dark spots upon its body, and that its axis is inclined to the earth's orbit'. This is precisely the case in the rotating variables discussed later in this book.

The idea that variability might be due to pulsation was raised (by A. Ritter) as early as 1873, but it was not until the observational studies by Harlow Shapley (1885–1972) and others around 1915, and the concurrent theoretical studies by Eddington, that the pulsational nature of the Cepheids, cluster type variables, and the long-period variables was established. The cause of the pulsation – the thermodynamic effects of hydrogen and helium in the outer layers of the stars – was not firmly established until after 1950.

1.7 Photoelectric photometry: the electronic revolution

Also around 1915, another important technique was added to the arsenal of the variable star astronomer: photoelectric photometry. At first, the technique was insensitive and cumbersome, and could only be applied to the brightest stars. However, through the careful efforts of its first practitioners, Joel Stebbins (1878–1966) and Paul Guthnick (1879–1947), this technique was successfully used to discover the shallow secondary eclipse of Algol (1910), the small light variations in the pulsating Beta Cephei stars (1913), and in the rotating magnetic stars such as α^2 Canum Venaticorum (1914). This technique has since revealed several other new classes of 'microvariables'. It has also enabled small colour variations to be observed. There is a lesson to be learned here: Stebbins and Guthnick were pioneers; they knew the limitations of their technique, and they took great pains to apply it carefully. Nowadays, photoelectric photometry is routine, and its practitioners often apply considerably less care than Stebbins and Guthnick did. The result is often mediocre, inferior data. On the other hand, photoelectric photometers are now simple and inexpensive enough for amateur use. With care, the amateur can now make important contributions to the photoelectric photometry of variable stars, particularly those with unpredictable, irregular or slow variations. This has included the observation of pulsating red giants, RS Canum Venaticorum binaries, and Be stars, as well as the timing of the minima of eclipsing binary stars.

1.8 Consolidation

During the twentieth century, the basic tools of variable star astronomy – visual, photographic and photoelectric photometry, spectroscopy and physical analysis – were gradually refined, and produced a steady stream of important results. To choose a few examples, we might arbitrarily turn to *A Source Book in Astronomy and Astrophysics, 1900–1975*, edited by K.R. Lang and O. Gingerich (Harvard University Press, 1979). It contains excerpts from the 'most important' astronomical papers published during that time. They include:

'The Measurement of the Light of Stars With a Selenium Photometer, With an Application to the Variations of Algol' by J. Stebbins (1910).
'Periods of 25 Variable Stars in the Small Magellanic Cloud' [discovery of the period luminosity relation] by Henrietta S. Leavitt (1912).
'Novae in Spiral Nebulae' [galaxies] by Heber D. Curtis, George W. Ritchey, and Harlow Shapley (independently in 1917).
'On the Pulsations of a Gaseous Star and the Problem of the Cepheid Variables' by Arthur S. Eddington (1918–1919).
'Cepheids in Spiral Nebulae' by Edwin P. Hubble (1925).
'Novae or Temporary Stars' by Edwin P. Hubble (1928).
'On Supernovae' by Walter Baade and Fritz Zwicky (1934).
'Rotation Effects, Interstellar Absorption and Certain Dynamical Constants of the Galaxy Determined from Cepheid Variables' by Alfred H. Joy (1939).
'Spectra of Supernovae' by Rudolph Minkowski (1941).
'The Crab Nebula' by Rudolph Minkowski (1942).
'T Tauri Variable Stars' by Alfred H. Joy (1945).
'The Spectra of Two Nebulous Objects near NGC 1999' [Herbig-Haro Objects] by George W. Herbig (1951).
'Studies of Extremely Young Clusters 1. NGC 2264' [Pre-Main Sequence Variables] by Merle F. Walker (1956).
'Binary Stars among Cataclysmic Variables III. Ten Old Novae' by Robert P. Kraft (1964).
'Observations of a Rapidly Pulsating Radio Source' [a pulsar] by Anthony Hewish *et al.* (1968).

This list, though somewhat arbitrary, touches on most of the key areas of variable star astronomy up to 1975, and mentions most of the outstanding contributors to this field. Note the predominance of North American astronomers. This is only partly due to bias: North American astronomers have indeed monopolized the field of observational stellar astronomy for most of the twentieth century. Sadly, the field of stellar astronomy has fallen somewhat out of favour in North America in the late twentieth century, in part because of the lure of fields such as observational and theoretical cosmology.

1.9 The modern age

The last entry on the list above is appropriate in that it introduces one of the many sophisticated new techniques which have been applied to variable stars – and all other fields of astronomy – in the last four decades. It is impossible to mention all the results of these techniques here; they will be discussed in detail later in this book.

The discovery of pulsars was probably the most unique and important contribution of radio astronomy to any field of stellar astronomy. Radio telescopes can also probe the interstellar matter around forming stars, as well as the matter ejected from supernovae, novae, and other eruptive stars. The most powerful radio telescope is the Very Large Array (VLA) in New Mexico, USA. An even more powerful facility – the Atacama Large Millimetre Array, to be located in Chile – is currently under construction as a multinational project.

Infrared (IR) astronomy began in the 1920s, using thermocouple detectors. Since then, detectors and telescopes have improved to the point where we can now observe IR radiation from stars such as Cepheids, even in external galaxies. IR telescopes are particularly useful for studying cool variables (which emit most of their energy in the IR) as well as for probing the cool gas and dust around stars, both old and young. IR detectors have become simple and cheap enough to be within the reach of the amateur, and there is particular need for systematic, long-term IR studies of variable stars – studies which can best be carried out at smaller, local observatories. The first major infrared satellite was the *Infrared Astronomical Satellite (IRAS)*, launched in 1983. The *Infrared Space Observatory (ISO)* was launched by the European Space Agency in 1995. NASA's *Space Infrared Telescope Facility (SIRTF)* was launched in 2003, and renamed the *Spitzer Space Telescope* after Lyman Spitzer, an early 'champion' of the concept of space telescopes.

Ultraviolet (UV) astronomy must be carried out from above the atmosphere, because the ozone layer absorbs all wavelengths shorter than 3000 Å. The first rocket observations were made in 1955, but UV astronomy truly blossomed with the launch of UV satellites: *Orbiting Astronomical Observatory 2* in 1968, *Orbiting Astronomical Observatory 3 ('Copernicus')* and the European Space Research Organization's *TDI* in 1972, and the *International Ultraviolet Explorer (IUE)* in 1978. The IUE was the 'workhorse' of UV astronomy, producing thousands of astronomical research papers in its long and productive life. The *Extreme Ultraviolet Explorer (EUVE)* was launched in 1992, and the *Far Ultraviolet Spectroscopic Explorer (FUSE)* in 1999. Two ultraviolet experiments – *ASTRO 1 and 2* – were flown on the Space Shuttle in 1990 and 1995. These UV satellites and experiments have been especially effective in studying hot stars (such as the Be stars) and hot gas (such as the chromospheres of 'active' stars and the discs of gas around the dense components of close binary systems).

X-ray astronomy must likewise be carried out from above the atmosphere, and has likewise been revolutionized by the launch of satellites, particularly *UHURU* in 1970 and the *High Energy Astrophysical Observatory B ('Einstein')* in 1978.

The European *Röntgen Satellite (ROSAT)* was launched in 1991, and the European *XMM-Newton* and NASA's *Chandra* satellite were both launched in 1999. Perhaps the most significant discoveries of X-ray astronomy have been the X-ray emissions from close binary systems with one normal component and one collapsed component: a white dwarf, neutron star, or black hole. In these systems, gas flowing from the normal component into the gravitational field of the collapsed component becomes heated to millions of K. Cataclysmic variables, which are systems of this kind, have been particularly well studied by this technique.

Gamma-ray astronomy began in the 1960s with a series of secret military VELA satellites, designed to detect clandestine nuclear tests. They detected occasional random bursts of gamma rays from space – gamma-ray bursters or GRBs. Then in 1991, NASA launched the *Compton Gamma-Ray Observatory*, which helped in proving that GRBs were super-energetic explosions, visible across the universe! The detection and study of GRB afterglows became a popular field for skilled amateur astronomers.

A noticeable trend in optical telescopes has been to larger and larger collecting areas – first to 4m-class telescopes, then to 8m, and now beyond. The largest telescopes (the Keck telescopes, for instance) now have segmented mirrors. There are even liquid (mercury) mirror telescopes, though these can only look straight up! There are multiple telescopes such as Keck I and II, and the European Southern Observatory's *Very Large Telescope (VLT)*. These multiple telescopes can be combined to form an *interferometer* with much greater resolving power than a single telescope.

There are, of course, advantages of putting optical telescopes into space: the superb resolution of the *Hubble Space Telescope (HST)* is well known (though *adaptive optics* is increasingly providing comparable resolving power in ground-based telescopes). Space telescopes can also provide high photometric accuracy, and a series of photometric satellites are due to be launched, starting with Canada's *MicroVariability and Oscillations of STars (MOST)* satellite, launched in 2003. The European *COROT* mission will follow soon after.

Finally, we should not forget another powerful tool of modern astronomy: the electronic computer. It allows us to deal with the millions of 'bits' of data sent back from space observatories, or generated by modern electronic detectors on ground-based telescopes. It also allows us to model the complex physical processes involved in stellar variability: pulsation and explosion, mass transfer in binary systems, and flares on stellar surfaces. This type of problem requires the world's most powerful computers, such as the CRAY-1 machines at the Los Alamos Scientific Laboratory in New Mexico in the 1970s and 1980s. In the early 1980s, however, evolution of computers began to move in reverse, towards smaller machines – personal computers and workstations. Thanks to Moore's

Law, which states that the power of computers increases exponentially with time, these have provided almost every professional and amateur astronomer with 'a CRAY on their lap'. Indeed, the most powerful scientific computers today are networks of individual processors or machines. Computers can automate virtually every aspect of the data acquisition and reduction process – even from CCDs, which generate gigabytes of data each night, in amateur as well as professional observatories. Technology has revolutionized amateur astronomy (and teaching observatories as well). For $US10 000, anyone can purchase a 0.4m-class telescope off-the-shelf, along with a CCD camera, and a PC. With this system, they can and do participate in forefront research on many classes of variable stars – high-speed photometry of cataclysmic variables, for instance. Small telescopes still play a major role in astronomy. Massive surveys such as the MACHO project were carried out using small telescopes. The first observation of a transit of an exoplanet was made with a small telescope. The study of bright stars continues to be important, because these are the stars which can be most intensively studied with other techniques such as spectroscopy.

And visual observation is not dead! Surprisingly, the demand for the AAVSO's visual observations (as measured by the number of requests for data and services) actually *increased* by a factor of 25 between 1970 (the beginning of the space age) and 2000. This is partly due to the important role which visual observers have played in monitoring variable stars during space-astronomy missions.

1.10 Variable stars: the present status

The study of variable stars is one of the most popular and dynamic areas of modern astronomical research. Variability is a property of most stars, and as such it has a great deal to contribute to our understanding of them. It provides us with additional parameters (time scales, amplitudes ...) which are not available for non-variable stars. These parameters can be used to deduce physical parameters of the stars (mass, radius, luminosity, rotation ...) or to compare with theoretical models. We can choose a 'model' of a given star by requiring that it reproduce the observed variability as well as the other observed characteristics of the star. From the model, we can then learn about the internal composition, structure, and physical processes in the star.

The study of variability also allows us to directly observe changes in the stars: both the rapid and sometimes violent changes associated especially with stellar birth and death, and also the slow changes associated with normal stellar evolution.

It is tempting to think that we have now discovered and studied all types of stellar variability, but history constantly proves us wrong. Even within the last few years, new types of variability have been discovered, usually (but not always) as a result of the development of new astronomical tools and techniques. Until astronomers have studied all types of stars, on all time scales from milliseconds to centuries, in all regions of the spectrum, the search will not be complete. Even then, it is unlikely that we shall understand all that we see.

As a branch of astronomical research, the study of variable stars has a unique breadth. While some astronomers, using the most sophisticated instruments of space and ground-based astronomy, continue to make important discoveries at the frontier of the science, other astronomers (professional and amateur), using much simpler techniques, make equally important discoveries. It is this breadth which makes this book so difficult – and so pleasant – to write.

Hoskin (1997) gives an excellent overview of the general history of astronomy. For further information on the history of variable star astronomy, see Hoffleit (1986), Hogg (1984), and Percy (1986).

2

Stars

Variable stars are stars, first of all. They can be studied and understood in the same way as non-variable stars, except that they have the advantage that they can 'talk' to us through their variability. If we can understand their language, then we can learn more about them than we can about non-variable stars. All of this information comes through 'the message of light' (and its relatives). We measure the direction, intensity, wavelength, and polarization of the light from the stars. In this chapter, we will describe how astronomers can understand stars through observation and theory. This is a collaboration between those who build the instruments that astronomers use, the observers who use them, the theoreticians who develop the laws of physics and astrophysics, and those who use modelling and simulation to apply those laws to real stars.

For an excellent introduction to all aspects of stars, see the monograph by Kaler (1997), and Kaler's website:
http://www.uiuc.edu/~kaler/sow/sow.html

Another option is to consult one of the many introductory astronomy textbooks which are designed for university students. Some of these textbooks are divided into two volumes, one being on stars and galaxies.

2.1 Positions

Stars seem fixed to the celestial sphere, as it appears to revolve around the earth in its daily motion. The position of a star is given by its declination and right ascension, which are analogous to latitude and longitude on earth. The *declination* is the angular distance of the star north or south of the *celestial equator*, which is the projection of the earth's equator on to the sky. The projections of the earth's axis of rotation on to the sky are the *north and south celestial poles.* The declination ranges from $+90^\circ$ to -90°, or 90°N to 90°S. The *right ascension* is measured eastward around the celestial equator from a reference point called the vernal equinox to the foot of the hour circle through the star. The *vernal equinox* is the point at which the sun crosses the celestial equator, moving from south to north, in its apparent annual motion around the sky. The *hour circle* is the circle that passes through the star and the celestial poles; it is analogous to a meridian on earth. The right ascension ranges from 0 to 24 hours, where one hour of angle corresponds to 15°. Depending on the latitude of the observer, the star may be observable throughout the year, or at certain times of the year, or not at all.

The position of a star places it in one of the 88 constellations (the same as the number of keys on a piano). This leads to some of the classical ways of naming stars: the Bayer (Greek letter) names in which a Greek letter (usually in order of decreasing brightness) is combined with the genitive of the (Latin) constellation name; and the Flamsteed numbers in which a number, assigned to each of the brighter stars across a constellation in order of increasing right ascension, is combined with the genitive of the constellation name. There is nothing magical about the 88 constellations that we use. Most were inherited from the Greeks and Romans, and augmented by some very utilitarian constellation names that European astronomers chose when they first mapped the southern skies. Other cultures have created completely different star patterns, but driven by the same curiosity about the sky, about nature, and about the forces that govern humankind.

Star positions are complicated by *precession*: a 26 000-year conical motion of the earth's rotation axis, caused by the gravitational pull of the moon and sun on the earth's equatorial bulge. This causes the celestial equator and the vernal equinox to move slowly with time. The position of the star must therefore be expressed for a specific date or *epoch* such as 2000.0. Precession may cause a star to stray from its original constellation boundaries, which may be confusing if the star name is based on the name of the original constellation!

2.2 Binary and multiple stars

The majority of stars in our galaxy are in binary or multiple star systems; the sun is an exception in this regard. The nearest star, α Centauri, is actually a pair of sunlike stars in mutual orbit, with a third, fainter star, orbiting much farther away. The brightest star, Sirius, is a pair consisting of a normal star and a white dwarf – a stellar corpse. And ϵ Lyrae is a pair of pairs. A *binary*

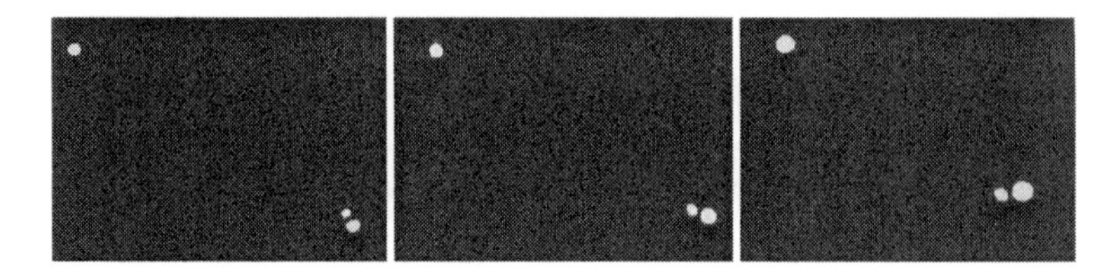

Figure 2.1 A visual binary star system Krüger 60, in Cepheus. The three images are taken over a 12-year period. The fainter component is a variable star – the flare star DO Cephei. The relative motion of the system can be seen; it can be used to determine the sum of the masses of the components. If the absolute motions of both components can be measured, then the individual masses of the components can be determined. (Sproul Observatory photograph.)

star system is a pair of stars that are gravitationally bound, and move together through space. They move in orbit around their mutual *centre of mass* (or gravity). Their distance from the centre of mass is inversely proportional to their mass. A *multiple star system* is three or more stars, gravitationally bound, and moving together. Multiple star systems are arranged hierarchically: a close pair and a single star in mutual orbit; two close pairs in mutual orbit etc. A non-hierarchical arrangement would not be stable over long periods of time. Binary systems may enable us to measure the masses of the stars, and are therefore of great importance (see section 2.9). In some binary systems, the stars are close enough to eclipse (see chapter 5), or to interact so as to influence each other's behaviour and evolution; these will be variable stars, so we will meet many examples of them later in this book.

Occasionally, a pair of stars may appear close in the sky, but be at very different distances. This is an *optical double*, and will not be considered further here. If the two stars appear close, and move together through space, they are presumably a binary system, even if their period is so long that there is no perceptible motion. If their period is a few hundred years or less, then we should expect to see orbital motion, and the system is a *visual binary* (figure 2.1). One of the components may be so faint as to not be visible, but its gravitational effect on the brighter (visible) star may be observable: it may cause the visible companion to move in a small orbit; this is called an *astrometric binary*.

Visual binaries are much easier to observe than astrometric binaries, because it is much easier to measure the position of a star, relative to a close companion, than to measure the position of a single star, relative to stars that do not appear close to it.

Even if the two stars are so close to each other that they appear as one, there are several ways in which we can detect and study them as a binary system. The spectrum of the 'star' may reveal the presence of two different sets of absorption lines; it is a *spectrum binary*. The two sets of absorption lines may exhibit varying Doppler shifts, as the two stars alternately approach and recede as they orbit around each other; this is called a *spectroscopic binary*. If the orbit plane is close to edge-on to us, then one star may pass in front of the other, and produce an eclipse – a decrease in the brightness of the star system; this is an *eclipsing binary*.

2.3 Star clusters

A *star cluster* is a system of up to a million stars, gravitationally bound, and moving together through space. Star clusters are particularly important to astronomers, because they provide a sample of stars with the same age and composition, but with different masses, born at the same time from the same material. So, as 'nature's experiment', they provide an excellent test of stellar evolution theories. The stars are also at the same distance, so their relative brightnesses reflect their true brightnesses. If the distance of any star or group of stars in the cluster can be measured, then the distance – hence the luminosity or power – of all the stars in the cluster is known. Star clusters are intriguing when seen through a small telescope, and may be spectacular when seen on a wide-field image. The nearest clusters are the Ursa Major cluster, at a distance of about 80 light years, and the Hyades, about 150 light years away. See section 2.7 for a definition of a light year.

There are two main types of star cluster. *Globular clusters* are systems of up to a million stars. There are about 150 globular clusters in our galaxy, forming a vast halo around it. They are among the oldest stars in our galaxy – over 10 billion years old – so they are 'fossils' that can tell us much about when and how our galaxy formed. They contain low-mass stars only; any high-mass stars have aged and died. They contain much less of the elements heavier than helium than in the sun. They contain several kinds of variable stars (figure 2.2 shows M14), including *RR Lyrae stars*, which are also sometimes called *cluster variables*.

Galactic or *open clusters* are systems of up to thousands of stars. There are thousands in our galaxy. The Ursa Major, Hyades, and Pleiades clusters are of this type. They inhabit the disc of our galaxy, and have ages from a few million to a billion years or more. They may appear sparse and ragged, and may be difficult to distinguish from the stellar background. Stars that appear in the direction of the cluster may actually be in front of or behind the cluster. It is important to check their motion or their distance to determine whether they are really a member. Galactic clusters eventually disperse, as a result of tidal interactions with stars and gas in the disc of our galaxy or of random exchanges of energy between stars in the cluster that causes some stars to escape or 'evaporate' from the cluster. Globular clusters may suffer some evaporation (indeed, there can

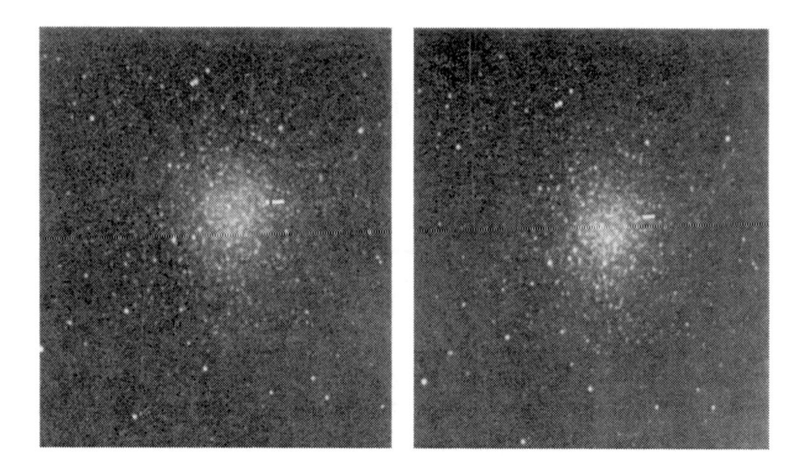

Figure 2.2 The globular cluster M14, showing a Population II Cepheid variable star at maximum (left) and minimum (right) brightness. (David Dunlap Observatory photograph.)

be some quite significant interactions between the stars within them, especially at their dense core), but they are relatively immune from the tidal interactions that affect clusters in the disc.

There are also *associations* of stars: very young groupings of ten to a few hundred stars that are expanding away from their birthplace in the interstellar gas and dust. They may or may not be gravitationally bound. The association of bright stars around the Orion nebula is a prominent example. About 70 associations are known within our galaxy. They may include O stars, that have lifetimes of only a few million years, and T Tauri variable stars, that are also known to be very young.

How does one distinguish a small cluster from a large multiple star system? Multiple star systems are hierarchically organized, in pairs of pairs. In star clusters, the stars orbit randomly around the centre of mass of the cluster.

2.4 Galaxies

Galaxies are families of 10^5 to 10^{12} stars that are bound by their mutual gravitation, and move together through space. Almost all stars are members of galaxies, though there are presumably some stars that have been ejected from galaxies, and are now wandering through intergalactic space. Galaxies range from *dwarf galaxies* that have low masses and luminosities, and are most common, to *supergiant galaxies* that have large masses and luminosities, and are very rare. Galaxies are also categorized by their morphology. The most conspicuous are *spiral galaxies*; they have a central *bulge*, a *disc* of stars, gas, and dust; and a spherical *halo* of old stars. In the disc, the gas and dust and young stars assume a pattern of *spiral arms*. In *barred spirals*, there is a bar-shaped 'formation' of stars that extends outward from the bulge; the spiral arms appear to be attached to the end of it. The spiral and bulge formations can appear and disappear over billions of years, so a galaxy can change from one type to another.

Our sun, and all the other visible stars in the sky are members of a spiral galaxy – the Milky Way galaxy. It is classified on Edwin Hubble's system as an Sb galaxy; the 'b' indicates that both the bulge and the spiral arms would be conspicuous, if seen from afar (the subclasses a, b, and c indicate the relative prominence of the bulge and disc). It also appears that the Milky Way may be a barred spiral galaxy, though the bar formation is largely hidden from us by absorbing dust in the plane of our galaxy. There also appears to be a 'thick disc' of older stars, whose orbits around the bulge have been 'puffed up' by ancient encounters with other galaxy fragments. And in the centre of the Milky Way, there is a dormant *super-massive black hole*. And surrounding our galaxy is a halo of *dark matter*.

The most numerous galaxies are *elliptical galaxies*, especially dwarf elliptical galaxies. They have a spheroidal shape, and contain little or no gas and dust – only old stars. There are also *irregular galaxies* that are also common; as the name suggests, they have no form or symmetry. Finally, there are rare *peculiar galaxies*, that appear to be spiral, elliptical, or irregular galaxies, which have been disrupted by some process such as a gravitational encounter with another galaxy, or a burst of star formation.

In galaxies, stars can form where there is gas and dust, so star formation is found in spiral and irregular galaxies where such gas and dust is present. Star formation can be triggered by *density waves* in the disc of a galaxy, or by the wind from a nearby massive star, or by the blast from a nearby supernova. Recently formed stars are referred to as *extreme Population I*. Less recently formed stars are normal *Population I*. Very old stars are *Population II*. Variable stars may be associated with Population I, II, or both. Pulsating variable stars have been especially useful in delineating the size and shape of our Milky Way galaxy, and the distances to other galaxies (figure 2.3).

2.4.1 *Groups and clusters of galaxies*

Large galaxies such as the Milky Way have smaller satellite galaxies; ours has at least half a dozen, including the *Large Magellanic Cloud* and the *Small Magellanic Cloud*. Their orbital periods are a billion years or more. Some galaxies are components of *binary galaxies* that are bound together, and move in slow mutual orbit. Our galaxy is a member of a small *group* of galaxies called *the Local Group*, that consists of one other large galaxy – M31 or the Andromeda

Figure 2.3 Cepheid variable stars discovered by E.P. Hubble (the numbered stars) in the nearby spiral galaxy M33. The stars marked by letters are comparison stars with known constant magnitudes. Palomar Observatory photograph taken by W. Baade, from an article by A. Sandage in the *Astronomical Journal.* (Print courtesy of Dr Sandage; from Kaler, 1997.)

Galaxy – and many smaller ones. There are also vast *clusters* of galaxies, containing up to many thousands of galaxies; the nearest large one is the Virgo Cluster.

The motions of galaxies in clusters first revealed one of the great mysteries of modern astronomy: over 90 per cent of the matter in galaxies and clusters is *dark matter* – it exerts gravitational force, but does not emit or absorb electromagnetic radiation. We now know that galaxies such as our own are embedded in a massive halo of dark matter, ten times as massive as the luminous matter that we see.

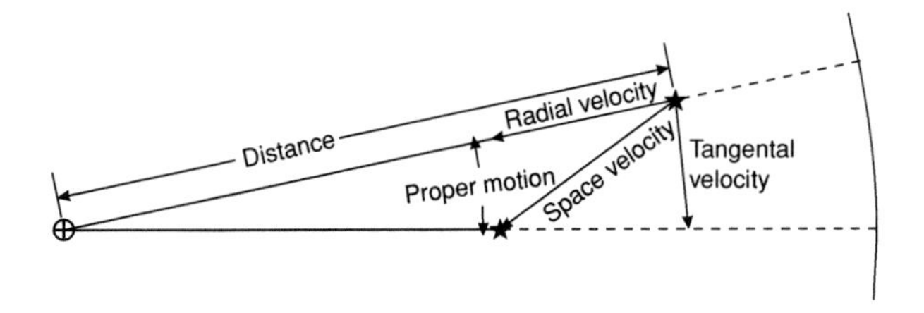

Figure 2.4 The proper motion is the annual angular (apparent) motion of a star across the sky. It is directly proportional to the tangential velocity of the star, and inversely proportional to its distance, and is measured in arc seconds per year. The space velocity also includes the radial component of the velocity, called the *radial velocity*, that is measured directly by the Doppler effect in km s^{-1}.

2.5 Motions of the stars

Although stars appear fixed in the sky, they are actually moving through space at tens of km s^{-1}. Only their great distances make them appear fixed. Their motions arise from two causes: the sun and its planetary system are moving, relative to their neighbour stars, at 19.4 km s^{-1} in the direction of RA 18 h, Dec. $+30^\circ$, so, on average, the neighbours seem to be drifting in the opposite direction; also, each neighbour has its own velocity, relative to the average. The velocities are generally 10–20 km s^{-1}, like the sun's, but the oldest stars in our galaxy – the so-called *Population II stars* – have much larger velocities, relative to the sun.

The motion of a star has three components, in three dimensions (figure 2.4). One component is the motion towards or away from us; that is called the *radial velocity*, and is measured by the *Doppler effect* (section 2.16). The other two components are the ones in the plane of the sky, across the line of sight; these are called the *proper motion*. The proper motion μ in arc sec per year is related to the tangential velocity T in km s^{-1} and the distance D in parsecs by

$$T = 4.74\mu D \tag{2.1}$$

The proper motion has two measurable components – in right ascension and in declination. The proper motion is determined by measuring the position of the star at two or more widely separated times, and noting the change. The proper motions tend to be much less than one arc sec a year, even for the nearest stars, so they can be determined only by careful measurement over a long time baseline.

Motions can be used to identify groups of stars such as clusters that are gravitationally bound, and are moving together through space. These will have similar motions. Motions can also be used to identify nearby stars: these will have large apparent motions, due to the sun's 19.4 km s^{-1} motion. A common analogy: when you are driving down the highway, objects nearby at roadside *appear* to sweep backwards, whereas more distant objects appear to move hardly at all. Motions can also be used to identify Population II stars, because these stars have relatively larger motions, even if they are not nearby.

An important class of 'moving stars' are *binary stars* – stars in mutual orbit. By measuring the motion of one star around the other, either visually or by the Doppler effect, astronomers can measure the most fundamental property of a star – its mass.

2.6 Apparent magnitude

The *apparent magnitude* is the astronomer's measure of the apparent brightness of a star or other celestial object. The magnitude system began with the Greek astronomer Hipparchus (190–120 BCE), who divided the stars into first, second, third, fourth, fifth, and sixth magnitude. Note that, in this system, a first-magnitude star is brighter than a sixth-magnitude star. Much later, it was realized that the magnitude system was a logarithmic or power-of-ten scale; a first-magnitude star is about 2.5 times brighter than a second-magnitude star which, in turn, is about 2.5 times brighter than a third-magnitude star and so on. In this system, a difference of five magnitudes corresponds to a ratio of $2.5^5 = 100$ in brightness. The magnitude system is now *defined* by this relationship

$$m_1 - m_2 = -2.5\log(B_1/B_2) \tag{2.2}$$

where m_1 and m_2 are the magnitudes of two stars and B_1 and B_2 are their physical brightnesses.

The apparent magnitude of a star also depends on the spectrum of wavelengths that it emits, and on the wavelength sensitivity of the detector. Historically, the first and most fundamental magnitude was the *visual magnitude* – the magnitude sensed by the average human eye. In the late nineteenth century, when astronomical photography was developed, the photographic emulsions were most sensitive to blue light, and the blue *photographic magnitude* was defined. With the development of accurate, sensitive photoelectric photometers, a standard photometric system was needed. The most widely used system is the UBV system developed by Harold Johnson and William Morgan (1953). It uses standard filters: *V* which is close to the visual magnitude, *B* which is close to the photographic magnitude, and *U* which is a violet-wavelength magnitude. This system has been extended into the infrared with RIJHKL magnitudes. The infrared JHK bands are ones at which the earth's atmosphere is relatively

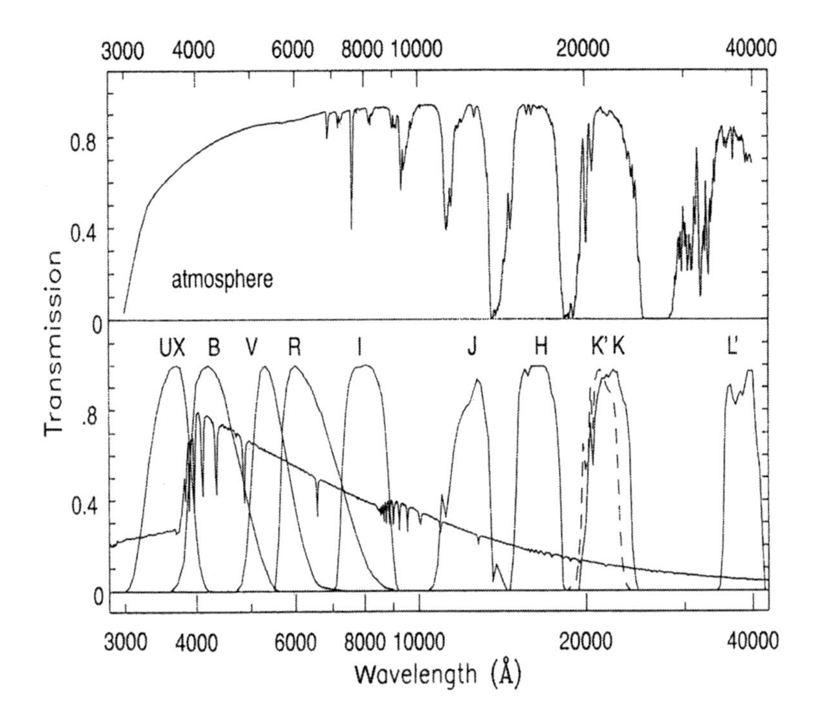

Figure 2.5 The passbands of the visual (UBV) and infrared filters (RIJHKL) in the UBV photometric system (lower panel), compared with the spectrum of the sun (lower panel), and the transparency of the earth's atmosphere (upper panel). The JHKL filters coincide with wavelength bands between the strong molecular absorption bands in the earth's atmosphere, at which the sky is relatively transparent. (From Sparke and Gallagher, 2000.)

transparent (figure 2.5). There are also 'narrow-band' photometric systems such as the Strömgren system, as well as others used for more specialized purposes.

2.7 Distance

The nearest star is the sun, 149 597 871 km from the earth, on average. The distance of the sun is determined by radar ranging of the planets such as Venus; this establishes a very precise scale of distances within our solar system.

The nearest other star is more than 200 000 times more distant, so astronomers have developed two more convenient units of distance. The *light year* is the distance that light travels in a year, at 299 792 458 km s^{-1}. A light year is 9.46×10^{12} km, or about 10^{13} km. The other unit, the parsec, is defined below. The distances to nearby stars are measured by parallax. *Parallax* is the apparent change in the position of an object, when viewed from two different locations (figure 2.6). The *parallax* of a star is defined as one-half the apparent

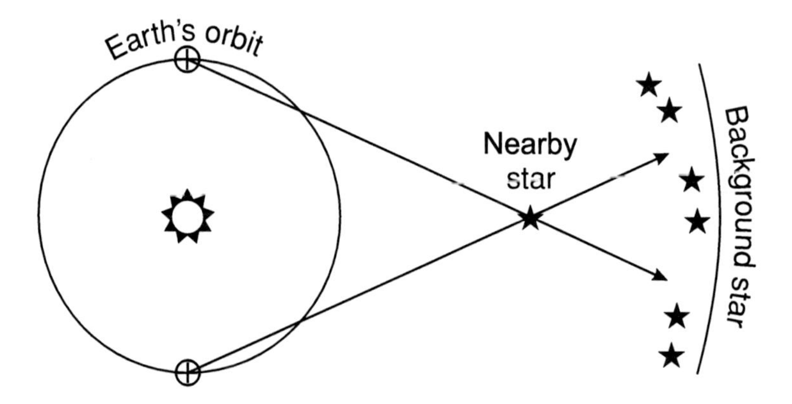

Figure 2.6 Parallax is the apparent change in the position of a star in the sky, as the earth revolves around the sun. The parallax is defined as half of the maximum apparent change in position, so it is also half the angle, subtended at the star by the diameter of the earth's orbit. It is determined by imaging the star at different times of year, from different points on the earth's orbit. It is measured in arc seconds, and is inversely proportional to distance in parsecs. (This figure is not to scale!)

annual shift in the position of a star when it is viewed from opposite sides of the earth's orbit. The parallax is measured in seconds of arc, or arc seconds; one arc sec is 1/3600 of a degree. The *parsec* is defined as the distance of a star that has a parallax of one arc sec; it is about 3.26 light years. The nearest star has a distance of slightly more than one parsec, and a parallax of slightly less than one arc sec. So

$$\text{parallax (arc sec)} = 1/\text{distance (parsecs)} \tag{2.3}$$

Because parallaxes are so small, they are difficult to measure, especially from the ground. The earth's atmosphere distorts sharp star images into trembling blurs. During its mission in 1989–93, the European Space Agency's *Hipparcos* satellite (an acronym for High Precision Parallax Collecting Satellite) measured the parallaxes of over 100 000 stars with an accuracy ten times better than before. It also measured accurate proper motions for hundreds of thousands of stars. GAIA (Global Astrometric Interferometer for Astrophysics), *Hipparcos'* successor, will measure many more stars, with much higher accuracy.

2.8 Absolute magnitude and luminosity

The apparent brightness of a star is determined by the power output of the star, and by its distance. Light travels in a straight line, so it passes through increasing areas with increasing distance from the source. This is expressed as the *inverse-square law of brightness* – the apparent brightness B is proportional to the absolute brightness or power P, and inversely proportional to the square of the distance D. As a formula

$$B = P/4\pi D^2 \quad (2.4)$$

The apparent brightness of a star, however, is measured by its (apparent) *magnitude* (section 2.6). The *absolute magnitude* of a star is arbitrarily defined as the apparent magnitude that it would have if it were at a distance of 10 parsecs (32.6 light years).

Just as the magnitude depends on the wavelength sensitivity of the detector, so does the absolute magnitude. The most commonly used absolute magnitude is the *absolute visual magnitude* M_v, that corresponds to the V magnitude. There is also the *absolute bolometric magnitude* M_{bol}, that corresponds to the magnitude measured by a detector that is perfectly sensitive to all wavelengths. These are related by

$$M_v = V + 5 - 5\log(D) \quad (2.5)$$

where D is the distance in parsecs; and

$$M_{bol} = M_v - BC \quad (2.6)$$

where BC is called the *bolometric correction*. The bolometric correction can be determined from models of the radiating layers of stars or, with difficulty, by directly observing the energy output of the star at all wavelengths. It depends primarily on the temperature of the star, and is largest for very hot or cool stars, for which much of the star's radiation falls outside the visible region of the spectrum.

A more physically meaningful quantity is the *luminosity* of a star, which is its power in Watts (its average power if it is a variable star) – the total energy that it emits each second.

2.9 Stellar masses

The mass of a star is its most fundamental physical property: it determines the star's evolution, lifetime, and fate. Stellar masses are determined most directly from *binary stars* (section 2.2), using Isaac Newton's extension of Johannes Kepler's Third Law

$$m_1 + m_2 = a^3/P^2 \quad (2.7)$$

where m_1 and m_2 are the masses of the components in solar units, a is the average distance between the components in astronomical units (the average distance of the earth from the sun), and P is the orbital period in years.

In a *visual binary*, the two components can be seen separately in the telescope, and their motion can be followed. Observation of their relative motion yields the sum of their masses $m = m_1 + m_2$. Observation of their absolute motion, relative to the background stars (which is much more difficult), yields m_1/m_2, hence m_1 and m_2. The modern techniques of *adaptive optics* and of *speckle interferometry* and the study of *lunar occultations* of stars enables some close binaries to be 'seen' as visual binaries. Even in an astrometric binary, in which the fainter component is not visible, some information about the masses can be obtained. One complication is that the orbit of the binary may be seen at an angle, rather than face-on, but there are techniques for deducing the orientation.

In a *spectroscopic binary*, the orbital motions of one or both of the stars can be measured from the spectrum, using the Doppler effect. If both spectra are visible, the masses m_1 and m_2 can be found, but they are multiplied by the (unknown) sine of the inclination i of the orbit to the plane of the sky.

In an *eclipsing binary* (chapter 5), the inclination is known; it must be close to 90° for an eclipse to occur. This removes the ambiguity in the velocities, sizes, and masses, if the eclipsing binary is also a spectroscopic binary.

2.10 Spectra

Most of what we know about stars comes from electromagnetic radiation – light and its relatives. We observe the direction of the radiation, the amount of energy as a function of wavelength, and the polarization, if any. The amount of energy as a function of wavelength is called the *spectrum*. The wavelength is measured in nanometres. (Die-hard astronomers such as I often use the more traditional *Angstrom unit:* Å, which is 0.1 nm.) If the wavelength falls between about 400 and 700 nm, then the radiation can be imaged and measured by the eye, and we call it light. Radiations with wavelengths shorter than 400 nm are called ultraviolet (UV), X-ray, and gamma-ray in order of decreasing wavelength; those with wavelengths longer than 700 nm are called infrared (IR), microwave, and radio in order of increasing wavelength. All of these electromagnetic radiations share the same properties, including their speed – they travel at the speed of light.

According to Kirchoff's laws that describe the spectra of electromagnetic radiation emitted by different types of sources: a hot solid, liquid, or dense gas emits a *continuous spectrum*; an energized low-density gas emits an *emission-line spectrum*; and a source of a continuous spectrum, seen through a cooler gas, produces

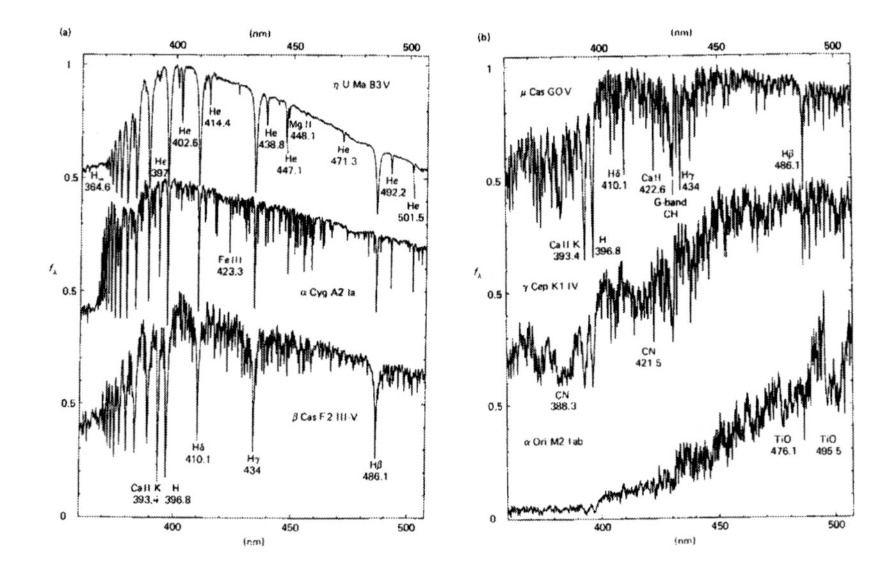

Figure 2.7 Spectra of six stars of different types: η U Ma (B3V), α Cyg (A2Ia), β Cas (F2III-V), μ Cas (G0V), γ Cep (K1IV), and α Ori (M2Iab). The lines show the flux of energy as a function of wavelength. The dips are the absorption lines, caused by different elements in the atmospheres of the stars. The wavelengths and origins (e.g. H = hydrogen) are marked for the stronger lines. The hotter (B, A, and F types) have more flux at short wavelengths; the cooler (G, K, and M types) have more flux at longer wavelengths. (From Walker, 1987.)

an *absorption-line spectrum.* Normal stars have an absorption-line spectrum; the deeper, denser layers emit a continuous spectrum, and the cooler atmosphere produces the absorption lines. Plotted as a graph, an absorption-line spectrum looks like a hill-shaped curve with sharp dips that represent the absorption lines (figure 2.7).

2.11 Colour

The colour of a star is determined primarily by its temperature: hotter stars produce relatively more blue light, and cooler stars produce relatively more red light. (This is contrary to some people's intuition, since it is opposite to the convention used on household taps!) The colour can be modified by 'reddening' by interstellar dust, just as the colour of the rising or setting sun or moon is affected by the reddening effect of particles in the earth's atmosphere; these particles scatter blue light out of the light path more effectively than they scatter red light.

The observant eye can perceive the colours of the stars. Betelgeuse has a reddish tint, Arcturus an orangish tint, and Rigel a bluish tint. But, because of the nature of colour perception by the eye, most stars appear almost white, even if their physical colours are quite different.

Star colours are measured with a photometer, with filters that transmit different colour bands in the spectrum. The most widely used colour system is the UBV system (though the colours measured with the filters in the *Sloan Digital Sky Survey* are now most numerous). The magnitude of a star is measured through each filter, and the differences in magnitude $(B - V)$ and $(U - B)$, which are called *colour indices*, are measures of the colour of the star. Any one colour such as $(B - V)$ is mainly a function of the temperature of the star, though the composition of the star, and its luminosity (actually the density of the atmosphere, which will be lowest for a high-luminosity star) have some effect, as does interstellar reddening. The colour index, corrected for interstellar reddening, is usually denoted $(B - V)_0$.

By measuring two different colours, we can separate the effect of the star's temperature from the effect of reddening by interstellar dust. By measuring three or four colour indices, as can be done in some other colour systems such as the Strömgren or Walraven systems, we can estimate the temperature, reddening, composition, and luminosity of the star. The colours can be measured to an accuracy of a few millimag, which corresponds to an accuracy in temperature of about 10K.

2.12 Temperature

The *temperature* of a star refers to the temperature of the *photosphere* – the average layer from which the starlight is emitted into space. The pressure, density, and temperature of the gas increase with depth. Below the photosphere, emitted photons would be absorbed by the dense stellar gas higher up before they could escape. From the photosphere and above, they can escape more or less freely. The level of the photosphere will depend on wavelength. At wavelengths at which the gas is more transparent, the photons can escape from lower levels. At wavelengths at which the gas is more opaque, the photons can only escape from higher levels. This range of levels is the star's *atmosphere*.

The temperature of the atmosphere can be determined in several ways. The hotter, denser, lower layers of the atmosphere emit a continuous spectrum of wavelengths. The shape of the spectrum is given by *Planck's law*, and the wavelength of the peak of the continuous spectrum is given by *Wien's law*, which states that the peak wavelength λ_{peak} (in metres) is given by

$$\lambda_{peak} = 0.0028978/T \qquad (2.8)$$

where T is the temperature on the Kelvin scale. The shape of the continuous spectrum can be measured through various calibrated intrinsic colour indices ($(B - V)$ being the most common), or more fundamentally by measuring the continuous spectrum at several wavelengths, and comparing the spectrum with the predictions of theoretical models of stellar atmospheres.

The temperature of the atmosphere can also be measured from the absorption-line spectrum of the star. Absorption lines occur when atoms in the atmosphere of the star absorb a photon, causing one of its electrons to move from a lower to a higher energy level. The atoms in the star's atmosphere are sensitive to, and therefore probes of, the physical conditions in the atmosphere – primarily the temperature, but also the density – and therefore the size of the star. At the lowest temperature, atoms can form molecules, so there will be molecular lines in the spectrum. At higher temperatures, the molecules *disassociate* into neutral atoms, with their electrons in low energy levels close to the nucleus. At higher temperatures still, the atoms are *excited*: the electrons are raised to higher orbits in the atom. At even higher temperatures, the atoms are *ionized*: the electrons are removed by energetic photons and by collisions. By measuring the degree of *dissociation*, *excitation*, and *ionization*, by noting which absorption lines are or are not present, and how strong they are, astronomers can measure the temperature to a relative accuracy of a few degrees.

One approach is *spectral classification* (Garrison, 1983). W. Morgan, P. Keenan, and E. Kellman (1953) devised the M – K system of classification, which defines *types* O, B, A, F, G, K, and M (and sub-types such as M0, M1, M2 etc.) and *classes* V, IV, III, II, and I. The types are those developed earlier at the Harvard College Observatory by a remarkable trio of 'Pickering's women' – Annie Cannon (1863–1941), Williamina Fleming (1857–1911), and Antonia Maury (1866–1952). Edward Pickering, Director of HCO, made a policy of hiring women research assistants because their salaries were significantly lower than men's – and history proved that they were extremely competent. Cannon classified the spectra of the over 200 000 stars in the *Henry Draper Catalogue*, and was recognized by an honorary doctorate from Oxford University.

More recently, two cooler types have been added to the sequence: L and T. These represent very faint, cool stars, and brown dwarfs, with temperatures in the range 2000 to 1000K.

Figure 2.8 show photographs of stellar spectra of different types. In figure 2.8 and later in this book, you may see reference to calcium H and K and sodium D absorption lines. The H, K, and D are not symbols for chemical elements; they are the designations that Fraunhofer gave to these absorption lines in the sun's spectrum. You may also see reference to HI or HeI (neutral hydrogen or helium) or HII (ionized hydrogen) or HeII (helium that has lost one electron).

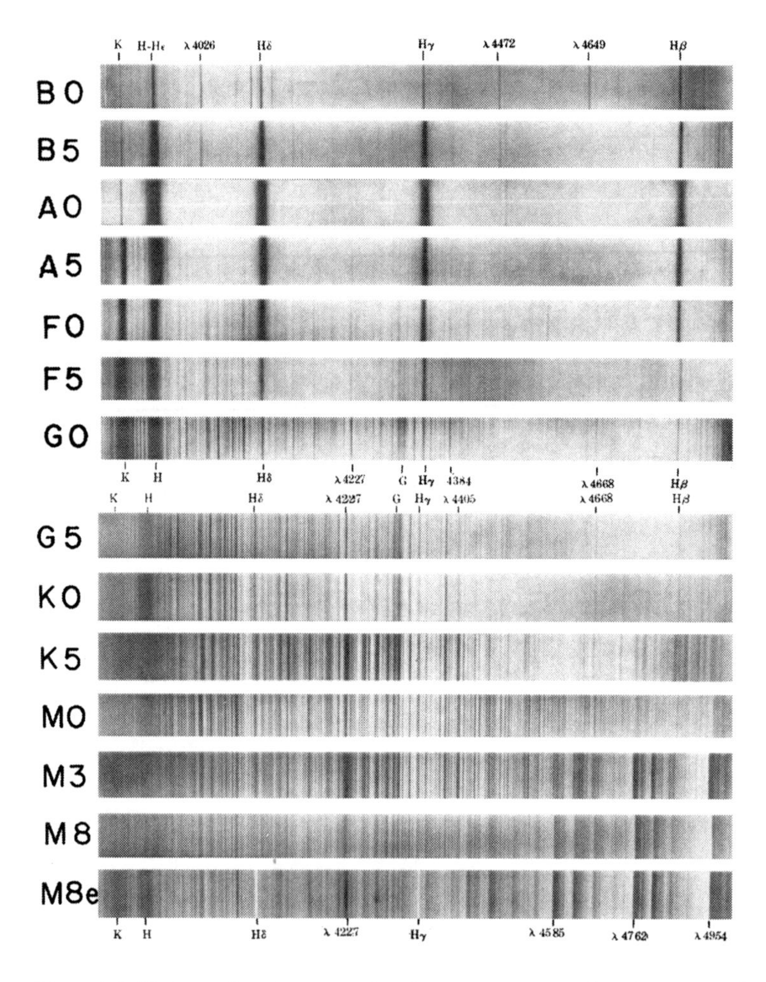

Figure 2.8 Photographs of spectra of main sequence stars of various types, from B0 to M8e. The spectra of stars are classified by comparison with spectra of standard stars, according to the relative strengths of the dark absorption lines, whose identities are marked. Of the stars shown here, the B0 stars are hottest, the M8e stars are coolest. (M8 stars are all variable stars, usually of the Mira type.) The designations H and K, under the strong lines at the left, are Joseph Fraunhofer's names for these absorption lines in the spectrum of the sun. The e in M8e denotes the presence of emission lines. (University of Toronto photographs.)

The M–K types and classes were empirical, but they turn out to be related to temperature and density, respectively. They can be calibrated by comparing the observed spectrum with the predictions of model stellar atmospheres, or by determining the temperature and density of the observed stars by other means. The luminosity classes V–I are important because, at any spectral *type*, they represent a sequence of increasing size and therefore luminosity: class I stars are supergiant stars. If these types and classes can be calibrated in terms of luminosity, then they can be used, along with the star's apparent brightness, to find its distance.

Spectral synthesis is a relatively new and very powerful tool for determining the temperature, density, and chemical composition of a star from the strength of specific lines in the spectrum. The observed lines in the spectrum are compared with the predictions of model atmospheres, making different assumptions about the temperature, density, and abundance of the element giving rise to that absorption line. The parameters of the best-fitting model are the ones adopted for the star.

2.13 Diameter

We see the sun as a finite disc in the sky, so we can measure its apparent (angular) diameter and, knowing the distance, calculate its actual diameter; it is about 1 390 000 km. More distant stars appear as points of light to the eye, and to most telescopes, so it is not easy to measure the apparent diameter.

There are several ways of measuring the apparent diameter of a star directly:

- Using a set of interconnected telescopes called an *interferometer*, it is possible to resolve the disc of some stars, and measure their apparent diameter.
- If the moon passes in front of the star, *high-speed photometry* can be used to measure the short but finite time required for the moon to extinguish the light of the star - a few milliseconds!
- For stars in an *eclipsing binary system* (chapter 5), the durations of the eclipses are related to the diameters of the stars. In the most favourable cases, the diameters can be measured to an accuracy of 1 or 2 per cent.
- Gravitational microlensing events, in which another object passes in front of a star; the bending of space by the gravity of the nearer object amplifies the light from the star behind.

Otherwise, a more indirect method must be used. Since the luminosity of the star is related to the radius R and temperature by the Stefan–Boltzmann law

$$L = 4\pi R^2 \sigma (T^4) \tag{2.9}$$

where σ is the Stefan–Boltzmann constant, the radius can be determined if the luminosity and the temperature are known, by the methods described above. This equation also defines the *effective temperature*, the temperature which would be calculated from the known luminosity and radius of the star.

In a sense, the diameter of a star is a nebulous quantity, since a star is a sphere of gas with a fuzzy edge. In the case of the sun, the atmosphere is only a few hundred km thick, a small fraction of the diameter. But stars like extreme red giants and supergiants may have a very extended atmosphere. And at wavelengths at which the gas is opaque, the star will look larger than at wavelengths at which the gas is more transparent.

2.14 Composition

The composition of a star can be inferred from the absorption lines in its spectrum; they are like a 'bar code' which identifies each element present. These lines are produced in the atmosphere of the star, so they reflect the composition of the atmosphere. The composition of the deep interior of the star might have been affected by nuclear processes going on there. The composition of the atmospheres of some stars may be 'polluted' by mixing of material from below, or by gravitational settling or radiative levitation of some elements. But, in most cases, the atmospheric composition reflects the initial mix of elements from which the star formed, and the composition of the bulk of the star.

Going from the absorption lines in the spectrum to an exact census of elements present, and their abundances, is not trivial. The strength of an absorption line depends, not only on the abundance of the element, but on the temperature and pressure in the atmosphere, and processes such as motions in the atmosphere which broaden the absorption lines. The element giving rise to that absorption line may be fully ionized, or in an energy state which does not give rise to that line. Astronomers must know the temperature and pressure of the atmosphere, and the properties of the atom in question, and build 'models' which incorporate the relevant physics.

The net result of such analysis is that stars are about three-quarters hydrogen, and one-quarter helium, by mass, with up to 2 per cent elements heavier than helium. By number of atoms, hydrogen makes up 0.91 and helium 0.09, with all the heavier elements making up less than 0.01. The 'heavy elements' are often called *metals*, even though most of them do not have metallic properties. The abundances of hydrogen, helium, and metals, by mass, are often denoted X, Y, and Z, respectively.

There are certain interesting characteristics to the observed abundances of the metals. The light elements lithium, beryllium, and boron, are practically non-existent; they are less than a billionth as abundant as hydrogen. As noted later in the book, they are nuclearly fragile enough to be destroyed by mixing, in the star, to deeper, hotter layers. Carbon, nitrogen, and oxygen are the next most abundant elements, after hydrogen and helium; that's good, since we are made primarily of those elements.

Elements with an even number of protons (or *atomic number*) are more abundant than those with an odd number. Heavier elements have much lower abundances: the abundances of the heaviest elements are a million times lower than carbon, for instance. Iron is noticeably more abundant than would be predicted from the general trend. These abundances can be explained by the particular properties of the nuclei, and the processes by which they formed.

2.15 Rotation

Everything spins – including stars. Our sun turns on its axis every 25–35 days, depending on latitude. (The sun is a fluid so, unlike the solid earth, different latitudes can and do rotate in different periods.) Among sunlike stars, young stars rotate an order of magnitude faster than the sun; they seem to slow down with age, as many of us do. Stars more massive than the sun can rotate in periods as fast as half a day; their equators are rotating at 500 km s^{-1}! This causes stars to have equatorial bulges, just as the earth and the author do.

We can measure the rotation of the sun by observing sunspots on its disc over periods of weeks to months. This is an active field of observational research for amateur astronomers. For other sunlike stars, we cannot see the spots on the disc, but we can observe their effect on the brightness of the star: as the star rotates, its brightness varies slightly, due to rotation. Such variable stars, and another kind of variable star with a patchy surface – peculiar A stars – are called *rotating variable stars* (chapter 4). Their period of variation is their period of rotation. We can also measure the rotation period spectroscopically: gases in the chromosphere, above sunspots, produce emission lines in the spectrum, so the spectrum also varies with time, with the rotation period. The *peculiar A stars* also have surface inhomogeneities in chemical composition that cause the brightness *and* spectrum to vary with the rotation period.

For stars without surface inhomogeneities, there is another way to measure the rotation from the spectrum. Since one side of the disc of the star will be approaching, and the other receding, the absorption lines in the spectrum of the star will be broadened by rotation by the Doppler effect (section 2.16). But the axis of rotation of the star may not be at right angles to our line of sight, so we may not measure the full effect of the rotation, only $v \sin i$, where i is the inclination of the rotation axis to the line of sight and v is the equatorial rotation velocity, in km s^{-1}. If the star is seen pole-on, the rotation will have no direct effect on the spectrum.

The sun rotates *differentially*: the rotation period is shorter near the equator, and longer near the poles. The MOST satellite has recently used starspots at different latitudes to detect differential rotation in a sunlike star – κ^1 Ceti.

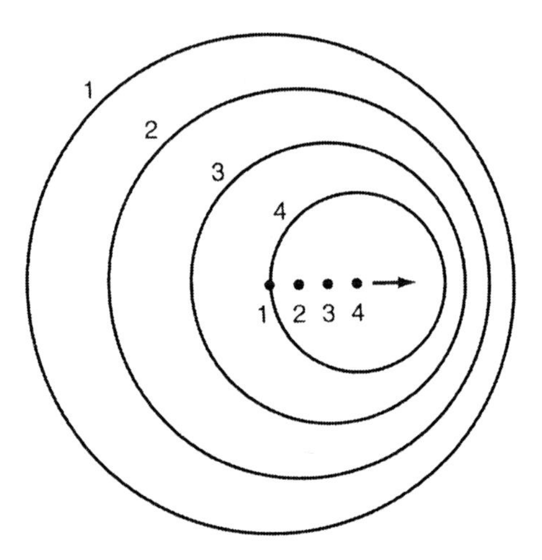

Figure 2.9 The principle of the Doppler effect. The source is moving from 1 to 4, emitting waves of a specific length. If the observer is on the right, they will perceive the wavelengths as being shortened; if they are on the left, they will perceive them as being lengthened. For light, shorter wavelengths correspond to bluer colours; longer wavelengths correspond to redder colours.

2.16 Radial velocity

The *radial velocity* of a star (or other object) is the speed at which it is approaching or receding, relative to the observer, i.e. its line-of-sight velocity. It is measured by the *Doppler effect*: if the source of waves is approaching, relative to the observer, then the apparent wavelength is compressed or shortened; if the source is receding, relative to the observer, then the apparent wavelength is stretched or lengthened (figure 2.9). In each case, the relative amount of the shift ([observed wavelength – true wavelength]/true wavelength) is equal to the speed of approach or recession (the radial velocity) divided by the speed of the wave

$$(\lambda_{obs} - \lambda_{true})/\lambda_{true} = v/c \quad (2.10)$$

where λ is the wavelength, v is the radial velocity, and c is the speed of light – 299 790 km s^{-1}.

The radial velocity is a key property of any star. It is even more important in the study of variable stars. In eclipsing variables, we can measure the orbital velocity of one or both stars in the binary system, that enables us to measure the masses of the stars. In pulsating stars, the near side of the star alternately approaches and recedes from the observer; from the radial velocity, we can confirm that the star is pulsating, and measure the actual velocity of pulsation

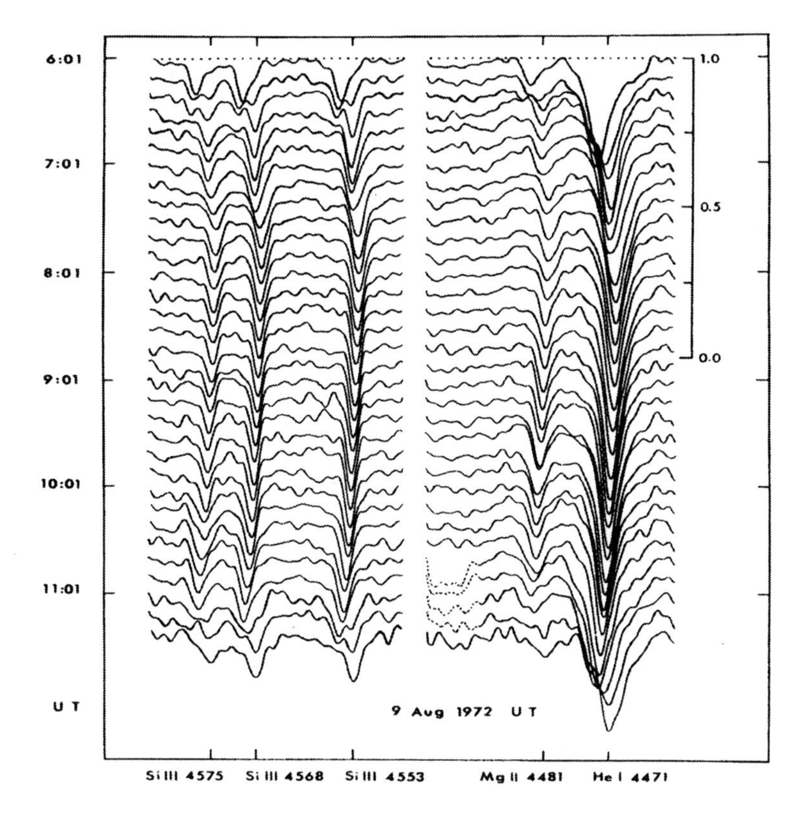

Figure 2.10 Radial velocity variations of BW Vulpeculae, a Beta Cephei pulsating star. In this series of spectra showing Si III, Mg II, and He I lines, the time increases downward (scale on left) and spans five hours. Note how the absorption lines move from the left (longer wavelengths) to the right (shorter wavelengths) and back, as the star expands and contracts. (From Goldberg, Walker, and Odgers, 1976.)

(figure 2.10). In novae and supernovae, the near side of the star approaches us at hundreds or thousands of km s^{-1} as the star erupts.

Prior to about 1970, radial velocities were measured by photographing the spectrum of the star, along with the spectrum of a light source – usually an iron arc, or a gas discharge tube which was attached to the telescope and therefore at rest, relative to the observer. The *spectrogram* was measured under a microscope: the positions of the absorption lines in the star spectrum were measured relative to the positions of the emission lines from the reference source. By the 1960s, this measurement process was being automated, by using a *microdensitometer* attached to a computer. By the 1970s, however, the whole process was revolutionized by the introduction of electronic detectors – initially *Reticons* and then *CCDs (charge-coupled devices)* – which were much more efficient than photographic plates, and responded linearly to light. The spectrum of the star was now imaged on to a detector which was thousands of 'pixels' (picture elements) long. Or, the spectrum could be 'folded' like the lines of words on this page with an *echelle spectrograph*, and imaged on to a detector which is effectively millions of pixels long.

For very high-precision radial velocity work, the light beam entering the telescope could be passed through a container of iodine or fluorine vapour, whose absorption lines could be used as a very stable reference source. Another development was *cross-correlation*: the whole spectrum of the star was digitally compared with the whole spectrum recorded at some initial time, and the average shift calculated. This means that every portion of the spectrum is used to measure the shift in wavelength. Yet another development was the *photoelectric radial velocity 'speedometer'*: the spectrum of the star was passed through a negative image of a spectrum of a similar star; the position of the negative image or 'mask' could be precisely shifted in wavelength to determine the *average* shift in wavelength between the star's spectrum and the reference mask. Radial velocities can now be measured to an accuracy of up to 1 m/sec – about the speed that a human can jog. This represents a Doppler shift of one part in a billion! Precision radial velocities have a variety of important applications to variable star astronomy, but the most interesting and important discovery, using precision radial velocities, has been *exoplanets* – planets around other stars. They reveal themselves by their gravitational effect on their parent star, causing the star to move in a small orbit at a speed of a few tens of m/sec. Exoplanets with masses as small as those of Uranus and Neptune have been discovered, but earth-mass planets would have too small a gravitational effect to be detected by the radial velocity technique. (See Hearnshaw and Scarfe (1999) for the proceedings of a recent conference on precision stellar radial velocities.)

2.17 The Hertzsprung–Russell (H–R) diagram

Early in the twentieth century, Ejnar Hertzsprung and Henry Norris Russell independently noted that a graph of a measure of the luminosity of a star (absolute magnitude), plotted against a measure of the temperature of a star (spectral type, or colour), gave an interesting result: the vast majority of stars lay along a band on this graph, extending from hot, high-luminosity stars, to cool, low-luminosity stars (figure 2.11) – the *main sequence*. The H–R diagram has subsequently become a cornerstone of both observational and theoretical stellar astrophysics. Observers tend to plot absolute magnitude against spectral type or against a colour such as $(B - V)$. Theoreticians tend to plot the logarithm of the luminosity against the logarithm of the effective temperature. For stars in a cluster, which are all at the same distance, it is possible to plot the *apparent* magnitude against some measure of the temperature, since the apparent and

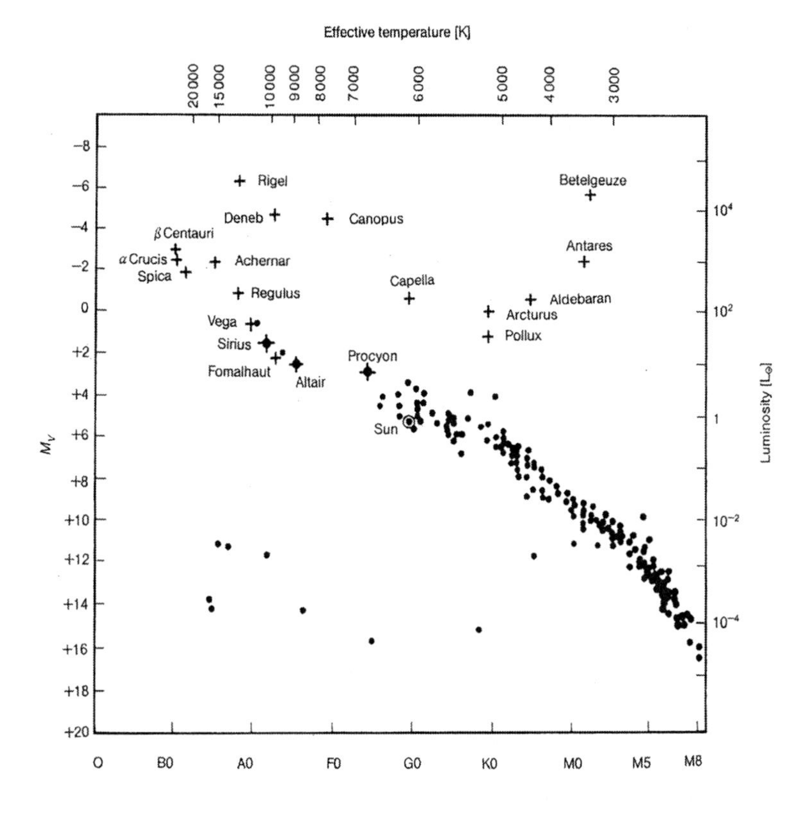

Figure 2.11 Schematic Hertzsprung–Russell (H–R) diagram, showing the locations of various types of stars: filled circles – the nearest stars (within 10 parsecs); crosses – the apparently brightest stars, identified by name. The distributions are different. The nearest stars are almost all intrinsically fainter than the sun; they are a more-or-less complete and unbiased sample. The apparently brightest stars are almost all intrinsically brighter than the sun, but this sample is naturally biased in favour of stars which are intrinsically bright. (From Karttunen *et al.*, 1996.)

absolute magnitudes will differ by the same amount for each star. For purely historical reasons, cooler stars are plotted on the right.

According to the Stefan–Boltzmann equation, the luminosity of a star is proportional to the square of its radius and to the fourth power of its effective temperature. Stars on the H–R diagram which are luminous and cool must therefore be very large – supergiants. Stars which are hot but not very luminous must therefore be very small – white dwarfs.

The H–R diagram has many applications in astronomy:

(1) It demonstrates a fundamental characteristic of stars: for over 90 per cent of a star's lifetime (the main sequence phase), its luminosity,

radius, and surface temperature are directly determined by its mass – the main sequence is a *mass sequence*; the star's luminosity and surface temperature are therefore correlated.

(2) It is a way of classifying and comparing populations of stars. For instance, the population of nearest stars to the sun almost all lie on the part of the main sequence below (fainter than) the sun, whereas the population of apparently brightest stars in the sky all lie above (brighter than) the sun (figure 2.11).

(3) It can be used as a distance-measuring tool: if a star is known to be a main-sequence star (and most of them are), then measuring its temperature (which is relatively easy) can lead to an estimate of its luminosity and hence, from its apparent brightness and the inverse-square law, its distance (which is relatively difficult).

(4) It can be used as a way of 'picturing' the evolution of stars: displaying the changes in luminosity and surface temperature which are predicted by theoretical models or simulations (figure 2.14) through *evolution tracks* on the H–R diagram, which can then be compared with the properties of observed stars. This technique is especially important for star clusters, since it can be used to determine their ages.

Another important trend is the *mass–luminosity relation*. For main sequence stars, the luminosity is proportional to $mass^x$, where x is between 3 and 4. This relation can be derived from observations of stars with known mass and luminosity, and can also be derived from the theory of stellar structure. The relationship between mass and luminosity can be seen in figure 2.14 which shows the masses of main-sequence stars and their luminosities.

2.18 Star structure

The thousands of stars which we see in the night sky – and the sun which we see in the day – are very stable, and remain so for up to billions of years. Many are variable, as we shall see, but this variability is a relatively minor perturbation on their long-term stability. Or the variability may be due to a geometrical process such as eclipse or rotation, not to any physical change in the star itself.

The first and most urgent form of stability for a star is *mechanical stability* or equilibrium. If the inward pull of gravity were not balanced, the sun would collapse in less than an hour. In actual fact, the inward gravity is exactly balanced by outward pressure. The pressure must increase with depth, so there is a net upward pressure force on any layer of the star, to balance the downward pull of gravity.

The pressure is produced by the thermal motion of the gas particles in the star. According to the equation of state for a 'perfect gas', it is directly proportional to the temperature, and to the number of particles per unit volume, which is related to the density. The temperature and density also increase with depth.

If the temperature increases with depth, then energy must flow out of the star, either by radiation, convection, or conduction, because energy flows from hot to less hot. Conduction occurs only in very dense regions, such as the cores of red giants, and in white dwarfs, where the electrons behave like a metal. Convection occurs in regions where the opacity of the gas is very high. Otherwise, the energy flows by radiation, the photons diffusing outward in a 'random walk' path, taking thousands of years to do so.

If energy flows out from the star, it follows that the centre of the star will cool, and the pressure which balances gravity will decrease, *unless* energy is created in the centre of the star. Indeed it is, as long as the star has a supply of thermonuclear fuel. For most of a star's lifetime, the fuel is hydrogen. Hydrogen is capable of transforming 0.007 of its mass into energy in *fusion reactions*, the net result of which is to convert four hydrogen nuclei into a helium nucleus. This can occur through the *proton–proton chain*, in which protons sequentially fuse through a chain leading to helium-4, or through the *CNO cycle*, in which protons are sequentially added to a carbon-12 nucleus, eventually yielding helium and carbon-12, which thereby serves as a 'catalyst' in this process. Later in the star's life, it may fuse the helium into carbon, but this transforms less than 0.001 of its mass into energy.

These general principles can be expressed as equations, and solved on a computer for a star of any particular assumed mass and composition, to give a numerical *model* of that star. The model gives the properties of the star, such as temperature and density, from the centre to the surface. This is one way in which astronomers can 'see' inside a star. Figure 2.12 shows a model of a star, determined in this way.

The structure of 'simple' stars is well understood, but the effects of rapid rotation, mass loss, and magnetic fields are less so. The presence of a close binary companion is another complication. The study of variable stars promises to help astronomers to understand 'real' stars in all their complexity.

2.19 Star formation

Stars form from *interstellar material* – loose unformed gas and dust between the stars. Some of this material is primordial; some has been recycled from previous generations of stars through stellar winds, and supernova explosions. In a galaxy like our own, about a tenth of the visible material is gas

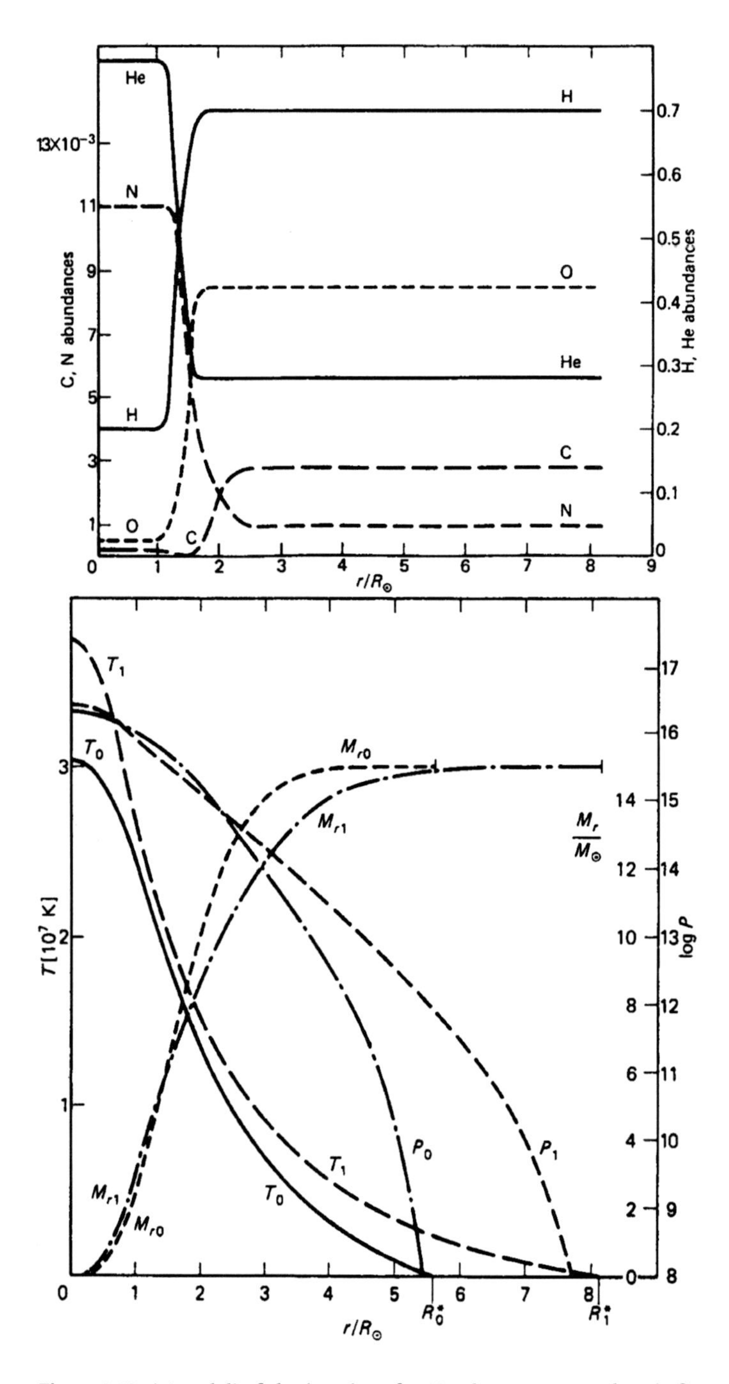

Figure 2.12 A 'model' of the interior of a 15 solar-mass star when it first reaches the main sequence (subscript 0) and 8.6 million years later (subscript 1). Top: the abundances of hydrogen, helium, carbon, and nitrogen are plotted as a function of r at age 8.6 million years. The evolved star would be a good model of a Beta Cephei variable star. Bottom: the temperature, pressure, and mass within a sphere of radius r – the distance from the centre – in solar units (Bohm-Vitense, 1992).

and dust; the rest is stars. (There is ten times as much invisible *dark matter* as visible matter; its nature is still not known.)

By mass, atoms and molecules of gas make up 99 per cent of the interstellar material; the dust makes up the rest. Together, the gas and dust are about three-quarters hydrogen, one-quarter helium, and 2 per cent everything else, by mass (section 2.14).

Atoms and molecules of gas are observable by the specific wavelengths of visible, infrared, or radio radiation, which they emit or absorb. Dust particles, which are a micron or less in size (the size of smoke particles), are detectable because they dim and redden the light which passes through them. If they are near a bright star, they may be visible by reflected light. If they are warm (typically a few to a few hundred K), they can be observed by the infrared radiation which they emit. Absorption and reddening by interstellar dust can be a significant problem when observing and studying stars – including variable stars. It modifies their observed properties, and may obscure them totally.

Clouds of interstellar material are normally stable to gravitational contraction. But if parts of the cloud are compressed by a *density wave* in the interstellar material in a spiral galaxy, or by the powerful wind of a nearby star, or by the blast from a nearby supernova, they may start to contract. As their density increases, they contract more quickly.

Stars may form in large numbers in *giant molecular clouds*. Within these, *pre-stellar cores* develop, hidden from view at visible wavelength by up to 50 magnitudes of absorption. The cores may occur in clusters, or in small isolated *Bok globules*, named after Bart Bok who studied some of them.

In the universe, though, everything spins, including the part of the interstellar cloud which will form a star. As it contracts, its rotation speeds up. The technical term for this is *conservation of angular momentum*, but most people have seen this phenomenon in everyday life when they watch a figure skater go into a spin by pulling their arms close to their body. Rotation then causes the contracting material to form an *accretion disc*; the star forms at the centre of the disc, and planets may form in the disc itself, as the dust accretes into *planetesimals* and the planetesimals into protoplanets. The discs can be detected and studied through the IR radiation from the dust in the disc, and through radio radiation from molecules in it. Figure 2.13 shows an HST image of a forming star, and protoplanetary disc, or *proplyd* – one of many that have been imaged in regions such as Orion.

The transition from pre-stellar disc, to star – and possibly planets – is complex and still poorly understood. There are complex processes in the compact magnetized region between the inner disc and the star which, among other things,

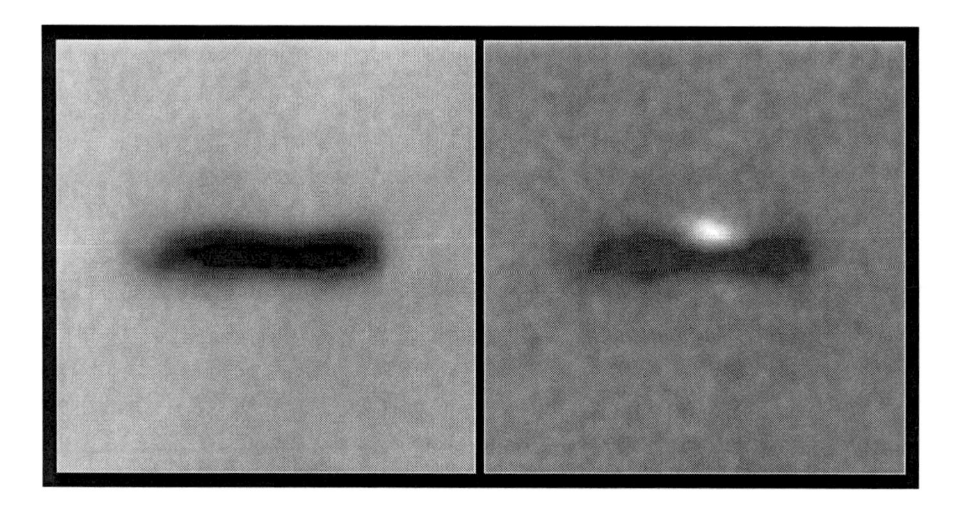

Figure 2.13 HST image of an edge-on protoplanetary disc in Orion. The star's light is mostly obscured by the disc, but is faintly visible in the right-hand exposure (M. McCaughrean, C. R. O'Dell, and NASA).

produce jets which are expelled and accelerated in a direction perpendicular to the disc. New technologies, such as adaptive optics, are enabling astronomers to observe and study these processes in detail. These processes have much in common with those that occur in the accretion discs in cataclysmic variables, and active galactic nuclei.

As long as the newly formed star has a mass greater than 0.08 solar masses, thermonuclear reactions will eventually be ignited in its core. The contraction will stop, and the star will be stable. The energy it radiates into space will be balanced by these thermonuclear reactions. By definition, the object is a star. If its mass is less than 0.08 solar masses, it will simply cool slowly with time; we call it a *brown dwarf*. What is the difference between a massive planet and a brown dwarf? By the usual definition, a planet forms in a disc around a star; a brown dwarf forms in the centre of a contracting cloud, like a star does. But the definition is still being considered.

Recent research shows that some young brown dwarfs have discs around them, and show other similarities to the coolest stars. Brown dwarfs may even have planets. But the study of brown dwarfs is even more challenging than the study of cool stars, and requires very large telescopes, and astute analysis and interpretation.

One of the mysteries that star formation must explain is the *initial mass function* – the observed distribution of masses of newly formed stars. There are many more low-mass stars than high-mass stars. And how to explain the formation of binary stars; why are over half of all stars born in binary systems?

Perhaps because the pre-stellar cores break into two or more fragments as a result of turbulence or rotation of the core.

The rapid rotation of the newly formed star generates a strong magnetic field by the *dynamo effect*. Stated simply, the dynamo effect is a process by which a magnetic field is produced by electrical charges in motion. In the sun, this involves a complex interaction between the sun's differential rotation, and the convective motions which occur in the outer layers of the sun. Magnetic processes involving the infalling material cause the star to be very *active*. Stellar activity includes spots, flares, and a hot chromosphere and corona. It also produces variability, which will be discussed later in this book (chapter 8).

2.20 Star evolution

Evolution means change. Gravity causes stars to shine. For most of their lives, stars make up this lost energy through thermonuclear fusion of hydrogen into helium, and helium into heavier elements. They change very little in structure, but their composition is slowly changing in their cores, as lighter elements are transmuted into heavier ones.

Models of stellar evolution can be calculated in the same way as for models of stellar structure. At each time step, the computer allows for the small changes in composition that have occurred. Figure 2.14 shows a grid of such models, for stars whose mass ranges from 0.8 to 120 in solar units. Note that the main sequence luminosity is a strong function of mass.

As the star's usable hydrogen becomes exhausted, its helium core shrinks. Hydrogen is fused to helium in a shell outside the core. The outer layers of the star expand, and the star becomes a large, cool, luminous *red giant star*. At some point, the centre of the core becomes hot enough so that helium can fuse into carbon. The star shrinks somewhat and, for a period of time, the star can balance gravity by fusing helium into carbon in its core. It becomes a *horizontal-branch star* on the H–R diagram. As the core becomes exhausted, helium fuses into carbon in a shell outside the carbon core. Again, the outer layers of the star expand, and it becomes a red giant once more, or more specifically an *asymptotic giant branch star*, so-called because of its position on the H–R diagram. At its largest and coolest, the star becomes unstable to pulsation, and becomes a *Mira star* (section 6.16). The pulsation drives off the outer layers of the star, revealing the hot core, which rapidly cools to become a *white dwarf*.

This picture applies to the vast majority of stars: those less massive than about ten solar masses. More massive stars evolve to become *red supergiants*. Their cores are hot enough to fuse carbon into oxygen, and oxygen into elements up to iron. But each successive element contains less nuclear energy, and satisfies the star's needs for a shorter period of time. When the core consists of iron, it can no

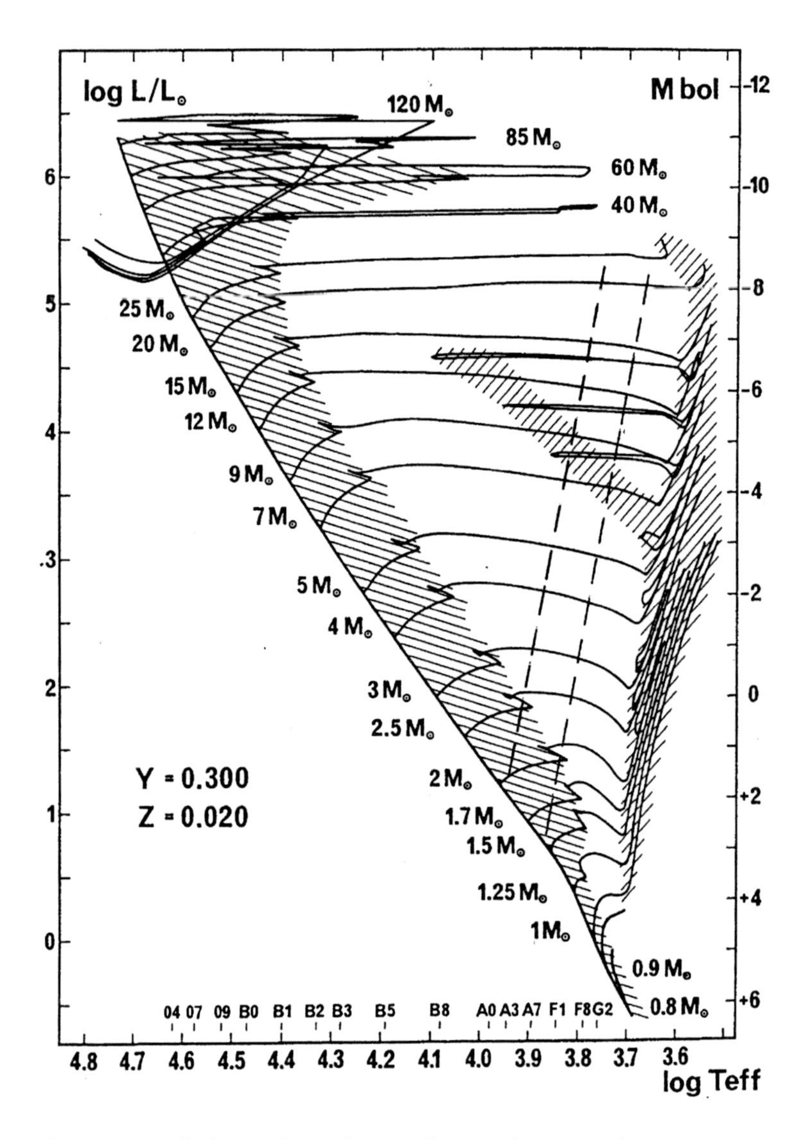

Figure 2.14 Evolution tracks on the H–R diagram, for stars with masses ranging from 0.8 to 120 times that of the sun, and chemical composition appropriate for that of our galaxy. The time scales for evolution, and therefore the lifetimes of the massive stars, are much shorter than for the less massive stars. The shaded regions are ones in which stars spend a significant period of time. The other regions are traversed very quickly. Evolution tracks show how stars reach the instability regions during their lifetimes; the dashed lines are the edges of the Cepheid instability strip. (From Schaller *et al.*, 1992.)

longer provide energy and pressure to resist gravity, and it collapses to form a neutron star; the released gravitational energy results in a *supernova explosion* of the star.

Stellar evolution is generally too slow to observe, but there are a very few cases in which a star has changed noticeably, due to evolution, in our lifetime. Since that, by definition, makes the star a variable star, these cases will be described later in this book. Variable stars provide another look at evolution in action: evolution causes the periods of pulsating stars to change and, since the effects are cumulative, they can eventually be observed. Star clusters provide another important test of stellar evolution theory, since they represent a sample of stars of different masses but similar age and chemical composition.

Note that *mass loss* can be an important process in the evolution of a star. The sun loses about 0.001 of its mass through thermonuclear fusion of hydrogen into helium; the 'lost' mass is transformed into the energy which has warmed the earth since its birth. During its main-sequence lifetime, the sun also loses about 0.001 of its mass through the *solar wind*. Red giants and supergiants lose mass through their dense *stellar winds* driven by pulsation; massive blue stars lose mass through winds driven by the pressure of their own radiation. Stars in close binary systems may lose mass to their companion star, as a result of its gravitational tidal effects, or the mass may be lost to the system entirely.

The total lifetime of a star depends on the total amount of available thermonuclear fuel, divided by the rate at which it uses it (its power), averaged over time. For the sun, this amounts to 10^{10} years; for more massive stars, it is less, and for less massive stars, it is more. As mentioned earlier, the luminosity is proportional to $mass^x$ so the lifetime T varies with the mass m as $T = m^{1-x}$, where x is between 3 and 4.

2.21 Star death

Star death is ultimately caused by gravity, 'the midwife and undertaker of the stars'. Gravity is an attracting force, which tends to contract or collapse every object. Small, cold bodies, like the earth, have enough rigidity to resist gravity. In a normal star like the sun, gravity is balanced by the outward pressure of hot gas, which is kept hot by thermonuclear reactions, primarily fusion of hydrogen into helium. When a star runs out of thermonuclear fuel, however, gravity will compress it into one of three forms:

- Most commonly, a *white dwarf*. A white dwarf can have a mass of up to 1.44 $M_\odot$, but can be descended from stars of up to 8 $M_\odot$, which have lost most of their mass as red giants. They have a radius of about 0.01 $R_\odot$ (i.e. a radius similar to that of the earth), and therefore have densities of approximately a million times that of water. The radius is a decreasing function of mass, approaching zero as the mass approaches 1.44 $M_\odot$. The outward pressure is provided by electrons, which have filled up all the available energy states within the star – a quantum-mechanical phenomenon called *electron degeneracy*. The electrons occupy all available states of position and energy; there is no room for further compression. For masses greater than 1.44 $M_\odot$ – the *Chandrasekhar limit* – electron degeneracy is insufficient to balance gravity, and the star collapses under its own weight.

White dwarfs are common: both Sirius and Procyon have white dwarf companions. The white dwarf companion of Sirius was discovered indirectly in 1844 from its gravitational pull on the visible star, and was first 'seen' in 1862. In 1914, its unusually small size was deduced from its luminosity and temperature. In 1931, Subrahmanyan Chandrasekhar was able to explain its nature, using the new theory of quantum mechanics. A white dwarf has no source of energy other than its stored heat. Its fate is to cool slowly, over billions of years, and to become a *black dwarf*.

- Rarely, a *neutron star*. Neutron stars have measured masses of 1.0 to 2.5 $M_\odot$, with most between 1.3 and 1.5 $M_\odot$; the average is 1.35. These masses are derived from pulsars which are in binary systems, and may not be the same as the masses of single neutron stars. They are normally descended from a star of more than 8 $M_\odot$, or possibly from a white dwarf which has exceeded the Chandrasekhar limit as a result of mass transfer from a binary companion. They have radii of about 10 km, and densities of 10^{15} gm/cm^3, well above the density of the atomic nucleus. Gravity is balanced by the outward pressure of neutrons which have filled up all the available energy states within the star – *neutron degeneracy*. The interior of a neutron star is a mixture of neutrons, some protons and electrons, and possibly stranger particles – hyperons, kaons, pions, or deconfined quarks. Since conditions in the deep interior are unlike anything on earth, the state of matter there is uncertain. Studies of the properties of neutron stars – their surface temperatures, for instance – can potentially provide information about the state of matter in these very extreme environments. The crust is mostly ultra-dense, ultra-strong iron, and the atmosphere may be only a few centimetres thick. Freshly formed neutron stars rotate rapidly; their rotation period is a fraction of a second. Their surface temperatures decrease rapidly, from billions of degrees when they are formed, to a million degrees or so after a thousand years. They are therefore visible with X-ray telescopes. The strong gravity at their surface causes a substantial *gravitational red shift* for any radiation emitted. Their magnetic fields are strong, and neutron stars were first observed in 1967 as *pulsars* (section 4.7), which emit rapid pulses of electromagnetic radiation as their magnetic fields are flung around, by the rotation, at close to the speed of light. The gravitational energy of a pulsar is about 3×10^{53} ergs, the magnetic energy about 1×10^{53} ergs, and the rotational energy about 2×10^{52} ergs. The 2004 discovery of a pulsar–pulsar binary promises to extend our understanding of neutron stars, as would neutrino observations of the next supernova in our galaxy (resulting from the formation of a neutron star), or observations of gravitational waves from the formation of a neutron star, or the merging of a neutron star binary. See Lattimer and Prakash (2004) for an excellent recent review.

- Even more rarely, a *black hole*. A black hole is an object which is so dense that its gravity prevents any mass or energy from leaving; the escape velocity exceeds the speed of light. Black holes result from the collapse of stellar cores more massive that 3 $M_{\odot}$, specifically in supernova explosions of stars with initial masses of more than 25 $M_{\odot}$. (It now seems possible, however, that some very massive stars can avoid becoming black holes, producing neutron stars instead.) The radius of their *event horizon* – the sphere on which the escape velocity is exactly equal to the speed of light – is about 10 km. Black holes have been identified in binary systems, by their gravitational effect on their visible companion, and by the X-rays which are emitted as gas spirals into the black hole through an accretion disc. From the spectroscopic orbit of the companion, the mass of the unseen companion – the black hole – can be deduced. The first identified black hole was the companion to HDE 226868. The system is an X-ray source, Cygnus X-1. The binary period is 5.6 days. A few dozen such stellar-mass black holes are known, and others are suspected.

  Supermassive black holes are found at the centres of many galaxies. Primordial black holes, with masses comparable to earth's, have also been hypothesized, but neither they nor their effects have yet been observed.

We should remember, however, that most stars are much less massive than the sun. They have lifetimes of tens of billion years or more and, for all practical purposes, live forever!

3

Variable stars

Variable stars are stars which change in brightness. The change may be as small as a few parts in a million, or it may be a factor of a thousand or more. It may occur in a second or less, or it may take years, decades, or centuries. These are extremes, but astronomers have developed an array of techniques for discovering, measuring, and analyzing the full range of possible variable stars. Why? Because the variations provide important and often-unique information about the nature and evolution of the stars. This information can be used to deduce even more fundamental knowledge about our universe in general.

The variations may be due to the rotation of a spotted star, or to an eclipse of a star by a companion star, or even by an unseen planet. The variations may be due to the vibrations of a star; if they are complex enough (as they are in our sun), they may provide an internal 'picture' of the star, like a CT scan. The variations may be due to eruptions on a star (flares), or an accretion disc (dwarf novae) or major explosions on a star (novae), or to the total disruption of a star in a supernova. Supernovae are the most violent events in our universe, yet we owe our existence to them, because they help to recycle the atoms, created in stars, into space, where some of them became part of our sun, our planet, and our biosphere. Also, the elements heavier than iron were mostly created in the supernova explosion. Supernovae may be dramatic and extreme, but they represent only one of the many roles that variable stars play in modern astrophysics, and in our understanding of the universe and the processes which govern it. In a sense, variable stars are 'speaking' to us. Variable star astronomers seek to learn their language, and understand what they are saying.

A variety of general resources on variable stars can be found in the 'Bibliography and Resources' at the end of this book.

3.1 Magnitude and Julian Date

The 'sound' of variable stars is their changing magnitude as a function of time. Of course, they do not really make a sound, but astronomers decode the *time series* of magnitudes m and times t, just as we decode the sound which a person makes as they speak. The time series is m_i, t_i, where t goes from 1 to N. N is the total number of measurements.

The concept of magnitude, and the many flavours of magnitude, are discussed in section 2.6. In variable star astronomy, visual magnitudes (m_v), photographic magnitudes (m_{pg}) and UBV photoelectric magnitudes are most commonly used. In astronomy, the time is traditionally given by the *Julian Date (JD)* – the interval of time in days since noon at Greenwich on 1 January 4713 BC. The choice of this date has a long and complicated history, which will not be given here. Suffice it to say that 4713 BC pre-dates all known astronomical observations. Note that the Julian Date begins at noon. The Julian Date avoids the complexities of the calendar, and is particularly appreciated by computers. The problem is that it is currently a rather large number: the Julian Date on 1 January, 2006 at 0h UT was 2453736.5. You may meet the quantity 'Julian Date – 2440000' or 'Julian Date – 2450000'. Beware of ambiguous Julian Dates, or ones which are in error by 0.5 day!

You may also encounter *heliocentric Julian Date*. As the earth orbits the sun, it may be closer to or farther from the variable star than the sun is. The light from the variable star may arrive earlier or later than it could arrive at the sun. This is a purely local effect; it has nothing to do with the behaviour of the variable star. So – especially for variable stars which are changing brightness rapidly – the Julian Date is corrected to what would be (hypothetically) recorded at the centre of the sun. In the case of very fast variable stars like pulsars, the Julian Date is actually corrected to the centre of mass of the solar system! (See: www.starlink.rl.ac.uk and go to 'slalib'.)

The heliocentric correction HC in days is given by

$$HC = -0.0057R\cos(\beta)\cos(L - \lambda) \tag{3.1}$$

where R is the distance between the earth and the sun in astronomical units, β and λ are the ecliptic latitude and longitude of the star, and L is the ecliptic longitude of the sun. For a discussion of Julian Date, and a Julian Date converter, see:
http://www.aavso.org/observing/aids/jdcalendar.shtml

3.2 Measurement of variable stars

Our personal view of the universe takes place through the complex process of *vision* – a process which is both physiological and psychological. Photons strike receptor cells in the retina, a biochemical reaction occurs, and triggers a signal to the brain. But there are two different kinds of receptor cells. *Cones* are colour sensitive, but have low brightness sensitivity; they are active at high light levels; and they are concentrated in the centre of the retinal field of view. *Rods* have higher brightness sensitivity, but little or no colour sensitivity; they dominate vision at low light levels. They are more abundant in the periphery of the retina, so one can often see fainter objects by *averted vision* – imaging the source at the edge of the field of view. Only a few photons are required to 'see' a faint source. The eye responds logarithmically to a light stimulus, hence the magnitude system.

3.2.1 *Photometry*

Hearnshaw (1996) has described the history of astronomical photometry over the past two centuries. The eye was the first photometer. Astronomers measured the brightness of a variable star, relative to one or more constant stars of known or assumed magnitude, by interpolation (or occasionally by extrapolation). Figure 3.1 shows a visual chart used by the AAVSO. The observer estimates the magnitude of the variable star, relative to the comparison stars. The magnitude of the variable star may be the same as that of one of the comparison stars. If not, it is estimated relative to the magnitudes of two or more comparison stars, which bracket it in brightness. For instance, if it is half way between the magnitudes of two comparison stars, then it is the average of those two magnitudes. The eye is reasonably good at making *comparisons* between the brightness of light sources. But the measurement is only as good as the eye that made it, the constancy of the comparison stars, and the validity of their 'known' magnitudes. Under the best conditions, the accuracy of a visual measurement is about 0.1 magnitude. If multiple independent measurements are made, and the errors are random, then the accuracy of the mean value increases as the square root of the number of measurements. But there are many possible sources of systematic error, especially if the colours of the variable star and the comparison stars are different. The *Purkinje effect* (named after Jan Purkinje (1787–1869)) is a change in the colour sensitivity of the eye that occurs at low light levels. The overall sensitivity increases, but the peak sensitivity moves towards the blue, and the eye becomes less sensitive to red light. This is because the cone cells, which function at higher light levels, are more red-sensitive than

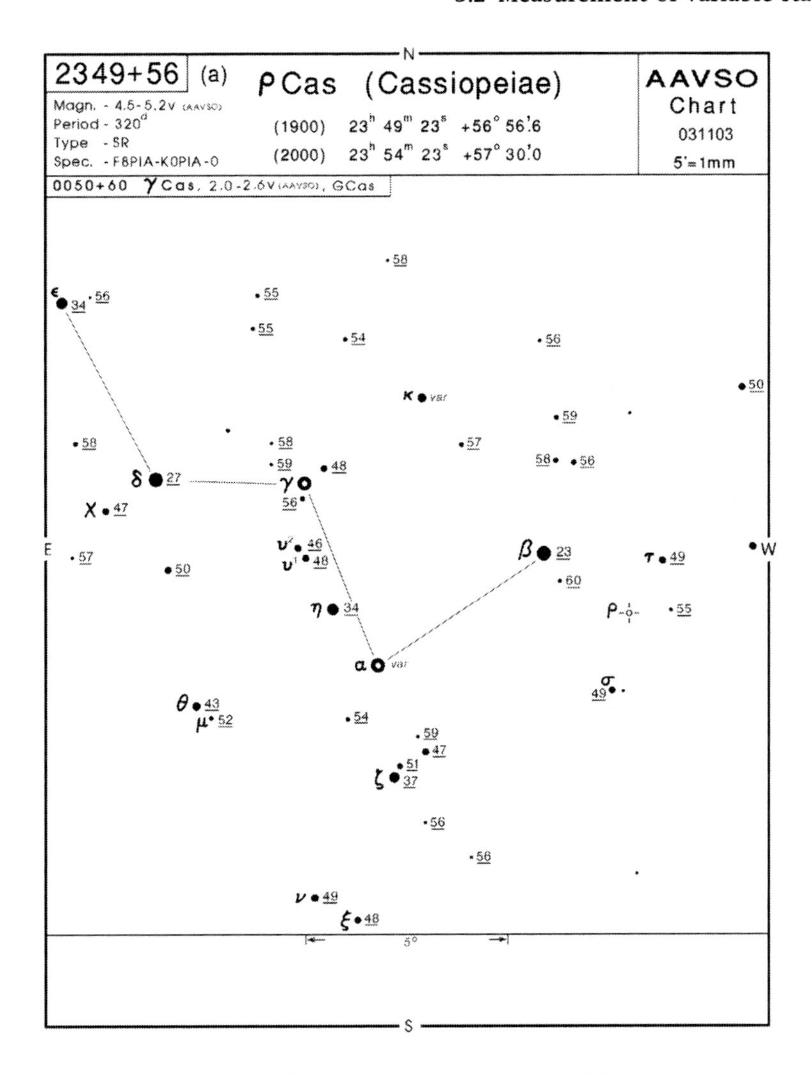

Figure 3.1 A chart for making visual observations of a variable star, ρ Cassiopeiae. The underlined numbers are the magnitudes of non-variable comparison stars, with the decimal points removed, in order not to confuse them with stars. (From AAVSO.)

the rod cells. Different eyes have different colour sensitivity. There are also small effects connected with the orientation of the variable star, relative to the comparison stars in the sky, which occur because of the uneven distribution of the rod cells on the retina. And the magnitudes of the comparison stars must, of course, be known.

The AAVSO is the largest organization to collect and archive visual observations of variable stars; their on-line database is a treasure trove. Remarkably, the demand for their data and services has increased dramatically in the last three decades, since the beginning of the space age. Almost 3000 requests for data were satisfied in 2004–05. The AAVSO publishes a *Manual for Visual Observing of Variable Stars*, which I recommend highly.

For continuity, the AAVSO has used the same charts and comparison stars for decades, and has used the same comparison star magnitudes, even though more accurate photoelectric magnitudes are available. Other variable star observing organizations may have used other charts and magnitudes. As of 2005, the AAVSO is engaged in a massive project to revise their charts and comparison star magnitudes, while making every effort to maintain the long-term continuity of their data.

There are other important variable star observing organizations. The Variable Star Section of the British Astronomical Association has carried out a variety of short-term and long-term studies of variable stars, and published them regularly in the *Journal of the BAA*. The Variable Star Section of the Royal Astronomical Society of New Zealand, led for decades by Frank Bateson until his retirement in 2004, has organized and carried out measurements of southern variable stars over many decades. Frank Bateson is certainly one of the foremost amateur astronomers of the twentieth century.

In the mid-nineteenth century, astronomical photography was developed. Photons strike silver halide in the emulsion, but the effect is latent; it has to be chemically developed and 'fixed'. The exposure of the photographic emulsion by the light from the star was a measure of its magnitude. Again, constant stars of known magnitude had to be available on the same photograph, preferably close to the position of the variable star. The darkening of the emulsion is an approximately linear function of the flux of the star at moderate illuminations, but not at high or low illuminations, so the exposure has to be calibrated with many standard stars.

Around 1900, physicists discovered the *photoelectric effect*: if a beam of photons shone on certain kinds of materials, the photons would liberate electrons from the material; the number of electrons was proportional to the number of photons. This effect could be used to create a current of electrons – *DC photometry* – or, with sufficiently sensitive and complex *photomultiplier tubes*, individual electrons could be counted through *photon-counting photometry*. Many amateur photometrists use photometers with *photo-diodes* – semi-conductors which are similar to solar-power cells. They are simple, but relatively insensitive. Figure 3.2 shows a photometer of this kind. Photomultiplier tubes have a series of diodes. Each one multiplies the electrons which fall on it until there is a measurable pulse of electrons; the pulses are then counted, to determine the number of

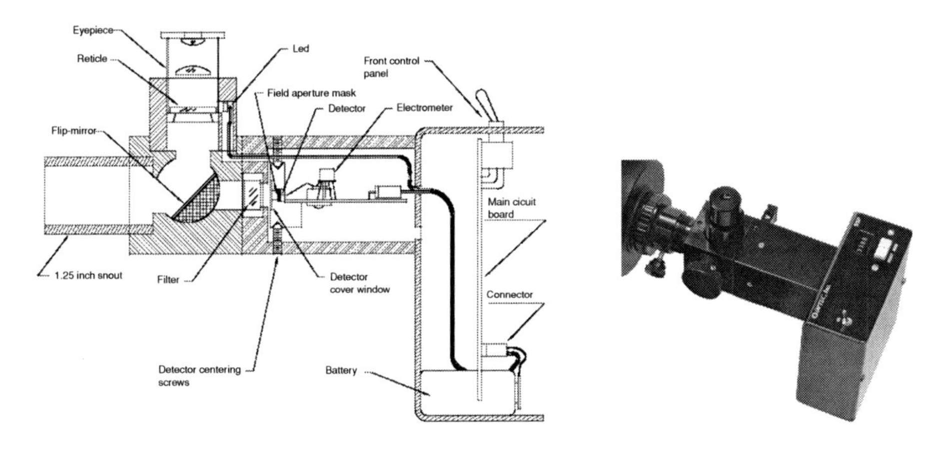

Figure 3.2 A photoelectric photometer. There is an aperture to isolate the star from other stars in the field, a set of filters to measure different wavelength bands, and a photo-diode detector. (From Optec Inc.)

incident photons. The observer uses a chart similar to figure 3.1, except that the magnitudes of the comparison stars are given to ±0.01 magnitude. The observer measures the current of photons from the variable star (including the sky around it), then from an equivalent region of blank sky; the sky current is subtracted from the star+sky current, and the result is converted into a magnitude. The same is done for a non-variable comparison and check star. The magnitude of the variable star is then expressed relative to that of the comparison star; this is called *differential photometry*. The magnitude of the check star is also expressed relative to that of the comparison star; if both are non-variable, the scatter should reflect measurement errors only. Finally, the differential magnitudes must be corrected for the small effect of *differential extinction* or dimming, if the stars are observed at slightly different altitudes above the horizon, and they must be *transformed* to the standard UBV system, to allow for the particular sensitivity of the observer's telescope and photometer system. Differential photometry is somewhat more accurate than absolute photometry – in which the stars being observed are situated all over the sky – because the variable, comparison, and check stars are being observed close in time, and close in position in the sky. The standard textbooks on photoelectric photometry are Hall and Genet (1988) and Henden and Kaitchuck (1982).

A standard photoelectric photometer measures one star at a time. In the 1970s, astronomers developed panoramic photoelectric detectors, or electronic cameras, notably *charge-coupled devices* or *CCDs* (figure 3.3). These consist of an array of typically 1000 x 1000 microscopic light-sensitive pixels. The camera is

Figure 3.3 A CCD chip. The chip has large numbers of microscopic detectors which form the pixels of the image. (From A. Henden.)

exposed to the sky; the electrons produced in each pixel are read out, and the number is stored in a computer. The numbers corresponding to each pixel must be corrected for the particular sensitivity and other characteristics of each pixel. These processes are referred to as subtracting the bias and dark frames, and flat-fielding; the images must also be inspected for defects such as cosmic ray events. Then the total number of electrons corresponding to each star must be added up and corrected in the same way as for single-star photoelectric photometry. CCD photometry is especially powerful if there are hundreds, thousands, or more variable stars on a single image or *frame*. Howell (2006) is an excellent guide to CCD observing, and Henden and Kaitchuck (2006) is forthcoming.

Recently, a powerful new method of CCD photometric analysis called *image subtraction* has been developed by Alard and others (e.g. Alard and Lupton, 1998; Olech *et al.*, 1999), which increases the photometric precision, especially in crowded fields. Alard's ISIS package is available on-line at:
www2.iap.fr/users/alard/package.html

Different kinds of detectors are required for IR and sub-millimetre-wave astronomy, and for the detection of UV, X-ray, and gamma-ray photons, which must be done from space, since these photons do not penetrate the earth's atmosphere.

See Sterken and Jaschek (1996: section 1.5) for a brief but excellent discussion of photometric systems and measurement.

3.2.2 *Spectroscopy*

Spectroscopy is another important tool for understanding variable stars, because information can be obtained from the full spectrum of wavelengths. But much more light is required to gain this information. This means larger telescopes and/or longer time exposures and/or brighter stars. The absorption lines provide a wealth of information about the composition and properties of the stars. Through the Doppler effect, their apparent wavelength provides information about pulsation, about the motion of binary and rotating stars, about the expansion of eruptive variables such as supernovas, and about the presence and properties of faint companion stars – and planets. In a spectrograph, the spectrum was originally produced by a prism, but is now almost always produced by a diffraction grating. The spectrum is imaged on a detector such as a CCD (figure 3.3).

3.2.3 *Polarimetry*

We normally do not think of starlight as being polarized, but it often is, especially if the star has a disc or wind or other non-spherical distribution of matter around it. If the starlight passes through interstellar material, then the dust particles can produce polarization, because they are non-spherical, and are partly aligned by the weak magnetic field of our galaxy. This component of the polarization tells us about the nature of the interstellar matter, but not about the star itself. A very few observatories are engaged in astronomical polarimetry, and this technique can provide useful information about the stars and their immediate surroundings. In a *polarimeter*, a polarizing filter is placed in the light path, to measure the amount and the orientation of the polarization.

See Kitchin (2003) for a comprehensive and excellent discussion of all aspects of astrophysical techniques.

3.3 Discovery and observation

Astronomy, like all sciences, is a human endeavour. Our understanding of variable stars has come about through the efforts of thousands of astronomers down through the ages: not only those outstanding individuals who tend to be named explicitly in books like this one, but also the multitude of others whose accumulated contribution is equally important. How exactly do these 'craftspeople of the stars' use the tools of their trade to turn unstudied stars into certified, well-studied, and well-understood variable stars?

Chapter 1 outlined the trends in the discovery and observation of variable stars, from earliest times to the massive automated surveys of today. At various times, there have been certain variable star 'factories' – observatories which have made a specialty of variable star discovery and research. Historically, these include Hamburg, Harvard, Johannesburg, Leiden, Moscow, Mt. Wilson, Sonneberg, and Yerkes. Some of these have since turned their attention to other fields, to be replaced by other observatories such as Bamberg (Germany) and Konkoly (Hungary). Some of these are still active; Moscow and Konkoly are still centres for study and documentation of variable stars. Others such as Harvard are no longer as active, but they still exert a strong influence through the well-trained apprentices who have gone from there to establish factories of their own. I was strongly influenced by Harvard-trained astronomer Helen Sawyer Hogg. The plate file of the Harvard College Observatory contains over 600 000 plates, and a massive project is currently underway to digitize and thus preserve these.

Until World War II, most astronomical research was carried out at observatories attached to universities or government laboratories. Their staff had regular access to the telescopes, often on every clear night, and were able to embark on projects lasting years or even decades. After World War II, however, several aspects of astronomy changed. Governments became heavily involved in the direct financial support of research. National observatories were established, beginning in 1958 with the Kitt Peak National Observatory in the USA. These had many advantages, of course; they could afford the biggest and best equipment, and were available to all qualified astronomers (and their students), including those at universities without observatories. Previously, some of the best astronomical facilities, such as the 5m telescope on Mount Palomar, were the private preserve of a few fortunate astronomers. The giant Keck telescopes are presently only available to Californian astronomers and their collaborators. Eventually, the concept of the national observatory was widened to include such unique facilities as astronomical satellites. In fact, the largest telescopes and the most sophisticated satellites are so expensive that they must be international projects; the Gemini 8m telescopes, and satellites such as the Hubble Space Telescope are examples. Any eligible, qualified astronomer may apply, but observing time on these facilities is in great demand, and is available only in small amounts on a highly competitive basis. Proposals must be made and submitted months in advance, though the director usually has a small amount of discretionary time available for 'targets of opportunity'. Proposals which are fashionable, or which produce quick, publishable results, are favoured, and long-term projects – which are so important to *all* of astronomy, including variable star astronomy – sometimes fall by the wayside. Variable star astronomy has suffered, to some extent, from this change of style. In North America, stellar astronomy is less in vogue; in Europe and some other places, it is still strong.

Long-term studies of novae, binaries, R Coronae Borealis, and Mira stars can best be carried out with small and medium-sized 'local' telescopes, such as the 1.88m telescope at my university's David Dunlap Observatory. Observing time can be assigned with this in mind. These observatories can react quickly to unexpected events such as novae or to desirable changes in an observing program. They also offer an obvious place to train graduate students, so that they can then make good use of national facilities. In the USA, the closure of small telescopes at the national observatories has to some extent been balanced by the funding of small local telescopes for undergraduate and graduate research. There is an important place for variable star astronomy at *any* small observatory – whether for amateur, student, or professional astronomer. There have been several important conferences on the roles of small, local telescopes, most recently *Small-Telescope Astronomy on Global Scales* (Chen, Lemme, and Paczyński, 2001).

Two very effective programs of long-term photometry of variable stars have both been carried out by Belgian and Dutch astronomers and their collaborators. The *Long-Term Photometry of Variables* project used a Dutch 0.91m telescope at the European Southern Observatory. Various observers co-operated in obtaining long-term photometric observations of a variety of types of variables for which this kind of observation was needed. The *Mercator* telescope is a 1.2m telescope on La Palma, which is also being used for a variety of projects for which long datasets are necessary. This includes observations of complex pulsating stars such as Slowly Pulsating B stars, Gamma Doradus stars, and their relatives.

Around 1990, astronomers began a series of projects to detect 'massive compact halo objects' or MACHOs by gravitational microlensing. This is a phenomenon wherein the brightness of a distant star is amplified by the curvature of space around a MACHO lying between the observer and the distant star. Such a phenomenon would be rare, because the MACHO and the star would have to be almost exactly aligned. Furthermore, the phenomenon is transient because, due to the motions of the three objects involved, the alignment lasts only a few days. The background star appears to brighten, then to fade. If hundreds of thousands of stars could be monitored every night for several years, however, a few microlensing events might be observed.

Astronomers therefore observed dense star fields in the bulge of the Milky Way, and in the Large and Small Magellanic Clouds, and in M31, with large-field CCD cameras. Each night, they processed the data, looking for the telltale signal of a MACHO. They found a few. But they also found thousands of background stars which varied in brightness for other reasons. They were variable stars – a MACHO survey by-product which has enriched and revitalized variable star astronomy. The largest surveys have been MACHO and OGLE (Optical Gravitational Lens Experiment). Figure 6.34 shows an example.

Then around 2000, a new series of surveys began, looking for transits of exoplanets (section 5.11). These were inspired by the discovery of the first exoplanet (51 Pegasi B) and the first exoplanet transit (HD 209458b). These surveys – of which there are about two dozen as of 2005 – also measure hundreds of thousands of stars repeatedly, looking for the small telltale fading caused by the transit. These surveys should also yield data on thousands of variable stars of all kinds. In particular, they provide statistics on what fraction of stars are significantly variable: about a half, when the photometric precision is 1–2 millimag.

There are also surveys such as the *Robotic Optical Transient Search Experiment* (ROTSE) which was designed to detect such things as the transient afterglows of gamma-ray bursts, but provides information about variable stars as a by-product; and *Hipparcos*, which was a satellite primarily designed to measure stellar parallaxes, provided photometry of thousands of variable stars, including newly discovered ones.

Deliberate surveys of variable stars are also being undertaken, thanks to the availability of wide-field CCD cameras with a million pixels or more, and powerful computers which can process the images almost in real time. Surveys of nearby galaxies are especially productive, since all the stars are at approximately the same distance; their apparent brightness reflects their true brightness. Amateur astronomers are deeply involved in some of these surveys, using everything from telephoto lenses to medium-sized telescopes. Two specific examples of such surveys are *The Amateur Sky Survey* (TASS) and the *All-Sky Automated Survey* (ASAS):

ASAS: http://www.astrouw.edu.pl/~gp/asas/

TASS: http://www.tass-survey.org

See Szabados and Kurtz (2000) for a recent conference proceedings on the impact of large-scale surveys on pulsating star research.

The relative numbers of variable stars of different kinds are rather misleading. The actual number in our galaxy will depend on how common that type of star is. The most common stars are cool red dwarfs. Most of these are flare stars, or rotating spotted stars. So these are probably the two most common types of variable.

White dwarfs are also common, and they can be pulsating variables – but only when their surface temperature lies within certain ranges as they cool. So the number of variables will also depend on how long a certain type of star remains in the stage of evolution in which it is variable.

Both red and white dwarfs are common, but they are also intrinsically very faint, so their apparent brightnesses will be very faint, even if they are relatively close. So very few red or white dwarfs have been studied for variability. Yet they are the most common type of variable star among the nearest stars.

And, if the amplitude of their variability is small, discovery will be difficult and therefore unlikely that they will be discovered. On the other hand, thousands of Mira variables, with amplitudes of up to ten magnitudes, have been discovered – even though they are in a relatively rare stage of evolution.

Among the *bright* stars in the sky – those in the *Yale Catalogue of Bright Stars*, for instance – the most common variables are the small-amplitude pulsating red giants, the Be stars (Gamma Cassiopeiae variables), eclipsing variables, supergiants of various temperatures including a few Cepheids, Delta Scuti stars, and rotating Ap and sunlike stars.

Remote/robotic telescopes are an interesting development of the last few decades. Space telescopes, of course, are to some extent remote and robotic, and some of the first developments of ground-based robotic telescopes were motivated by the need to develop space observatories. Amateur astronomers such as Russell Genet developed and used robotic telescopes, and made them available to professional astronomers through schemes such as the *Automatic Photometric Telescope Service*, more recently maintained by Lou Boyd and Michael Seeds. There was, of course, a move by professional astronomers to build telescopes in remote locations. Now, much of the observing at such sites is done in the 'service mode', with the observer not actually present – except perhaps via the Internet. Robotic telescopes proliferated in the 1970s and 1980s, with the development of computer power. Sometimes, their use was motivated by the sheer joy of the technology, but they are ideal tools for both science and education.

The next step in the study of variable stars is to go into space, where photometry can be done with a precision of a few parts in a million. Satellites such as *Hipparcos* and the Hubble Space Telescope have already made important measurements of variable stars. In 2003, the first 'dedicated' variable star satellite – Canada's MOST (Microvariability and Oscillation of STars) – was launched; it will soon be followed by the European satellites *COROT, GAIA*, the successor to *Hipparcos*, planned for launch in 2011 at the earliest, and the NASA satellite *Kepler*, planned for launch in 2007–8. The satellite *WIRE (Wide-Field Infra-Red Explorer)* turned to asteroseismology after its main instrument failed, but its star-tracking camera was able to make the first dedicated photometry of variable stars from space.

This is not to say that the traditional approach to variable stars is the only approach. New and better techniques, often involving new regions of the electromagnetic spectrum, have brought new practitioners into the field. Radio studies of flare stars, infrared studies of novae, ultraviolet studies of Be stars, and X-ray studies of active sunlike stars have all shed 'light' on the visual behaviour of these stars. Physicists have become aware of the fascinating complexity of astronomical objects, and have begun to take up astronomy. The relationship between them and the traditional astronomer is often an uneasy one, as is often the case with individuals who are different but equal, but the relationship is to the eventual benefit of both. Most branches of astronomy have reached the stage where more physical analysis is desirable; that is the reason why astronomers today should and frequently do have a strong background in physics. As an example, the modelling of pulsating and exploding stars requires an intimate knowledge of hydrodynamics, plasma physics, nuclear physics, and radiative processes, as well as numerical analysis. An unfortunate consequence is that, in the USA, astronomers trained in this field were in great demand in thermonuclear weapons research, and could (and did) command high salaries in defense laboratories!

My department was formerly a Department of Astronomy. We considered becoming a Department of Astrophysics. I was personally concerned that this would drive away the thousands of non-science students who take our introductory astronomy courses to satisfy the requirement that they take at least one science course. We are now a Department of Astronomy and Astrophysics.

Variable star research is not carried out in a vacuum. The astronomer's work springs from his/her training and interests, pushed by curiosity, guided by interactions with the literature, facilities, and collaborators. Email and the worldwide web have amplified this situation. There is much to be gained from collaboration and coordination of observing programs. A few carefully chosen stars may be studied using different techniques and different facilities in what is known as a *campaign*. One or more stars may be studied with the same technique from different longitudes, so that 24-hour coverage is possible – a *multi-longitude campaign* (figure 6.21). This is especially useful when the time scale of variation is only a few hours. Astronomers may choose to study a unique event such as a nova, or the eclipse of a very long-period binary. A carefully chosen sample of stars may be agreed upon for a long-term study, so as to get a complete and unbiased picture of their behaviour. Multi-wavelength, multi-technique observations have been especially useful. Not only do campaigns make efficient use of astronomers' limited time and facilities, but they also promote healthy collaboration and friendship between astronomers in different countries and areas of specialization. On the other hand, I know from personal experience that these campaigns are a challenge to organize, from the planning to the publication of the results.

The International Astronomical Union (IAU) does much to facilitate interaction between astronomers. This organization, with over 9100 members in 70 adhering countries, brings together almost all those who are professionally involved in astronomy. The general work of the IAU is carried out by its Executive Committee; its specific work is carried out by 40 interest groups or Commissions, organized into 12 divisions, each with their own executive. Some commissions deal with general matters such as nomenclature and documentation, education and development; others deal with specific areas of astronomy. One commission publishes the *IAU Circulars*, which electronically announce such discoveries as supernovae. Commission 27 deals with variable stars; Commission 42 deals with close binaries. These two commissions, which make up the Division on Variable Stars, advise the editors of the *General Catalogue of Variable Stars* (GCVS), sponsor the useful *IAU Information Bulletin on Variable Stars* (IBVS),

prepare triennial reports on progress in the field, and organize conferences on both general and specific aspects of variable stars. The *Triennial Reports* were once quite extensive, and were major reviews of the field; now, they are somewhat briefer. Most of all, the Commissions bring together, in their membership, the hundreds of astronomers from all over the world who share an interest in understanding variable stars better. Unfortunately there is not an equivalent to the IAU for amateur astronomers, though the AAVSO has fulfilled some of the role through its international meetings, its international database, and website full of tools, data, and information. See the websites of IAU Commissions 27 and 42:
http://www.konkoly.hu/IAUC27/
http://www.konkoly.hu/IAUC42/index.html

There are many meetings, or series of meetings on variable stars, many of them sponsored by the IAU. For instance, the IAU holds a General Assembly every three years, and Commissions 27 and 42 hold business and scientific sessions. There are also several-day symposia and colloquia on specific aspects of variable stars. Those of us working on pulsating variable stars, for instance, organize a meeting every two years, to present the latest research results, and to develop new projects for the future. The proceedings of those meetings (e.g. Kurtz and Pollard, 2004) are a valuable source of current information and research. André Heck has edited a number of interesting (albeit expensive) books on issues of information and organization in astronomy.

3.4 Analysis of variable star data

Variable star measurements consist of values of magnitude, colour, radial velocity, or some other property, along with the Julian Date – the time at which they were made. We shall assume that the Julian Date has been corrected to the sun, or to the centre of mass of the solar system if necessary.

We begin by asking an important question: is the variable star which we are analyzing actually variable? If its range of variation is much larger than the observational error, then there may be no doubt. If the variability is based on a small number of measurements which differ from the others, could they be due to misidentification of the variable, or to clouds? In sky surveys, variables are identified by comparing their scatter (as measured by the standard deviation of their measurements, for instance) to the scatter of the survey measurements as a whole, using a statistical criterion. But there will always be a few low-amplitude variables lost in the 'noise', and a few 'noisy' constant stars which are misidentified as variables.

Another thing to remember is that photometric measurements are normally expressed as magnitudes, and these are a *logarithmic* function of the physical quantity, flux. For instance:

- The output of theoretical models of variable stars is normally given as flux.
- In calculating the average observed magnitude or colour, the light or colour curve should be flux-averaged, especially if the amplitude is large.
- The adding of two sinusoidal components should be done as fluxes; likewise, Fourier analysis of light curves should ideally be done on flux curves, if valid amplitudes are to be obtained.
- Some variable stars (such as Mira) have an unresolved companion of constant brightness; the decomposition of the combined light curve should be done as flux, to allow for the constant contribution of flux from the companion.
- A linearly declining light curve represents an exponential decrease in flux; this is seen in the declining phases of the light curves of supernovae which are being powered by exponential radioactive decay.
- Imagine the *flux* curve for a Mira star with a sinusoidal light curve with a total range of ten magnitudes – a factor of 10 000 in flux!

For most purposes, however, it is sufficient (and much easier) to work with the magnitude curve.

3.4.1 *Light curves*

The simplest form of variable star analysis is inspection of the shape of the light curve (figure 3.4), and the time and magnitude of maximum and minimum. The term *range* is usually used to denote the difference between maximum and minimum. The term *full amplitude* can also be used, but *amplitude* by itself is often used to denote the half-range, as in the coefficient of a sine or cosine function. It is best not to be ambiguous.

3.4.2 *Times and magnitudes of maximum and minimum*

Assuming that there are sufficient measurements around the time of maximum (or minimum), the historical method for determining the time and magnitude of maximum is Pogson's method of bisected chords: (i) draw a smooth curve through the points on the light curve going up to and down from maximum; (ii) draw a series of horizontal lines at equal increments below maximum, going from the ascending branch to the descending branch; (iii) bisect each line, between the ascending and descending branch; (iv) draw a smooth curve through

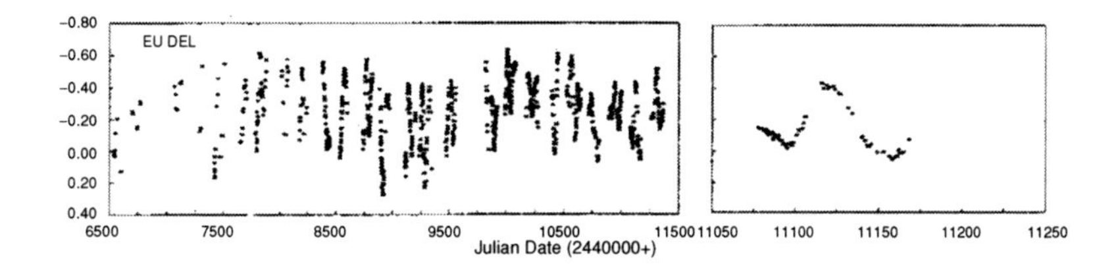

Figure 3.4 The light curve of the pulsating red giant EU Delphini on a 5000-day scale (left) showing the slow variability of the star, and on a 200-day scale (right) showing the 63-day period of the star. Light curves, on time scales of hours to decades, can provide a visual impression of the star's variability on a range of time scales. (From Percy, Wilson, and Henry, 2001).

the midpoints, and extend it upward to intersect the smooth curve through the points at maximum – the intersection point gives the time and magnitude of maximum.

Nowadays, with the availability of computers, it is more customary to fit a low-order polynomial through the points around maximum, and to determine the time and maximum value of that polynomial. But the human eye and brain are still a powerful tool for making sure that this has been done sensibly, so make sure that the polynomial actually appears to fit the observed points!

3.4.3 *Periods, and period determination*

If it is known or suspected that the variability of the star is periodic, then it is possible to determine the period. Many types of eclipsing, rotating, and pulsating stars are periodic, but some of them tend to be semiregular, and others tend to be multiperiodic. We will assume, for the time being, that the variability is singly and perfectly periodic.

If the period of a star is known, and constant, it is possible to define a quantity called *phase* – the fraction of the star's variability cycle which has elapsed. The phase is the decimal portion of

$$(t - t_0)/P \qquad (3.2)$$

where t is the time of the measurement of the star, t_0 is an arbitrary *epoch* – usually a time of maximum or minimum brightness, such as the time of the first observed maximum – and P is the *period* of the star. The integral part of this quantity is the *cycle number* N – the number of cycles elapsed between t_0 and t.

If t_i and m_i are the times and magnitudes observed, and ϕ_i is the corresponding phase, then a graph of m_i versus ϕ_i is called a *phase diagram*. Phase goes from

0.0 to 1.0, but usually the phase diagram is plotted from 0.0 to 2.0, or from −0.5 to +1.5 to show the full behaviour of the star during its cycle.

The phase diagram shows the average shape of the light curve. It is possible to quantify the shape through *Fourier decomposition*, a technique developed and applied by Norman Simon and his collaborators (e.g. Simon and Schmidt, 1976). The phase curve is represented by a series of sine functions with periods of P, $P/2$, $P/3$, $P/4$ etc. The amplitudes of the sine functions, and the phases of the harmonics relative to that of the fundamental, are determined. These can then be compared with the same parameters of theoretical light curves, determined from non-linear models. Siobahn Morgan has created a World-Wide Web database of variable star Fourier coefficients at:
http://nitro9.earth.uni.edu/fourier/

The phase diagram provides one tool for period determination: if the star is singly periodic, then the correct period should produce a phase diagram in which the scatter is due only to observational error. With the advent of computers, several methods were developed which automatically tested trial periods. These more sophisticated methods of studying the periodicity of variable phenomena are called *time series analysis*. See Templeton (2004) for a good introduction to the topic. Periods are tested between some pre-determined lower and upper limits, which are set by the length and spacing of the data, and by the period which is expected in the star. The longest period tested should be several times shorter than the length of the dataset; the shortest period tested should be significantly longer than the typical separation of the measurements. The spacing ΔP between the trial periods is optimized: if it is too large, the correct period might be missed; if it is too small, the phase diagram will not change significantly from one trial period to the next.

One class of methods tested the phase diagram by calculating the length of the 'string' which would connect adjacent points on the phase diagram. The shortest string length should correspond to the 'best' period. These are called *string length methods*; Dworetsky (1983) describes an elegant version, and gives a review of earlier methods of this kind.

Another class are *bin methods*. In these cases, the phase diagram is divided into intervals or 'bins' of phase, the average scatter within bins is determined, and the lowest average scatter corresponds to the 'best' period. The method of Stellingwerf (1978) is the best known of this class. These methods are part of a more general class of methods known as ANOVA or analysis-of-variance methods.

Yet another large class of methods are *Fourier methods*. In these methods, the dataset is multiplied by a sine curve corresponding to the trial period. If there is a signal of this period in the data, then the product of the data and the

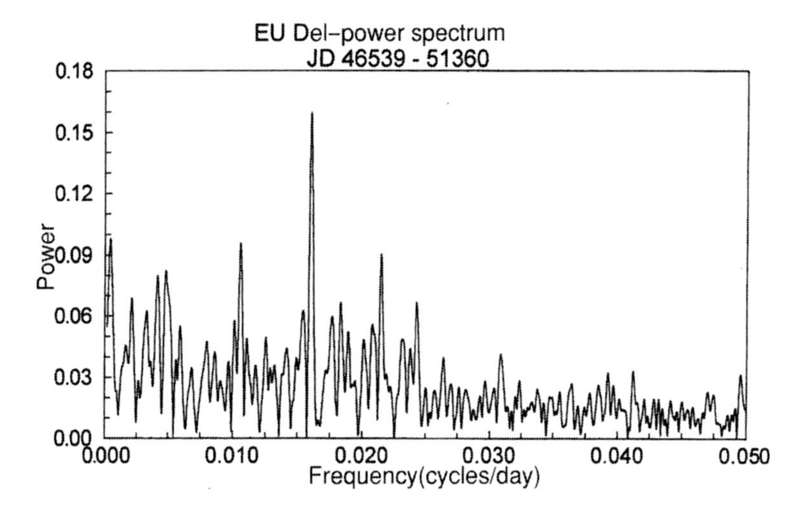

Figure 3.5 The power spectrum of the measurements of EU Delphini. The highest peak is at the 63-day pulsation period. The lower peaks are either alias periods (those separated from the highest peak by a multiple of 0.00274 cycles/day) due to the periodic seasonal gaps in the data, or possibly additional low-amplitude pulsation modes, or noise. (From Percy, Wilson, and Henry, 2001.)

sine curve will be non-zero. Otherwise, the product will be approximately zero, because the data will be randomly multiplied by positive and negative parts of the sine curve. A graph of the product *vs* trial period is called a *periodogram* or, if the square of the amplitude of the signal is used, a *power spectrum*. A commonly used version of this method is the *Lomb–Scarge* periodogram (Scargle, 1982); *TS* which contains a hybrid Fourier-CLEAN method known as CLEANest (Foster, 1995) is available on the website of the AAVSO. The Lomb–Scarge algorithm provides a measure of the statistical significance of any peak. The CLEAN algorithm can, under favourable circumstances, remove alias peaks from the power spectrum.

Figure 3.5 shows the power spectrum of EU Delphini. There is a 63-day peak due to the main pulsation period, along with alias peaks which are due to the seasonal gaps in the data. There are low-frequency peaks which may be due to the slow variations in the star.

If there are regularly spaced gaps in the data, then a problem may develop with these methods: *alias periods*. (These should really be called *pseudo-alias periods*; alias periods arise because the measurements themselves are made at equal intervals, and lead to alias periods if the actual periods are of the same order as the interval between the measurements.) These are periods which may fit the data, but which are actually spurious. The alias periods P_{alias} are related to the true period P_{true} and the regular spacing of the data T by

$$1/P_{alias} = N/P_{true} \pm 1/T \tag{3.3}$$

where N is an integer. Common spacings may be one day, if the observations are made from one site only, and/or one year, since most stars are not observable at certain seasons of the year. As a result, the power spectrum may display not one but many peaks. The peaks corresponding to the alias periods are equally spaced in frequency on either side of the true peak; the spacing of the peaks is the spacing of the gaps in the data, expressed as a frequency. For instance, in figure 3.5, there are alias peaks which are separated from the main peak by multiples of 0.00274 cycles/day – one cycle per year.

There may also be peaks in the spectrum which are purely due to the 'noise' in the data. How can one tell which peaks, if any, are real, that is statistically significant? Two elegant discussions of this issue are by Scargle (1982) and by Horne and Baliunas (1986), who present formulas for the *false alarm probability* for any peak.

A rather different technique is *auto-correlation*. It does not assume that the light curve is strictly periodic. Rather, it asks how the light curve compares with itself at times separated by some interval Δt. This technique, developed extensively by Box and Jenkins, is widely used in many fields of science and social science.

There is a simple version of this, which, in some circumstances, can be very useful – *self-correlation*, e.g. Percy, Ralli, and Sen (1993). Among other things, it can often avoid the problem of alias periods. This method shows the cycle-to-cycle behaviour of the star, averaged over the dataset. For all pairs of measurements, the difference in magnitude Δmag and the difference in time Δt are calculated. Δmag is plotted against Δt from zero up to some appropriate upper limit, which depends on the time scales to be expected in the star and the total length of the dataset. The $\Delta mags$ are binned in equal bins of Δt, so that, if possible, there are at least a few values in each bin. The average Δmag in each bin is plotted against Δt in a *self-correlation diagram* (figure 3.6). The average Δmag will be a minimum at multiples of the period or time scale present, if any. The level of the minima will depend on the magnitude measurement error, and on the irregularity of the variations. The difference in level between the maxima and the minima is approximately equal to the half-range of the periodicity. Figure 3.6 shows the self-correlation diagram for EU Delphini. There are minima at multiples of the 63-day period.

Another widely used class of methods are *least-squares methods*, which find the best-fitting function – the sum of a set of sine curves – which best fit the data.

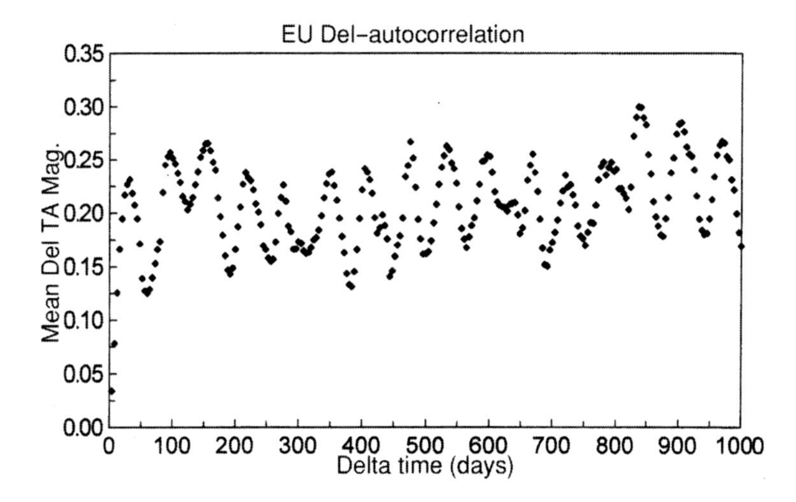

Figure 3.6 The self-correlation diagram of the measurements of EU Delphini. Minima occur at multiples of the 63-day period. The fact that they persist to large values of Δt means that the star is highly periodic. The height of the minima above the baseline is normally a measure of the observational error; in this star, it is much higher (about 0.15) than the measurement error of about 0.03, because of the irregularity of the star. (From Percy, Wilson, and Henry, 2001.)

The user specifies the number of separate periods to fit; the method gives the periods, amplitudes, and phases of the sine curves which minimize the sum of the squares of the differences between the observed magnitudes and the magnitudes predicted by the function at the time of observation. This methodology has been extensively developed for variable star astronomy by Michel Breger, and his group at the University of Vienna. They have developed publicly available software, the latest version of which is *Period04*; see:
http://www.astro.univie.ac.at/tops/Period04/index.html

These methods assume that the period(s) and amplitudes are constant through the dataset. *Wavelet analysis* is a powerful method which can track the changing period and amplitude through the dataset. Wavelet software is publicly available on the website of the AAVSO (Foster, 1996); a new version *WinWWZ* was made available in 2005. It can be used, for instance, to study variables whose periods change, or come and go, or vary in amplitude. AAVSO wavelet software, and other useful time-series analysis and other software, is available at:
http://www.aavso.org/data/software/

Figure 3.7 shows the wavelet plot for the Mira-like variable R Doradus. There are two periods, corresponding to two pulsation modes, which alternate. The wavelet plot deals effectively with this complication.

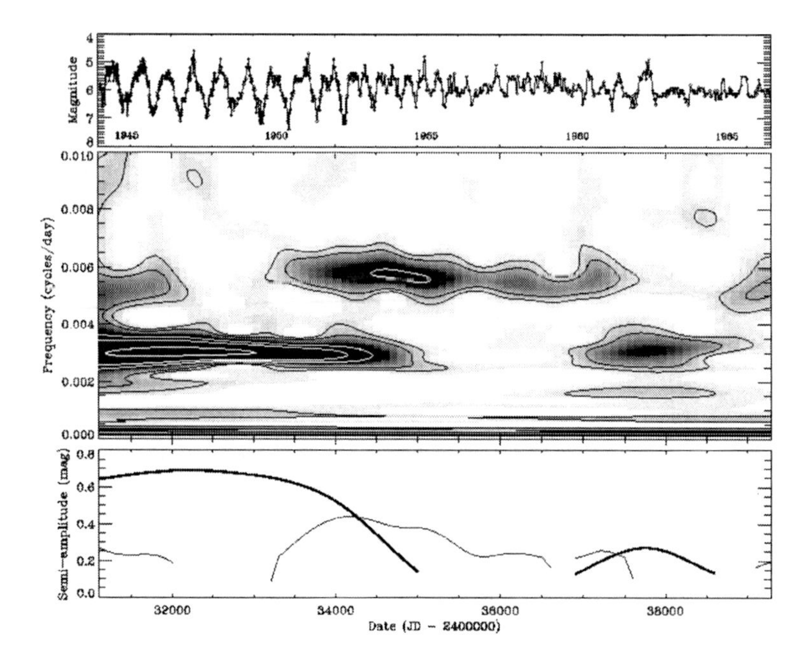

Figure 3.7 Wavelet plot for the variable star R Doradus. Top: the 20-year light curve, based on visual observations by Albert Jones. Middle: the power as a function of frequency and date. Bottom: the amplitudes of the 332-day and 175-day periods, as a function of date. The two periods, representing two different radial pulsation modes, appear and disappear on a time scale of years. (From Bedding *et al.*, 1988.)

Finally, there is the problem of variable stars which are not strictly periodic. Some are multiperiodic; the individual periods can be determined by Fourier or least-squares analysis, as long as there is a sufficient number and timespan of observations. Others are semiregular; there is some form of irregularity superimposed in the periodicity. Some types of variable stars, mostly pulsating variable stars, exhibit *chaos*, a form of irregularity which can arise from a simple dynamical system behaving in a non-linear way (see, for example, Buchler *et al.*, 2004).

3.4.4 *Period changes in stars*

Many variable stars are periodic: their light curve repeats regularly, with a well-defined period. The period of an eclipsing variable depends on the masses of the components, and on their average separation. The period of a rotating variable is the rotation period of the star. The period of a pulsating variable depends on the radius, mass, and structure of the star, and the mode in which it is pulsating. Normally, these properties of stars change very slowly over time, so the period changes very slowly also. But the changes in period produce a *cumulative* effect, which, over a period of time, can be measured for some types of variable stars. This can provide important information about evolution and other processes in stars.

Consider the following analogy. Suppose you have three clocks. Clock A is a perfect clock: it always ticks 86 400 seconds in a day. Clock B keeps the same period at all times, but it ticks only 86 399 seconds in a day. Clock C is slowly running down; it ticks 86 400 seconds in the first day, 86 399 in the second, 86 398 in the third, 86 397 in the fourth, and so on. If you compare clock B with the perfect clock A, you will notice an error of 1 second the first day, 2 seconds the second day, 3 seconds the third day, and so on. The cumulative error is a *linear* function of time. If you compare clock C with the perfect clock A, you will notice no error the first day, 1 second the second day, $1 + 2 = 3$ seconds the third day, $3 + 3 = 6$ seconds the fourth day, $6 + 4 = 10$ seconds the fifth day, and so on. The error of clock C which is slowing down accumulates as the *square* of the elapsed time. After a week or so, its error is quite obvious.

Astronomers measure period changes in stars by comparing the observed time of maximum or minimum brightness O with the calculated time C according to an ephemeris which assumes a known, constant period: $t_{max} = t_{max,0} + P.N$, where N is the number of cycles elapsed between $t_{max,0}$ and t_{max}. The quantity $(O - C)$ is plotted against time t or cycle number N. The so-called $(O - C)$ *diagram* may take one of several forms:

- If the period of the star is constant, and equal to the assumed period P, then the diagram is a horizontal straight line.
- If the period of the star is constant, but is slightly greater than or less than the assumed period P, then the diagram is a straight line, with a slope which is equal to the error ΔP in period.
- If the period of the star is constant, abruptly changes to another constant period, then the diagram is a broken straight line; the change in slope is the change in period.
- If the period of the star is changing at a uniform rate, then the diagram is a parabola, opening upward if the period is increasing, and opening downward if the period is decreasing. The coefficient of the t^2 term is $(\beta/2P)$, where β is the rate of period change in days per day.

Figure 3.8 shows the $(O - C)$ diagrams for two RR Lyrae stars, one showing evidence for a uniformly changing period, and the other showing evidence for an abruptly changing period. Several other $(O - C)$ diagrams are shown later in this book, figure 6.12 and figure 6.31 for instance. Both of these show uniformly changing periods, which provide evidence of the existence and rate of evolution of the star.

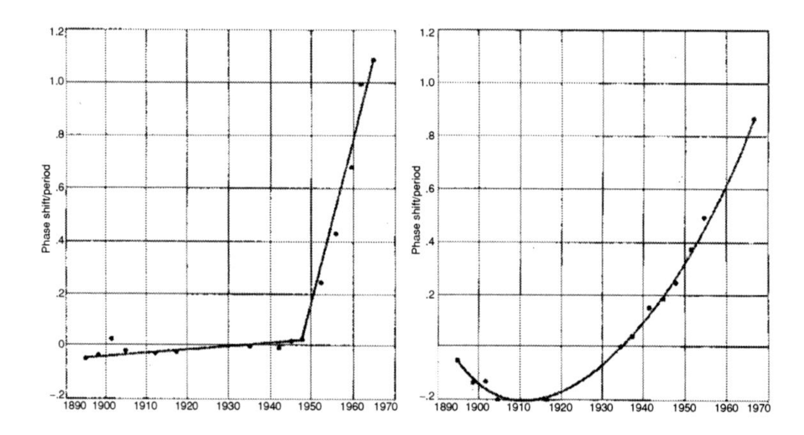

Figure 3.8 $(O - C)$ diagrams for two RR Lyrae stars in the globular cluster M5. The period of Variable 70 (right) is lengthening uniformly; the $(O - C)$ diagram is parabolic. The period of variable 19 (left) changed abruptly in 1947. (From Percy, 1975, based on the work of H. S. Hogg and C. M. Clement.)

It is important to note whether $(O - C)$ is being plotted in days or in cycles, and whether it is being plotted against days or cycles, see Belserene (1986, 1989) for a discussion of the $(O - C)$ method, and for useful algorithms. The most recent and comprehensive discussion of the $(O - C)$ method is the edited book by Sterken (2005), especially Sterken's introductory review.

In certain pulsating variables such as Mira and RV Tauri stars, there are random cycle-to-cycle period fluctuations which tend to produce irregular wave-like patterns in the $(O - C)$ diagram. Eddington and Plakidis (1929) developed a formalism for identifying this effect and measuring the size of the fluctuations. This effect makes it very difficult to measure evolutionary period changes in these stars. No systematic study of this phenomenon has been carried out, but it appears most conspicuous in stars with a large radius/mass ratio. One example is shown in figure 3.9.

If the variable star is a member of a binary system with a large orbit and a long period, another process can affect the $(O - C)$ diagram – light-time effects. When the variable star is on the far side of its orbit, it takes extra time for its light to reach us, and the observed maximum or minimum appears late. When the variable star is on the near side of its orbit, it takes less time for the light to reach us, and the observed maximum or minimum appears early. This produces a periodic cyclic pattern in the $(O - C)$ diagram, with a period equal to the binary period. Normally, the binary orbit must be very large, with a period of years or decades, for this effect to be observable, so it requires careful, long-term monitoring. A notable example is the high-amplitude Delta Scuti star SZ Lyncis. Its pulsation period is 0.12053 day. The $(O - C)$ diagram has a periodic component with a period of 1178 days and an amplitude of 0.006 day, which is

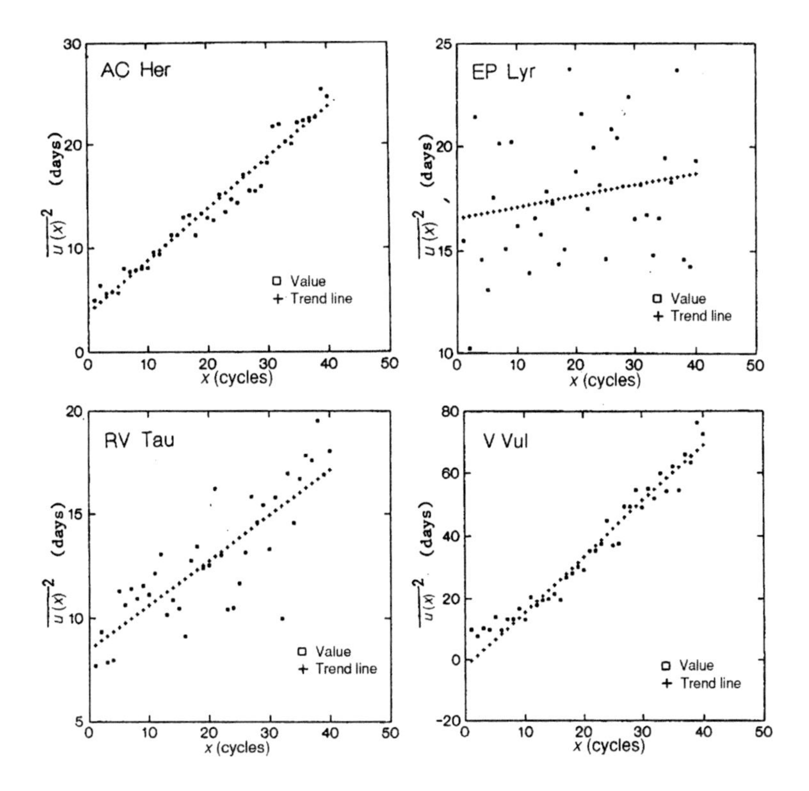

Figure 3.9 Graph showing the Eddington–Plakidis analysis for random cycle-to-cycle fluctuations in period, for four RV Tauri stars. If there are such fluctuations, the graph shows a linear trend with a finite positive slope, as it does in three of the stars. The situation with EP Lyrae is less clear. (From Percy *et al.*, 1997.)

due to a binary companion whose mass is similar to the sun's. Radial velocity observations confirm this interpretation.

If there are two very close periods present, with similar amplitudes, the interference between the periods will produce positive or negative $(O - C)$s, with a 'phase shift' of half a cycle when the periods cancel each other out.

Because of these processes, and other unknown ones (e.g. Sterken, 1993), the interpretation of $(O - C)$ diagrams is something of a 'black art'. Waves in the $(O - C)$ diagram which are due to random effects may be interpreted as light-time effects, especially if only one or two cycles have been observed. Or they may be interpreted as a result of several discrete changes in period.

3.5 Classification

An ideal classification system for variable stars uses only directly observed quantities, distinguishes between physically different systems, and groups together similar objects. To begin with, however, it is not often obvious which observable quantities will accomplish the other goals. We need to understand the objects to be sure that the classification is good; but, usually, classification is the first step required before we can understand what we are looking at. The constant tension between a clean, purely observational classification and a physically meaningful one is behind most of the apparent confusion in the classification of variable stars.

The light curve provides an obvious starting point, because it is easily and directly observed. Is it strictly periodic, strictly irregular, or something in between? What is the characteristic period or time scale? Its amplitude? Its detailed shape? Each of these parameters can be represented by a number to produce a classification scheme with potentially infinite detail. Consider the eruptive variables, for instance those characterized by rapid increases in brightness, followed by slower decay. The time scale, amplitude, and shape of light curve can distinguish between supernovae, novae, dwarf novae, and flare stars, and can subclassify these almost *ad infinitum*.

The colour and velocity curves provide evidence of temperature changes or motion, such as might be produced by various physical mechanisms including pulsation, eruption, rotation, or orbital motion. Thus they also aid in classification, both by providing more numbers, and by providing clues to the physical nature of the variability.

Beginning in the late nineteenth century, it became possible, in some cases, to make a *physical* classification of variable (and non-variable) stars. Astronomical spectroscopy enabled astronomers to measure the temperature of a star, and later the luminosity as well. This led to one of the most fundamental concepts of astronomy: the two-dimensional classification of stars according to temperature and luminosity, and its graphical representation on the Hertzsprung–Russell diagram (H–R diagram: section 2.17).

In its original form, the H–R diagram was a plot of absolute magnitude (usually M_v) vs. spectral type. Since these quantities can be determined for many variable stars, such stars can be plotted on the H–R diagram. When this is done, it is found that, for many types, there is a correlation between the properties of their light curve and their position on the H–R diagram. Thus, different 'light curve' classes of variables inhabit different regions of the H–R diagram. Since M_v and spectral type are related to luminosity and temperature, this means that variability properties are correlated with – and perhaps even determined by – the luminosity and temperature.

For some variables, however, properties other than these may be important: age, mass, chemical composition, and internal structure, for instance. It is possible for different types of variables to have the same position on the H–R diagram. For instance, red giants of a variety of masses have about the same luminosity.

And it is difficult to distinguish Population I and II Cepheids on the basis of light curves and position on the H–R diagram alone. Fortunately, these additional properties can often be lumped together under the concept of *stellar populations*, a term invented by Walter Baade. Stars of greater age tend, on the average, to have lower masses, lower abundances of the elements heavier than helium, lesser concentration to the plane of the Milky Way, and higher velocities relative to the average stars in the solar neighbourhood. Of these properties, abundances and high or low relative velocity can be used to classify most individual stars as to their population type, and the *mean* velocity and concentration towards the galactic plane can be established for *classes* of variables which contain sufficient members. Together, the light curve, temperature, luminosity, and population type are sufficient to form the basis for the 'official' classification system for variable stars, as used in the fourth edition of the GCVS. A discussion of this system can be found at CDS:
http://vizier.u-strasbg.fr/viz-bin/getCatFile?II/214A/./vartype.txt

Although this system is official, it is far from perfect. Since it is often based on quantitative properties of the stars, it invites sub-classification, and astronomers seldom resist the temptation to seek homogeneity by devising smaller and smaller groups of stars, often with uncertain affiliations. One tends to 'lose sight of the forest and see only the trees'. Often, classes or groups may be named after a prototype which later proves to be unique or atypical – β Lyrae, for instance. There may be groups with subtle distinctions whose names become interchangeable – Mira stars and long-period variables, for instance. There may be groups with unrelated names but great similarities – Delta Scuti stars and dwarf Cepheids, for instance. Also, since there are many variables for which we have incomplete information, a too-detailed classification scheme makes misclassification more likely.

Another complication is that, over the last two or three decades, astronomers have become aware that *binarity* has a profound effect on stellar variability – not just eclipsing variables, but cataclysmic variables, and even, to some extent, rotating variables. So we must also classify the star *system*.

Classification systems reflect our understanding and interpretation of variable stars, and can therefore affect it. A cumbersome classification system is more dangerous than a cumbersome nomenclature system.

As our understanding of the mechanism of variability has grown, it has become increasingly popular to try to classify variable stars according to the physical mechanisms of variability. The appeal of this scheme is that it is logical and fundamental, and helps to focus theoretical efforts on similar physical situations. Not the least of its advantages is the fact that some stars may show several types of variability due to different causes. In RS Canum Venaticorum and BY Draconis stars, for instance, eclipses, rotational, and flare-type variability may occur in the same star. The sun is a pulsating, flaring, rotating variable – and an eclipsing variable during a transit of Mercury or Venus! Conversely, similar physical processes may occur in apparently dissimilar types of variables. To some extent, this book adopts the 'mechanisms' approach. However, there are two serious disadvantages of this approach to classification: (1) We cannot immediately classify a new variable according to this scheme, and (2) we may have to reclassify some variables as we revise our ideas regarding their nature.

3.6 Certification

Once a variable star has been discovered, it is considered for official 'certification'. Prior to World War II, this was done by the Commission on Variable Stars of the Astronomische Gesellschaft in Germany. From then until 1952, it was done by a subcommission of the International Astronomical Union (IAU). Since 1952, it has been done by the compilers of the *General Catalogue of Variable Stars* (GCVS) in the USSR and now in Russia.

The criteria for certification are quite strict in principle: there must be definite evidence of variability; the brightness must be determined reliably; an accurate position must be given; and the type of variability must be known. In practice, certification is perhaps done too casually, especially for small-amplitude variables. This is not entirely the fault of the certifiers. They must base their certification on the information and interpretation given in the discovery paper, and these are not always accurate. As a general rule, the variability of *any* star should be confirmed by an independent observer, especially when the amplitude is small. There are far too many variables whose properties are poorly known, and far too few which could be described as 'well studied'. Unfortunately, not much credit is given for confirming someone else's discovery – even though this is an essential part of the scientific process.

3.7 Nomenclature

After a variable star is certified, it is given an official name. Name lists of new variable stars are included at regular intervals in the *Information Bulletin on Variable Stars*, published by the Konkoly Observatory in Hungary, on behalf of the IAU Division of Variable Stars. The name lists give alternate designations of the stars, and references to the discovery papers.

The standard system of variable star nomenclature was developed by Argelander, and is as follows; see Hoffleit (1987) for a brief history. Stars which have a Greek-letter name (such as δ Cephei) or a small-Roman-letter name (such as g Herculis) keep that name, and are listed at the *end* of the list of variable stars in their constellation in the GCVS. All other stars are designated by one or two capital letters, followed by the genitive case of the constellation's Latin name (such as Cephei or Cygni). The letters are assigned in the order in which the variables are *discovered*, starting with R, S, ... Z, RR, RS ... RZ, SS, ST ... SZ through to ZZ, then AA ... AZ, through to QZ, omitting J. (The letters A through Q had already been used in the naming of non-variable stars in the newly charted southern constellations.) This system provides 334 designations per constellation; subsequent variables are called V335, V336 etc., followed by the genitive form of the constellation name. In order to provide for the unambiguous assignment of every star to a constellation, the definition of the constellation boundaries was fixed by a committee of the IAU in 1935 in 'Delimitation Scientifiques des Constellations'; they are line segments of constant right ascension or declination. Precession of the earth's rotation axis, which gradually changes the right ascension and declination of a star, occasionally leads to a star's moving from one constellation to another.

There are many other designations which can be assigned to a variable star, both before and after certification. Variables discovered in large-scale surveys are often designated according to the observatory at which they were discovered. HV, for instance, means 'Harvard variable', and BV means 'Bamberg variable'. The star may already be listed in the *Henry Draper Catalogue* (HD) or *Bright Star Catalogue* (BS); stars in the latter catalogue are usually referred to by their HR (Harvard Revised) number. The variable may be designated by its position at a given epoch, either accurately or in a shorthand notation. For example, the Harvard system of designations, used by the AAVSO in its database, is based on the right ascension and declination of the star for the epoch 1900. A finder chart for the variable can be provided; this is especially useful for faint stars in crowded fields such as clusters, the Milky Way, and other galaxies. Novae initially may be given a special designation, before being given their variable star name, consisting of the constellation and year of discovery – Nova Delphini 1967 (HR Delphini), for instance. There is a special system of nomenclature for supernovae.

The system of variable star nomenclature is cumbersome, but it is well-established, and has no negative influence on our understanding of variable stars, since no assumptions about the nature of the stars are built into it. On the other hand, it *can* lead to confusion if the reader is unfamiliar with the system. It can also lead to confusion if a typographical error or other ambiguity occurs. For instance, u Herculis and U Herculis are two different stars, as are mu Centauri and MU Centauri. Every year, a few 'misidentifications' of variable stars are discovered and published. In order to avoid these and similar problems, the IAU strongly recommends that at least two different designations (e.g. variable star name and HD number) be given for any variable star which is referred to in a publication.

The system of variable star nomenclature is now facing a crisis, as thousands of new variables are being discovered each year in large-scale sky surveys, such as those for gravitational microlenses, and exoplanets. Eventually (in my opinion), all star names will have to be based on a precise position at a specific epoch – language which computers can easily understand.

3.8 Bibliography

Once a variable has been certified and named, it is included in the next edition or supplement to the *General Catalogue of Variable Stars* – the 'bible' of variable star astronomy. The GCVS is prepared by the Astronomical Council of the Russian Academy of Sciences, at the Sternberg Astronomical Institute of the Moscow State University, under the auspices of the IAU Division of Variable Stars. The fourth edition of the GCVS was published in 1985. It is not clear whether further formal editions will or should be printed. The current version of the GCVS (GCVS4, Kholopov *et al.*, 1998), or the more recent *Combined General Catalogue of Variable Stars* (Samus *et al.*, 2004), can be accessed and searched on-line through CDS (see below). Catalogues of specific types of variable stars can be generated from the on-line GCVS, though it should be realized that the classifications in the GCVS are not necessarily complete, especially for stars with complex variability of multiple types. The search capability is quite sophisticated: it can generate lists, based on criteria such as brightness, type, and period.

The GCVS contains extensive explanatory information and remarks (in Russian and English), as well as the name, position, classification, maximum and minimum magnitude, time of maximum or minimum brightness, period, spectral type, and references to the discovery and finder chart. It should be realized that this information may be incomplete or incorrect, especially if it was incomplete in the discovery paper.

Stars which were not considered by the editors of the GCVS to be definitely variable may be included in the *Catalogue of Suspected Variable Stars* (CSV) or the *New Catalogue of Suspected Variable Stars* (NSV), the latter containing 14 810 objects. This listing is highly tentative, and at least one astronomer irreverently refers to it as 'the suspect catalogue of variable stars'. Stars in this listing are good candidates for confirmation or rejection by other observers. Unfortunately, there is more glory in discovering new variable stars than in confirming or rejecting them. Thus, the list of suspected variables gradually increases in length.

Early editions of the GCVS did not include most variables in globular clusters, the Magellanic Clouds, and other external galaxies. An extremely useful *Catalogue of Variable Stars in Globular Clusters* was compiled by Helen Sawyer Hogg, and a new version is maintained on-line by Christine Clement (see Resources).

The publications mentioned above provide basic information on certified variables. For the majority of variables, this is the only information known. For the rest, it at least provides a starting point for a literature search. For the serious investigator, a problem arises for the so-called 'well-studied' variables, for which there may be thousands of observations and hundreds of results scattered through catalogues and research papers. The astronomer *must* deal with this information; the alternative is duplicated effort, and sometimes embarassment. Astronomical research papers proliferate like rabbits, and in a diversity of journals, conference proceedings, observatory publications, and web sites, and – especially in the early years of astronomy – in a diversity of languages. Proliferation is not necessarily a bad thing; it reflects a healthy growth of our science.

Fortunately, automated computer search procedures have been developed, and these are becoming more powerful and user-friendly all the time. One of the most important sources (at least for me) is SIMBAD, a database maintained at the Centre de Données astronomiques de Strasbourg (CDS). Entering the name of any star, in any acceptable designation, immediately gives access to all other designations, all measurements, and all recent references. It also gives access to catalogue data for the star, from the GCVS, for instance:
SIMBAD: http://simbad.u-strasbg.fr/simbad
The GCVS4 can also be searched, by star designation, at GCVS HQ in Moscow:
GCVS: http://www.sai.msu.su/groups/cluster/gcvs/
Until about 1970, information on specific variable stars was found in the *Geschichte und Literatur der Veranderliche Sterne*, published at intervals from 1920 to 1963, by which time it had fallen so far behind the current literature that it was of limited value. About this time, the *Jahresbericht* was replaced by the semi-annual *Astronomy and Astrophysics Abstracts*, which 'held the fort' until quite recently. It had the advantage of semi-annual and cumulative indexes by topic and author, and was extremely useful at the time. The *Geschichte* was replaced by the predecessor of SIMBAD at Strasbourg. Now, the literature can be searched, either for references on a particular star, or on a particular topic, by systems such as SIMBAD.

There are still gaps in the system. Many early, foreign, or otherwise obscure references are not yet entered into bibliographic databases, and many of these publications are not easily available, even in the libraries of the largest professional institutes. For people like me, who are often interested in the long-term behaviour of variable stars, that can be a problem. And *Astronomy and Astrophysics*

Figure 3.10 The *Microvariability and Oscillations of STars* (MOST) satellite, and the Project Scientist Jaymie Matthews, in the 'clean room' of the Institute for Aerospace Studies, University of Toronto. The satellite is about the size of a large suitcase and, as the Project Scientist has pointed out, it weighs less than he does. (From MOST Science Team.)

Abstracts has not entirely been superseded by SIMBAD; there are some references in the former that are not in the latter.

A related question of bibliography is: how should variable star observations be published? This is part of a larger question of how *any* scientific information should be published. At one time, IAU Commission 27 maintained an *Archives of Unpublished Photoelectric Observations of Variable Stars*, in the form of paper files; such information is now usually published as an electronic supplement to a journal paper. With the proliferation of information in general, and research papers in particular, and with the increasing expense of printing and mailing, there is a move to electronic publishing. But the complications of electronic publication are not trivial, and it has required the wisdom and patience of astronomers, publishers, and librarians to oversee the successful transition from paper to electrons, at least for major journals. Smaller journals such as the *Journal of the AAVSO* are currently facing the challenge of converting to electronic publication; the *JAAVSO* is successfully on-line as of 2005. The AAVSO is also (successfully) meeting the challenge of publishing all of its *data* on-line. Universities such as my own are setting up systems so that some or all of my on-line data and papers could be published and archived there. But many problems remain. There is a large volume of spectroscopic information on stars which exists on photographic plates that are no longer cared for, and that may suffer from damage, or worse. Much of this information is about stars with variable spectra, and is therefore of interest and importance. And the collections of photographic plates of star fields, accumulated by observatories such as Harvard for over a century or more, are subject to the same dangers and neglect.

Sterken and Duerbeck (2005) is a book which contains some very thought-provoking historical papers on the topic of archiving astronomical information through the ages.

Box 3.1 MOST Satellite (Microvariability and Oscillations of STars)

Canada's MOST (Microvariability and Oscillations of STars) satellite (figure 3.10) is the first to be devoted entirely to the measurement of variable stars. Its 'claim to fame' is its ability to make ultra-precise photometric measurements of stars of magnitude 6 and brighter. With this precision, it is studying the low-degree acoustic oscillations of sunlike stars, and rapidly pulsating Ap (roAp) stars, both with periods of minutes, to probe their structure by *asteroseismology*. It is searching for reflected light from giant, short-period exoplanets – planets around other sunlike stars – as they exhibit phases as they orbit their star. It is studying the turbulent variability in the dense winds of Wolf–Rayet stars. It will provide new information about the structure, evolution, age, and processes in the stars.

This 'Humble Space Telescope' is equipped with a 15-cm telescope and CCD camera; this and all its other components are housed in a satellite the size of a suitcase. It features two innovations: a new type of attitude control system, using reaction wheels and magnetotorquers, which were designed to keep it pointed to better than 10-arc sec accuracy (but achieved close to 1 arc sec), and a system of microlenses which project a stable image of the star on the science CCD which can detect photometric variations as small as one part in a million, or better if the signal is periodic. Other than the reaction wheels, it has one moving part – a sun shield over the telescope diagonal. The total cost of the mission is about $US 5 million – humble, indeed.

MOST was launched from Russia on 30 June, 2003 with a former Soviet ICBM. After a short engineering test phase, it began collecting data, and has already provided exciting new results on several stars. In particular, it observed the sunlike star κ^1 Ceti, and deduced the presence of two spots at different latitudes, rotating with different periods because the rotation is differential. Even more important: it has shown that photometric variations, previously 'observed' in Procyon and thought to be due to pulsation, are actually an artifact. Asteroseismology is actually even more difficult than had been believed! The website of MOST contains its latest results:
http://www.astro.ubc.ca/MOST/index.html

4

Rotating variable stars

Only in theory are stars bland, featureless spheres. In the case of the sun, we can directly observe its complex, ever-changing surface – sunspots, flares, magnetic fields, low-level pulsation, and seething convection cells – bubbles of rising and falling gas. On other stars, we cannot 'see' the appearance, but we can deduce it.

As the star rotates, regions of different brightness (and colour, and magnetic field strength, and other properties) will pass into the observer's field of view, and these properties will appear to vary. The period of variation is the period of rotation. Since all stars rotate, any star with a patchy surface will be a rotating variable, unless the patches are symmetrical about the axis of rotation, or unless the axis of rotation points at the observer. The variations will be both photometric and spectroscopic.

What causes a star to have a patchy surface? Rotation itself causes a star to be slightly oblate, and to be slightly cooler at the equator and hotter at the poles. But this effect is symmetric about the axis of rotation. A binary companion can cause patches: it distorts the star, and causes 'gravity darkening' and a reflection effect; this is discussed in section 5.4.6. *Ellipsoidal variables* are sometimes considered to be rotating variables. In this case, a star is distorted into an egg shape by the tidal gravitational effect of a companion star. The axis of the 'egg' points to the companion, and appears to rotate as the star orbits the centre of mass of the system.

Large convection cells may also cause patches; on the sun, the convection cells are small and numerous, and their effects on the brightness of the sun cancel out. On red giants and supergiants, the convection cells may be much larger, and may explain some of the complex variability of these stars.

But the usual cause of patches are magnetic fields, either global or localized.

Whatever the cause of the patches, the period P of the variability (the period of rotation) will be related to the equatorial rotation velocity v and the radius of the star R by $P = 2\pi R/v$ or

$$v = 50.6(R/R_{\odot})/P(day)\,\text{km s}^{-1} \tag{4.1}$$

Although v cannot be measured directly, its projection along the line of sight $v\sin i$ (where i is the inclination of the axis of rotation to the line of sight) can be measured from the broadening of the absorption lines in the spectrum of the star, as discussed in section 2.15. This provides a test of rotational variability in a group of stars of similar radius: if $v\sin i$ is plotted against P, the upper envelope or boundary of the graph should be a hyperbola whose equation is

$$v\sin i = 50.6(R/R_{\odot})(\sin i)/P \tag{4.2}$$

From the equation for the upper envelope, the average radius of the group of stars can be determined. Alternatively, if the radius of a rotating variable star is known from its other properties, and if its period and its $v\sin i$ can be measured, the inclination of its axis i can be estimated.

There are two especially important and interesting groups of rotating variable stars: peculiar A (Ap) stars and their relatives, and sunlike stars with starspots. The Ap stars have global, approximately dipole magnetic fields, like the earth. Sunlike stars have magnetic fields which are concentrated primarily in small regions called *starspots*, though, as noted later, the sun does have a weak global magnetic field of about one Gauss.

4.1 The spotted sun

Sunspots are regions of the sun's photosphere which are cooler (3800K) and therefore darker than their surroundings (5800K). Sunspots tend to be localized in one or a few areas at moderate solar latitudes (figure 4.1). Often, the spots or regions occur in pairs. The number of sunspots rises and falls in an 11-year *sunspot cycle*. The cycle is not strictly periodic: the lengths of the cycles vary between 10 and 12 years, and the number of sunspots at maximum varies between about 50 and 200. In fact, there was an interval between about 1645 and 1715 (the *Maunder minimum*) when there were very few sunspots, even at sunspot maximum. Early in a cycle, spots tend to appear at moderate solar latitudes; later in a cycle, they tend to appear much closer to the solar equator.

The number of sunspots detected will, of course, depend on the observer's equipment, sky conditions, and experience. In deriving the sunspot numbers mentioned above, statistical techniques are used to allow for these factors. The

Table 4.1. *Bright and/or interesting rotating variable stars*

Star	HD	V Range	Period (d)	Spectrum	Comments
α And	358	2.02–2.06	0.966222	B8IVp(Hg-Mn)	2nd-brightest ACV in *GCVS4*
ϵ Eri	22049	3.73–3.78	11.10	K2V	brightest BY Dra in *GCVS4*
α Dor	29305	3.26–3.30	2.95	A0IIIp(Si)	–
μ Lep	33904	2.97–3.41	2:	B9IV	bright ACV
θ Aur	40312	2.62–2.70	1.3735	B9.5p(Si)	bright ACV
σ Gem	62044	4.13–4.29	19.423	K1III	2nd-brightest RS CVn in *GCVS4*
V645 Mon	65953	4.68–4.70	0.207878	K4III	brightest FK Com in *GCVS4*
V816 Cen	101065	8.00–8.02	long	F8p	roAp; Przybylski's star
ϵ UMa	112185	1.76–1.78	5.0887	A0p(Cr-Eu)	brightest ACV in *GCVS4*
α^2 CVn	112413	2.90–2.98	5.47	A0p(SiEuHg)	prototype
RS CVn	114519	7.93–9.14	4.797887	F4IV-V + K0IIIe	prototype RS CVn
FK Com	117555	8.14–8.33	2.4	G2eapnIII + K3V	prototype FK Com
BY Dra	234677	8.04–8.48	3.813	K3V + MVe	prototype BY Dra
λ And	222107	3.69–3.97	54.2	G8III-IV	brightest RS CVn in *GCVS4*
ι Cap	203387	4.27–4.33	4.5–5.0:	G8III	2nd-brightest BY Dra in *GCVS4*

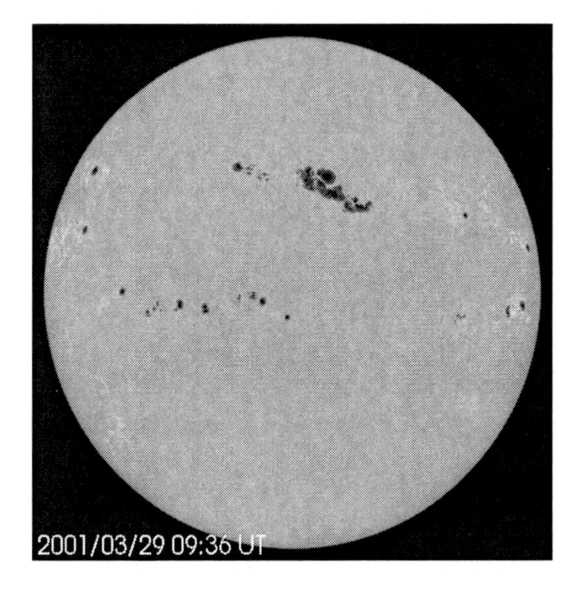

Figure 4.1 A white-light image of the sun, showing sunspots. As the sun rotates, these would cause the brightness of the sun to change slightly, if it were viewed from a distant star. (ESA/NASA.)

AAVSO has had an active sunspot-observing program since World War II, when sunspot observations, previously obtained from Europe, were no longer available. The AAVSO also monitors Sudden Ionospheric Disturbances (SIDs) which are caused by solar flares. See the AAVSO website for detailed information about how the measurements are made, analyzed, and used.

The sun has a global magnetic field of about one Gauss, similar in strength to the earth's. In sunspots, the magnetic field is concentrated and strong, up to several thousand Gauss. This is comparable with the strength of a good laboratory or industrial magnet! Sunspots in pairs have opposite magnetic polarity. Sunspots are darker because the magnetic field impedes the flow of energy from below, reducing the temperature. Energy is present, in these regions, in forms connected with the magnetic field. The magnetic field results in other forms of *solar activity* near sunspots, including flares (section 7.2), clouds of gas above the photosphere called *prominences*, and also coronal heating. The temperature of the corona is about a million degrees, so the sun's corona (and the coronas of other sunlike stars) are sources of X-ray emission. In alternate solar cycles, the polarity of the general magnetic field and of the sunspot pairs reverses, so the 11-year solar cycle is actually a 22-year cycle.

On average, the sun rotates once in about 30 days, but it is a fluid, and does not rotate as a solid object: its rotation period is 26.93 days at the equator, and 35 days near the poles. The *total solar irradiance*, the sun's total brightness as seen from earth, varies by a fraction of a percent as sunspots are carried around the sun by its rotation.

How do other sunlike stars compare with the sun in their spottedness and other forms of activity? What are their rotation periods? Is the rotation differential? Do they have activity cycles? Studying the variability of other sunlike stars helps to answer questions like these, and also helps in understanding how the sun's activity may have changed over time. See Golub and Pasachoff (2002) for an up-to-date account of our sun.

4.2 Sunlike stars

What would the sun look like from a great distance? It would look like a point of light, but presumably its brightness, and colour, and spectrum might change as starspots rotated on and off the side of the star which faces us. The normal absorption-line spectrum of the star would include strong, dark absorption lines of elements such as hydrogen, sodium, and ionized calcium, and, in the cores of these lines (especially ionized calcium), there would be emission peaks due to the chromospheric gases associated with the starspots. The strength of these emission peaks would vary with time, as the starspot regions

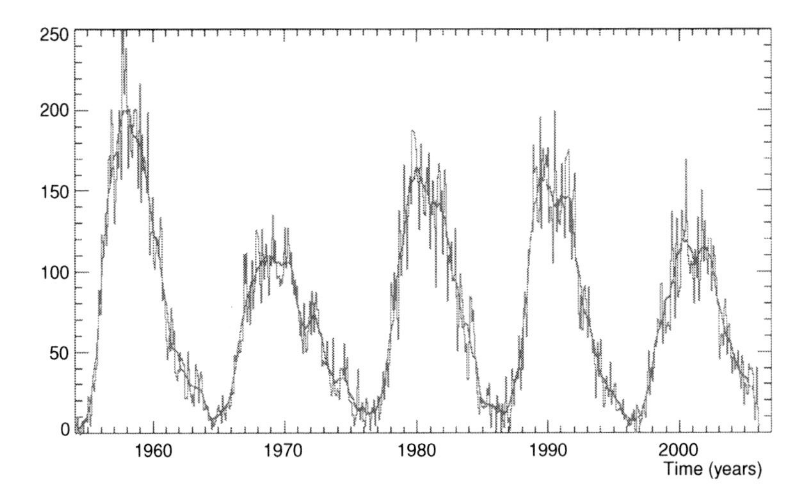

Figure 4.2 The sunspot cycle: the changing number of sunspots with time. At present (2006), we are beginning a new cycle. Neither the length of the cycles, nor the height of the maxima are constant from cycle to cycle. (From World Data Center for the Sunspot Index.)

rotated in and out of the field of view. Thus, starspot activity, and the rotation period, can be measured by monitoring the brightness and spectrum over time scales of days, weeks, and months. The sun would be microvariable in brightness, with a range of about 0.01 in V, and a period of about 30 days. The amplitude would change with time, as major sunspot groups formed and decayed. There would be little variability at sunspot minimum, when sunspots were rare. The period of variability would vary slightly, depending on the rotation period at the latitude where the major sunspot groups were situated.

Long-term spectroscopic monitoring of sunlike stars has been done especially by George Preston, Arthur Vaughan, and Olin Wilson. They have found that the Ca II emission does indeed vary on two different time scales: days or weeks, a time scale which they attribute to the rotation of a star whose activity is concentrated in active regions (as with the sun); and years, a time scale which they attribute to activity cycles similar to the 11-year activity cycle of the sun. Rotation rates have been measured for about 1500 young sunlike stars in clusters and associations of a variety of ages. These studies provide important information about the rotation at any age, and its relationship to the spectral type of the star, and about the change of rotation rate with time. Figure 4.3 shows the photometric variability of a typical young sun-like star.

These and other studies have now revealed a number of interesting and important facts about rotation and activity on sunlike stars:

- In general, the amount of stellar activity is correlated with the speed of stellar rotation, because it is produced by the magnetic field, which, in turn, is produced by the rotation of the star.
- Sunlike stars are born with relatively rapid rotation but, for any kind of star, there can be a range of initial rotation rates. This is because of the varying conditions under which stars form.
- As the newborn star shrinks toward the main sequence state, its rotation will increase, due to the *conservation of angular momentum*.
- As sunlike stars grow older, they rotate more slowly. For the first few million years or so, the decrease is quite rapid, and is due to braking by the remains of the accretion disc, which is still coupled to the star through the magnetic field. Later, their angular momentum is carried off as a result of their stellar wind.
- Stellar activity decreases with age; as a sunlike star grows older, it shows less activity. In 1972, Skumanich proposed that the rotation and activity decreased with the square root of the age, but this is a simplification: stars of a given mass and age can have a range of different rotation rates. This is because rotational braking depends on the rotation rate (at high rotation rates, the braking mechanism is less effective). It also depends on the mass of the star; less massive stars are more convective, and this affects the nature and strength of their magnetic field.
- The sun is a middle-aged, relatively slowly rotating, inactive star. But presumably it rotated much faster, and was much more active, when it was young.
- Stars with rotation periods slower than about 20 days show activity cycles similar to that of the sun. The length of the cycle is typically about ten years, relatively independent of any other property of the star.
- Stars with more rapid rotation periods tend to have more chaotic, irregular activity.

Since 1990, Edward F. Guinan and his collaborators have been carrying out *The Sun in Time* project; see Messina and Guinan (2002, 2003), for instance. This long-term monitoring project has studied starspot cycles in single sunlike stars. Young, rapidly rotating stars tend to have slightly longer cycles, but stars of the sun's age and rotation speed tend to have cycles like that of the sun. This project also determines rotation periods, and even finds evidence of differential rotation, as in the sun. High-resolution spectroscopy can also be used, to

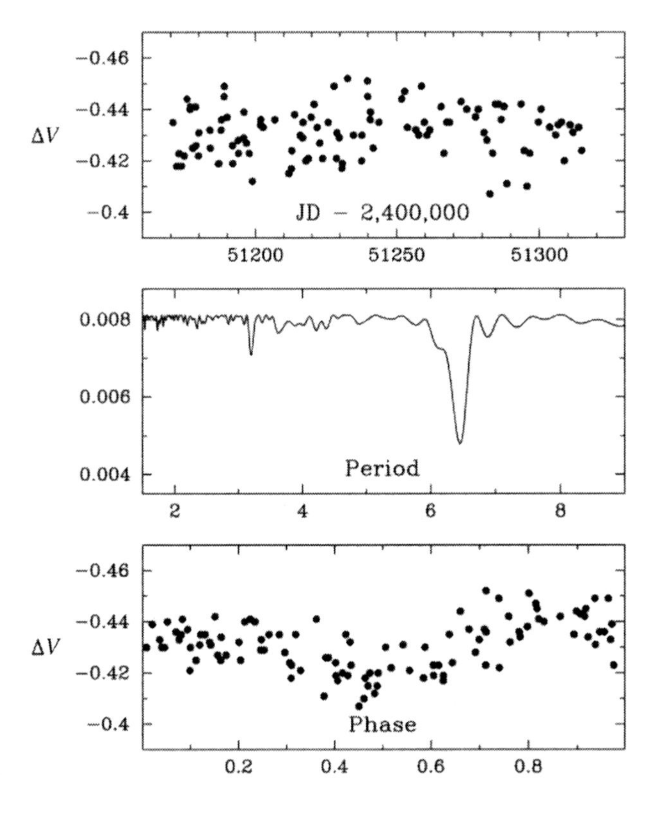

Figure 4.3 For the young sunlike star HD63433: the V light curve during one season (top), the period analysis using phase-dispersion minimization (middle), and the phase curve using the period 6.46 ± 0.01 days (bottom). The amplitude is larger than for the sun. (From Gaidos, Henry, and Henry, 2000.)

Doppler-image the starspot distribution as a function of latitude and longitude on the star.

As the name of Guinan's project suggests, the study of sunlike stars helps astronomers to understand more about the nature and evolution of our own star's rotation and activity.

Rotation is an extremely important process in any star. Many studies of the structure and evolution of stars have not considered the effect of their rotation, because it complicates the calculations. Recently, extensive sets of models of rotating stars have been calculated, by Meynet and Maeder (1997 and many subsequent papers), for instance.

In addition to its own effect on the star, rotation may produce magnetic fields, and mass loss, which have their own effects on the star. These effects have rarely been included in stellar structure calculations, except with very simple approximations. There is still much to be done in understanding the complex topic of stellar structure and evolution!

4.3 FK Comae stars

In the 1980s, partly as a result of X-ray astronomy, astronomers became aware of – and interested in – sunlike stars with a high level of activity, including X-ray emission. The X-ray emission comes from the hot corona, which is one consequence of stellar activity. The RS Canum Venaticorum stars, mentioned below, fell into this category. A wide assortment of stars were found to have such high activity, and it became tempting to subclassify them. FK Comae stars were initially defined as late-type *giants*, which were not binaries (as RS Canum Venaticorum stars were), but which showed high activity. The GCVS now includes both single and binary stars in this class. These stars are rotating so rapidly that astronomers believe that they may be the end product of a merged W Ursae Majoris binary system, rather than being very young stars with unusually high rotation at birth. Notable members of the class include FK Comae Berenices (G5II) with a period of only 2.4 days, and UZ Librae (K2III) with orbital and photometric periods near 4.75 days (the rotation and revolution are presumably synchronized). The V ranges are up to 0.5, though are more typically 0.2. Stars with smaller ranges would probably be classed as 'ordinary' sunlike stars, though it is not clear where the boundary lies.

4.4 RS Canum Venaticorum stars

The RS Canum Venaticorum stars have been defined as 'binaries with orbital periods between one and 14 days, with the hotter component F-G IV-V and with strong H and K line emission seen in the spectrum outside eclipse' (Hall, 1976). There are similar stars with orbital periods less than a day ('the short-period group') and greater than 14 days ('the long-period group'). It is probably appropriate to lump all these stars together and call them 'stars showing the RS Canum Venaticorum phenomenon'.

Individual members of this group have been known and studied for decades, but they first began to attract wider attention in the mid-1960s on account of their unusual photometric variability outside eclipse. In the mid-1970s, even more exotic properties were discovered: radio emission and flaring, thermal X-ray emission indicating temperatures of 10^7 K, and strong and variable Ca II H and K, hydrogen, and magnesium II h and k line emission. It is now accepted that these properties are due to stellar activity – starspot groups, a thick chromosphere, and coronal magnetic loops. In many ways, they are like solar activity, but on a grander scale.

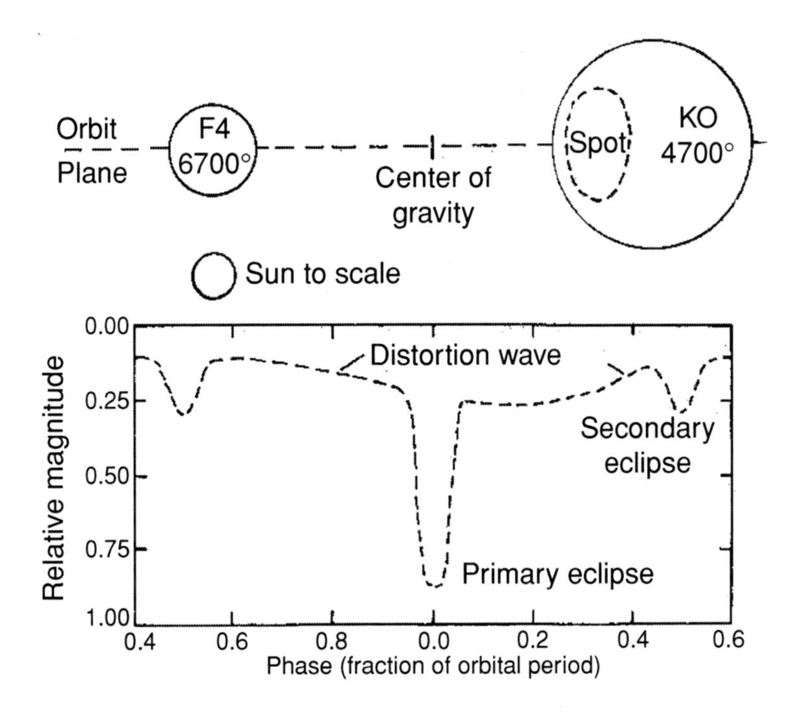

Figure 4.4 Light curve and schematic model of RS Canum Venaticorum, with the sun drawn to scale. Below: the light curve of the system; the primary and secondary eclipses are shown, as is the 'distortion wave' which extends throughout the entire light curve. The distortion wave is due to the presence of the starspots. These may 'migrate' in longitude, causing the distortion wave to migrate, relative to the eclipses, in the phase curve (Based on a light curve by E.W. Ludington.)

The RS Canum Venaticorum stars show photometric variability other than that which may be due to eclipses. The variability consists of a more-or-less sinusoidal 'distortion wave' whose amplitude and phase (relative to the orbital period) slowly vary. The phase variability causes the distortion wave to migrate relative to the orbital period or – if the system eclipses – relative to the eclipse light curve. The distortion wave is due to the starspot groups: darker regions near the equator of the star. The rotation of the star, which is usually synchronous with the orbital revolution, then produces the distortion wave, as the darker regions are turned toward or away from the observer. The variability of the *amplitude* of the distortion wave is due to the changing area of the spots; this measure of the stellar activity changes in long-term cycles, as it does on the sun and on other sunlike stars. The variability of the *phase* of the distortion wave is due to the differential rotation of the star, which affects the position of the starspot region, relative to the companion star and to the orbital phase. Figure 4.4 shows a schematic diagram and light curve of RS Canum Venaticorum itself.

Why is the level of activity so high on RS Canum Venaticorum stars? Because stellar activity is caused by magnetic fields on a star, and magnetic fields are produced by the star's rotation. RS Canum Venaticorum stars rotate rapidly because tidal interactions with their companions have 'spun them up' to rotation rates several times faster than 'normal'.

Because of the complex photometric variability of RS Canum Venaticorum stars, it is desirable and necessary to monitor these stars systematically over many consecutive observing seasons. This requires extensive observing, and Douglas S. Hall in particular recruited the help of dozens of amateur astronomers with photoelectric equipment. Most RS Canum Venaticorum stars are bright. Their colours are moderate. The amplitudes can be of 0.2 magnitude or more. They are ideally suited for photoelectric observation by amateur astronomers. Results were obtained and published relatively quickly, providing prompt feedback and satisfaction for the observers.

This period, around 1980, marked the birth of large-scale participation in photoelectric photometry by amateur astronomers, and undergraduate research students. Simple, relatively inexpensive solid-state photometers became commercially available; users no longer had to build their own. The International Amateur-Professional Photoelectric Photometry organization came into being, with its meetings, *Communications*, and various 'wings' around the world. Books were published about how amateurs and students could contribute to science through photoelectric photometry (Hall and Genet, 1988; Henden and Kaitchuck, 1982). The AAVSO photoelectric photometry program was started. There was – and still is – a lot of good science which can be done, in this field, by amateurs and students with small telescopes.

Box 4.1 Star sample – IM Pegasi

IM Pegasi [HR 8703, HD 216489, K1.5II-IIIe, V = 5.892] is an RS Canum Venaticorum star with a photometric period of 24.73 days and an orbital period of 24.65 days. Doug Welch (then an undergraduate research student) and I observed it in 1979, and published the results in Percy and Welch (1982); I immediately forgot about the star until quite recently, when it re-surfaced in my life as the guide star for the *Gravity Probe B* satellite. The GPB mission scientists requested monitoring of this star by the AAVSO photoelectric photometry program, which I co-ordinate. Here's why.

GPB is a very sophisticated satellite, many years in development, whose purpose is to test aspects of the General Theory of Relativity which had heretofore not been well tested. The satellite required a guide star whose position could be measured and monitored to an accuracy of better than 0.001 arc sec. IM Pegasi satisfies that requirement because it is also a *radio source*, and the positions of point radio sources can be measured, with radio interferometers, to high accuracy. The radio emission comes from around the *active regions* on the K star, which are produced by the star's magnetic field, which is generated by its above-average rotation, which in turn is caused by tidal interaction with its binary companion. The radio position measurements are so precise that they reveal the star's motion in its 24.65-day binary orbit! The spots on the active star have been extensively studied by *Doppler imaging* (Berdyugina *et al.*, 2000); they evolve in a 6.5-year activity cycle.

The role of the AAVSO photoelectric observers, most of whom are amateur astronomers, is to monitor the presence of flares on the active star. If there were a flare, then it would slightly change the 'average' position of the two stars in the binary star system, as seen by the GPB star tracker. That would be enough to affect the experiment.

(For updates on GPB, see:
http://einstein.stanford.edu)

4.5 BY Draconis stars

BY Draconis stars are defined as *cool* (K and M type) stars, as compared with G type for sunlike stars, with noticeable variability due to cool spots and rotation. They are also recognizable by the emission lines in their spectrum due to chromospheric activity. They can be single or binary, so their relatively high rotation may be genetic, or induced by a close orbiting companion. Many also show flares, and are therefore UV Ceti stars (flare stars). The variability of these stars has been difficult to study, because cool main sequence stars are intrinsically faint; the brightest member is well beyond the limit of the unaided eye.

BY Draconis itself consists of a K4V star and a K7.5 star, with a 5.975 day orbital period. A more famous example is YY Geminorum (Castor C), discovered as an eclipsing binary in 1926; the components are M1Ve and M2Ve, and the orbital period is only 0.814 day. Periods between one and ten days are typical of these stars. The V ranges are up to 0.3, but are typically 0.1. The range, phase, and shape of light curve can change slowly with time, as starspots grow and decay.

An interesting question is: to what extent does solar-type variability persist to M stars and cooler – even to brown dwarfs? Unlike the sun, M stars are completely convective, so the dynamo that produced their activity and variability would have to be of a different type. But these cool stars *do* seem to have magnetic fields, and *do* seem to be active and variable.

I feel that it would be reasonable to group all of these cool, active rotating variables together. The over-riding principle is that the activity is generated by rotation, and the rotation can be genetic, or induced by a close orbiting companion star.

On the other hand, the Ap stars described in the following section are quite different. They are much hotter, and their magnetic fields are global, not localized.

4.6 Peculiar A (Ap) stars

Over 90 per cent of stars lie on the main sequence in the H–R diagram. For these (and indeed for almost all stars), the spectrum is primarily determined by the star's temperature, secondarily by its metal abundance, and very slightly by its luminosity – luminous stars are less dense, and that affects the appearance of the spectrum. But there is a small group of main sequence B8 to F2 stars whose spectra are quite bizarre: the apparent abundance of certain elements is thousands of times different than in the sun. For instance, there are stars with enormous excesses of the rare-earth elements (those with atomic numbers between 58 and 71), others with an over-abundance of silicon, others with an over-abundance of mercury, or manganese. These are the *peculiar A stars*, or Ap stars. About 5 per cent of mid-B to mid-F stars are Ap stars. The proportion is highest at B8, where it is about 25 per cent.

The Ap stars can be divided into a number of classes, which form a temperature sequence; in order of decreasing temperature, they are Si-4200, Hg-Mn, Si, Si-Eu-Cr, Eu-Cr, Eu-Cr-Sr, and Sr. More generally, three groups are sufficient: Hg-Mn, Si, and Eu-Sr-Cr (figure 4.5). (For those of you who have forgotten: Cr = chromium, Eu = europium, Hg = mercury, Mn = manganese, and Si = silicon.) There are also helium-strong and helium-weak stars, which are hotter, but related. The GCVS4, for instance, includes a class of SX Arietis (SXARI) stars, which are main sequence B0p-B9p stars with variable He I and Si III lines, and magnetic fields. The most famous example is σ Orionis E (V1030 Orionis). As we shall see, all of these types are rotational variable stars. For a comprehensive discussion of Ap, Am, and other A-type stars discussed in this book, see Zverko (2005).

There are also *metallic-line A (Am) stars* which are a spectroscopically defined group of non-magnetic, chemically peculiar, population I, main sequence A and early F stars that display an apparent surface under-abundance of calcium and scandium, and an apparent over-abundance of the iron-group elements (Kaye

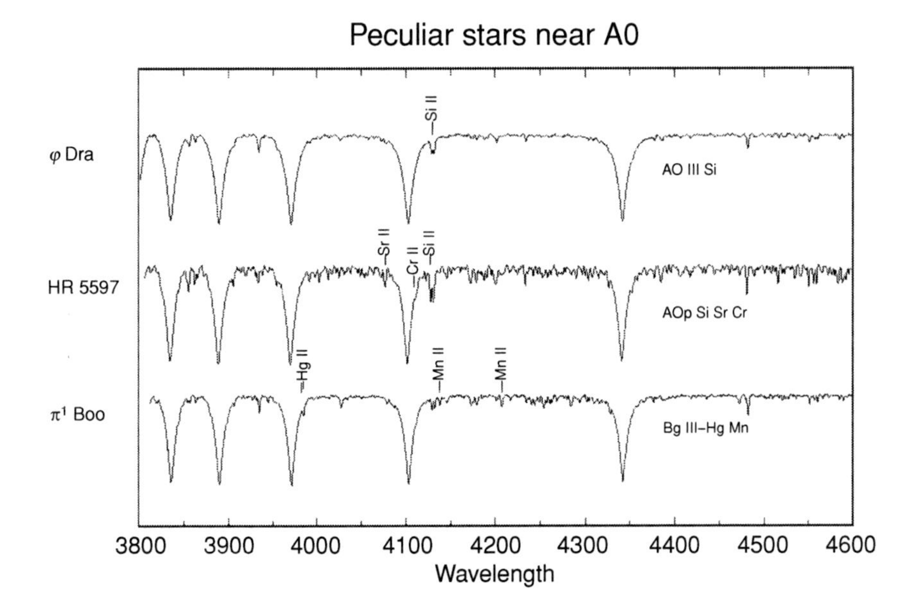

Figure 4.5 The spectra of three Ap stars, of different types: Si (silicon), SiSrCr (silicon–strontium–chromium), and HgMn (mercury–manganese). The strong absorption lines in each spectrum are of hydrogen. The wavelength is in Angstroms. (From Richard O. Gray, private communication.)

et al. 2004). And there are Lambda Bootis stars which are milder versions of the Am stars. Neither of *these* types are rotational variables. But we can still ask – what causes the peculiar abundances in these stars?

With the development of high-resolution spectroscopy, early in the twentieth century, astronomers discovered that the spectra of these stars actually varied periodically with time! The periods were typically one to ten days, or a few weeks at most. They were spectrum variables! Photoelectric photometry, developed by Joel Stebbins in the US and Paul Guthnick in Europe, revealed that the brightness and colour of these stars also varied, with the same period as the spectrum. They became known as α^2 Canum Venaticorum stars, after a bright prototype. Over a hundred Ap stars have been carefully monitored, photometrically. Their periods are typically a few days. The amplitudes in V are generally less than 0.05, with 0.02–0.03 being most common, but a few have amplitudes of up to 0.3. The shapes of the light curves are generally close to sinusoidal, and are strictly periodic (figure 4.6). They are among the most common variables in any magnitude-limited sample or survey of variable stars. But what causes their periodic variations?

The next piece of the puzzle appeared in 1946, when Horace Babcock discovered a strong magnetic field in the Ap star 78 Virginis. Magnetic fields in stars

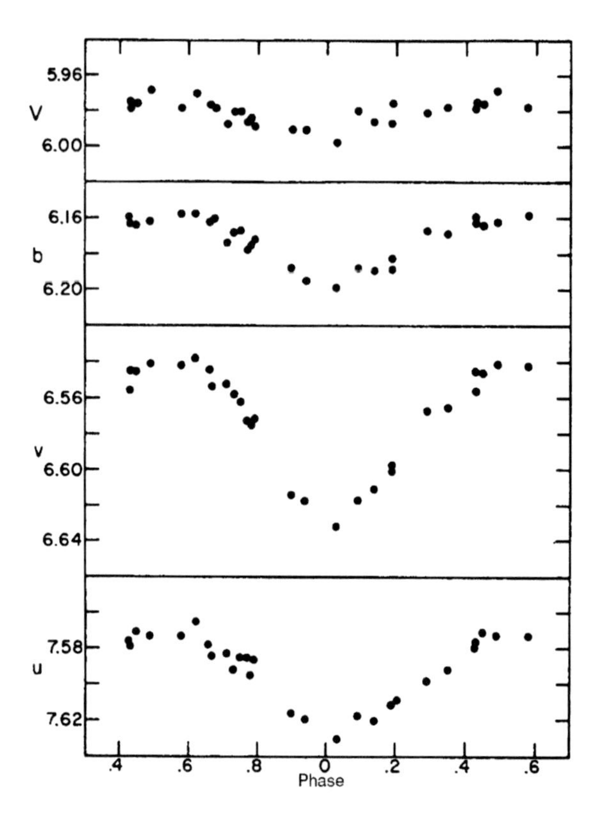

Figure 4.6 The variations in brightness, in the Stromgren uvby system, for the Ap star HD 24712 (HR 1217, DO Eridani). Phase 0.0 is when the spectral lines of europium are at maximum strength. The star is faintest at this phase. The opacity of the visible part of the star's atmosphere is such that less radiation is being emitted in the visible wavelengths. See the Star Sample description of DO Eridani in section 6.13. (From Wolff and Morrison, 1973.)

are detected and measured by the *Zeeman effect*, a splitting of the absorption lines into two or more components by the effects of the magnetic field on the energy levels of the electrons in the absorbing atoms. These components are linearly and/or circularly polarized, and stellar magnetic fields are now normally measured by polarimetry.

By 1960, 70 magnetic Ap stars were known, with *global* magnetic fields of typically 1000 Gauss but, in some cases, up to 10 000 Gauss. The sun has magnetic fields this strong, but they are localized in sunspots; the global magnetic field is about one Gauss, comparable with that of the earth. Imagine a star with a *global* magnetic field which is as strong as that of an industrial or laboratory magnet! What could their origin be?

Furthermore, the apparent strength of the magnetic field of the Ap stars was variable, with the same period as the spectroscopic and photometric variations. During the 1950s, D. W. N. Stibbs and Armin Deutsch developed a successful model for the variability of the Ap stars – the *oblique rotator* model. The stars' magnetic fields were primarily dipole fields (like the earth's), inclined to the axis of rotation (like the earth's). As the star rotated, the apparent orientation and strength of the magnetic field changed with time. Furthermore, the magnetic field somehow localized the distribution of the elements with peculiar abundances into 'patches', so that, as the star rotated, the apparent abundance changed with time. These abundance regions had slightly different brightnesses and colours, hence the photometric variations. By making measurements of the strength and radial velocity of the absorption lines of each element, over time, it is possible to construct maps of the element distribution over the visible parts of the star (figure 4.7).

But what causes the peculiar abundances? In the 1970s, Georges Michaud and others showed that, if the outer layers of an A-type star were very stable, with little or no convection, rotation, mass loss, or other kinds of mixing, then certain elements would slowly rise to the top of a star's atmosphere, and other elements would sink downward, relative to these. Depending on whether the upward force of photons, absorbed by the atom from the radiation field, was greater or less than the downward force of gravity, the atom would rise or fall. Those that rose to the surface would be visible in the spectrum of the star. Depending on the temperature of the star, it turned out that this effect could levitate rare-earth elements, silicon, or mercury and manganese, as observed. The magnetic field of the star provides the stabilizing effect necessary for this diffusion to occur. (In the Am stars, there is no significant magnetic field. Diffusion occurs because the stars are slowly rotating, compared with A stars in general, and there is little or no turbulence, circulation and mixing in their outer layers. But the chemical peculiarities are uniformly distributed on their surfaces, so there is no photometric or spectroscopic variability as the star rotates.)

But where does the magnetic field come from? In sunlike stars, the magnetic field is produced by a dynamo effect in the convective outer layers of the star, by its rotation. In the A stars, the outer layers are radiative, not convective. And the strength of the observed magnetic field is not correlated with the rotation period of the star. And the strength of the magnetic field does not seem to change with time as it does on the sun – at least at the present level of detectability.

So is the magnetic field primordial, coming from the weak magnetic field in the interstellar material from which the star formed, or from a dynamo effect in the star when it was forming? Or is it somehow generated in the star's convective core by the dynamo effect, by the star's rotation? Recent surveys have

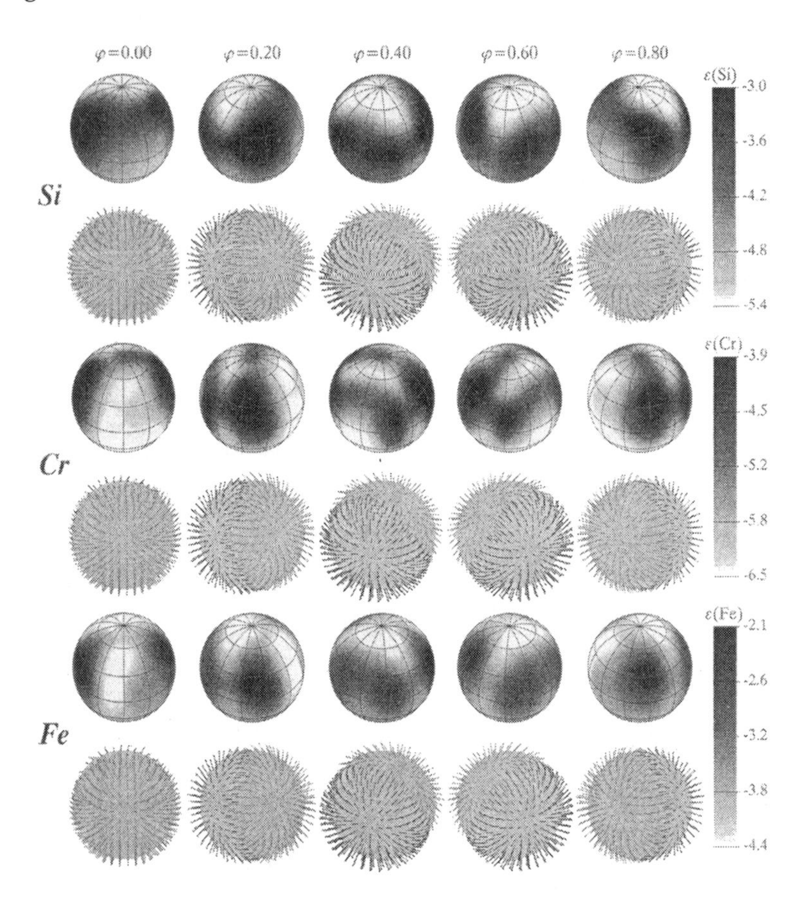

Figure 4.7 A map of the distribution of the chemical elements and the magnetic field on the surface of the Ap star α^2 Canum Venaticorum, derived from high-resolution spectropolarimetry. The magnetic field is primarily dipolar, with a small quadrupole component. The surface distributions of the chemical elements form symmetric patterns, which closely follow the magnetic field geometry. The scales at right are logarithmic abundances. (From Kochukhov *et al.*, 2002.)

detected magnetic fields in a few A and B stars which are in the last stages of formation – the Herbig Ae/Be stars (chapter 9). Most likely then, the magnetic field is a 'fossil' field, produced during the formation of the star, and 'frozen' into the highly conducting stellar plasma.

4.7 Pulsars

If Ap stars seem extreme, pulsars are even more so. In almost every respect, they belong in the *Guinness Book of Records*.

Box 4.2 Star sample – ϵ UMa (a peculiar A star)

Epsilon UMa (A0pIV(Cr-Eu), HR 4905, HD 112185) is notable in several respects. By virtue of belonging to the Big Dipper, it is probably the best-known star which is an Ap star, at least for northern observers. It is the brightest Ap star, at $V = 1.76$, and is probably the most extensively studied. At 81 light years, it is probably the closest. It belongs to the Ursa Major moving cluster, with an age of 500 million years.

The photometric variability was discovered in 1934 by Guthnick, one of the pioneers of photoelectric photometry; he derived a period of 5.0887 days. In 1947, Deutsch discovered spectrum variability with this same period. The large $v \sin i$ of 50 km s^{-1}, together with the relatively long period, suggests that this star is a bit larger than the average Ap star, and this is consistent with the luminosity class IV assigned to the star.

As of 2000, it had the *weakest* magnetic field ever measured in an Ap star; the longitudinal field varies between -10 and $+100$ Gauss, on the rotational period of 5.0887 days. Nevertheless, the star has a photometric amplitude of 0.02 in V, which is quite healthy for an Ap star. Using Doppler imaging, several chemical elements have been mapped on the surface of this star; see figure 4.2. It is possible that a stellar interferometric telescope could resolve the 'patches' on the surface.

The star tracker on the WIRE satellite was used to search for roAp pulsations in this star, but they were less than 75 parts per million. So not all Ap stars are roAp variables!

Pulsars emit pulses of (mostly) radio radiation, with periods of 0.001 to 10 seconds, with exquisite precision. They are rapidly rotating neutron stars with super-strong magnetic fields. Their pulse period is their rotation period. They are not *pulsating variable stars*, they are *rotating variable stars*. (See Kramer *et al.*, 2000, and *Science*, **304**, 532, and especially Manchester, 2004 and Stairs, 2004 for excellent recent reviews. See also the web site:
http://www.atnf.csiro.au/research/pulsar/
for a catalogue, and extensive information; click on 'tutorial' for a good introduction.)

Pulsars were discovered in 1967 by Jocelyn Bell (now Bell Burnell), working with Antony Hewish at Cambridge University (figure 4.8). Hewish (but not Bell) subsequently shared a Nobel Prize in Physics for the discovery.

But how could a celestial object vary, like clockwork, with a period of one second? There are three possible processes which can produce periodic behaviour

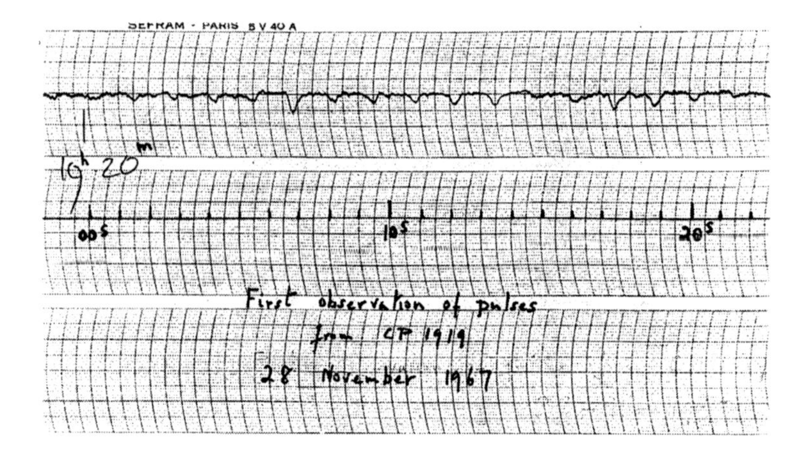

Figure 4.8 First observation of radio pulses from a pulsar, CP1919, by Jocelyn Bell and Anthony Hewish. On this now-old-fashioned recorder chart, time increases to the right; the intensity of radio emission increases downward. Note the regular 'blips' in the top panel, separated by about a second.

in stars: rotation, revolution, and pulsation. A normal star, or even a white dwarf cannot rotate, revolve, or pulsate so fast. Only the *rotation of a neutron star* could explain the observations. But how to prove it?

Neutron stars are produced by supernovae, in the collapse of a massive star at the end of its nuclear-burning lifetime; the internal pressure in the star is insufficient to balance gravity. In 1968, a pulsar was discovered at the heart of the Crab Nebula (figure 4.9), the remnant of a supernova which was recorded in 1054 AD. Its period was only 0.033 second. As with other pulsars, the period could be measured with exquisite accuracy, and so could its rate of change. It was slowing down at a perceptible rate. In fact, the rate of rotational energy loss by the slowing-down pulsar was approximately equal to the rate at which the Crab Nebula radiated energy! This supported the model, first proposed by Thomas Gold, that pulsars were rotating, magnetic neutron stars; their magnetic fields carried rotational energy into the nebula, energizing the nebula, and also braking the rotation of the neutron star.

About 1500 pulsars were known as of mid-2004, and the number should rise rapidly as a result of new surveys which are about to start. There are many more pulsars which are undetectable because their beams of radiation do not sweep across our location in space. About 50 pulsars are known to be in binary systems, which enable the mass of the neutron star and/or the companion to be determined; the masses of the neutron stars are typically 1.3 to 1.7 $m_\odot$; the average is about 1.35. The companions may be neutron stars, white dwarfs, or normal stars. Two pulsars (B1913+12 and B1620−26) have planet-mass companions.

Figure 4.9 Probing the mysterious heart of the Crab Nebula, the tattered remains of an exploding star, astronomers have found this object to be even more dynamic than previously understood. The pulsar is in the centre; the tangles of nebulosity and magnetic field are a result of the rotation of the pulsar. These findings are based on a cosmic 'movie' assembled from a series of Hubble telescope observations. (HST/NASA.)

Pulsar period changes not only led to the confirmation of Gold's theory, but they have also revealed 'glitches' due to starquakes in the neutron star's crust, the presence of companion stars (and even planets), and the existence of the long-predicted *gravitational radiation*. Pulsar B1913+16, with a period of 0.059 second, was found in 1974 to have a second neutron star as a companion, in an eccentric eight-hour orbit. Observations of the slowdown of its orbital motion, using the Doppler effect on its pulses, and the $(O - C)$ method, led to precise tests of the General Theory of Relativity – including the detection of gravitational radiation – and to a Nobel Prize in Physics for its discoverers Russell Hulse and Joseph Taylor.

In 1982, the first 'millisecond pulsar' was discovered, spinning 642 times a second – comparable with a kitchen blender, but 20 km in diameter. At their equators, the surface velocity is over 10 per cent of the velocity of light. 'Normal' pulsars with periods of seconds spin rapidly because they are the collapsed corpses of large, slowly rotating stars; conservation of angular momentum amplifies the spin. Millisecond pulsars are 'spun up' by mass and angular momentum transferred from a companion star. They have been 'recycled' to a short period. Millisecond pulsars are particularly abundant in some globular clusters; the cluster Terzan 5 has at least 20.

Pulsars were initially detected at radio wavelengths, but they are increasingly studied with X-ray satellites such as *Chandra*. Accreting neutron stars often show coherent modulations called *burst oscillations* during their X-ray bursts. These may be driven by non-radial pulsation of the neutron star surface. If this is the case, these oscillations can provide important information about the neutron stars themselves.

One millisecond pulsar – B1957+20 – has been dubbed 'the black widow pulsar' because its intense, high-energy radiation is slowly evaporating its unfortunate companion star.

In 1992, three planets were found, using the Doppler effect and the $(O - C)$ method, orbiting the millisecond pulsar B1257+12. Did they survive a supernova explosion, or were they somehow formed after the supernova? A pulsar B1620–26 in the globular cluster M4 is orbited by a white dwarf, and also by a planet!

In 2003, a binary pulsar in which relativistic effects are even stronger than in B1913+16 was found, and, in January 2004, a binary system of *two* pulsars was discovered. It has the potential to provide detailed understanding of relativistic orbital motion, beaming of radiation by a strong magnetic field, and internal structure of a neutron star.

In 1979, an even more bizarre subset of pulsars was discovered: a gamma-ray burst named SGR0526-66, coming from the supernova remnant N49 in the Large Magellanic Cloud, was observed to show 8-second pulses. In 1981, an 'anomalous X-ray pulsar' (AXP) was discovered. Both turn out to be neutron stars with *ultra-strong* magnetic fields, now called *magnetars*. 'Normal' pulsars have magnetic fields of 10^{12} to 10^{14} Gauss. (Recall that the earth's magnetic field is about one Gauss.) These enormous 'normal' fields can be explained by the collapse of a normal star with a normal magnetic field. But magnetars have magnetic fields of 10^{15} Gauss! Unlike normal pulsars, magnetars are powered by the decay of their magnetic fields, not by rotation or accretion. A few AXPs have been detected at visible and IR wavelengths, and this helps to define their spectra – thermal radiation from their hot surfaces, modified by its passage through their ultra-strong magnetospheres, and further modified by absorption by interstellar gas and dust.

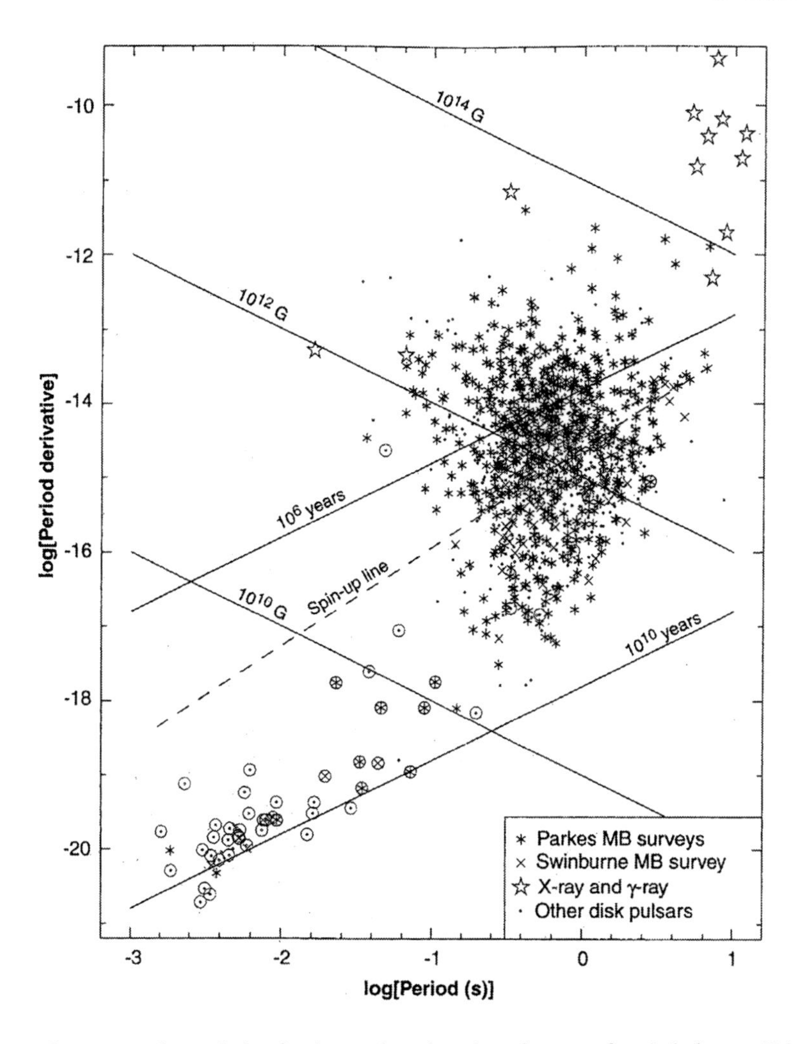

Figure 4.10 The periods of pulsars, plotted against the rate of period change. Using theoretical models, the age and the strength of the magnetic fields of these pulsars can be deduced. The star symbols in the upper right are *magnetars* with ultra-strong magnetic fields. The circled dot symbols in the lower left are millisecond pulsars that have been 'spun up' by accretion. Normally, pulsars evolve from the upper left to the lower right, as their rotation slows down. (From CSIRO Australia.)

On 27 December 2004, scientists detected a giant flare from the surface of the magnetar SGR 1806-20. Though the flare energy was mostly in the form of gamma rays, and not visible at optical wavelengths, the total apparent brightness actually exceeded that of the full moon! This magnetar is 50 000 light years away. Imagine the effect on the earth if the magnetar had been nearby!

Like pulsars in general, AXPs test the very limits of modern-day physical theory. Their observed properties give astronomers information about the nature and properties of matter under the most extreme possible conditions.

The key properties of pulsars are shown together on a graph which relates their rate of period change to their periods (figure 4.10), which shows their estimated ages and magnetic fields as well.

Pulsars connect with about two dozen major fields of physics and astronomy, and are therefore one of the most exciting objects in the 'celestial zoo'.

5

Eclipsing variable stars

Although a substantial fraction of stars belong to binary or multiple star systems, most of these are *wide binaries*; the average separation of the stars is tens or hundreds of times greater than the average distance of the earth from the sun. These are usually seen as *visual binaries*, if they are nearby. They are not usually variable in brightness for geometric reasons, because, with such large separations, they are unlikely to eclipse as seen from the earth, or to influence each other. *Close binary stars*, however, are usually variable in some way. That is because the probability of an eclipse is greater, for geometric reasons. And the degree of physical interaction is much greater.

Normally, close binaries will not be visual binaries, but techniques such as optical interferometry and adaptive optics will gradually enable astronomers to 'see' the two components of some close binaries. Visual binaries, spectroscopic binaries, and eclipsing binaries will increasingly overlap.

For an excellent introduction to close binary stars, see Hilditch (2001).

5.1 Overview

Eclipsing variables (figure 5.1) are binary stars in which the observer sees the orbit almost edge-on. One star periodically eclipses the other and, at these times, the total brightness of the pair decreases. The brightness change is a geometrical effect; there is not necessarily any physical or intrinsic change in the stars. (In *close binary stars*, however, there may be.)

Binary stars are of interest for several reasons. First of all, at least half of all stars are binary or multiple stars, so they are a normal occurrence in our universe. At least 3 per cent of bright stars are eclipsing variables, based on the sample of the 300 brightest stars.

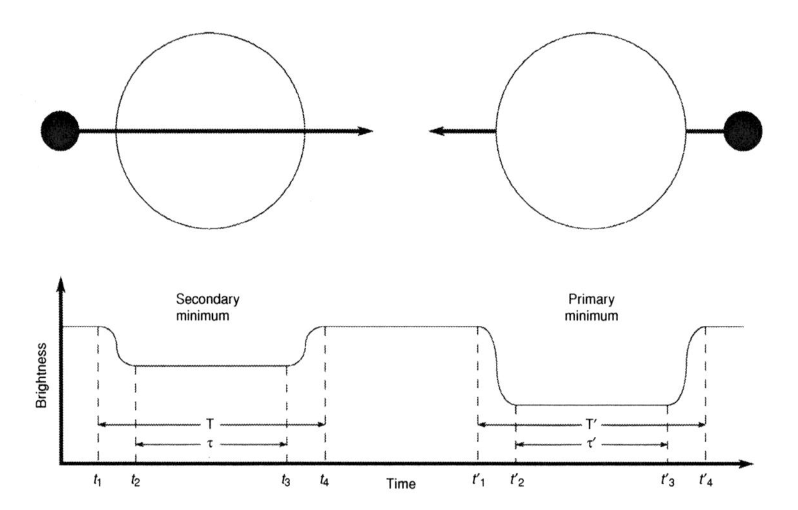

Figure 5.1 Schematic picture of a simple eclipsing binary, and its light curve. In the light curve below, time increases to the right. As the smaller, brighter star (in black in this diagram) passes in front of the larger, fainter star (left), there is a secondary minimum. t_1, t_2, t_3, and t_4 are the times of the four *contacts*. As the brighter star passes behind the fainter one (right), there is a primary minimum.

The mode of origin of binary stars is still unclear. Close binaries were once thought to result from the fission of a single rapidly rotating star, but this is no longer a viable hypothesis. Wide binaries probably result from two independent condensation events in the star-forming process. A few – especially in dense stellar environments – may be gravitational captures or exchanges.

Binary stars tell us much about the physical properties of stars, as was discussed in section 2.9. If an eclipsing variable is also a spectroscopic binary, many of its properties can be deduced. For instance, the spectroscopic observations give $m_1 \sin i^3$ and $m_2 \sin i^3$ which then give m_1 and m_2 if i is known, which it will be if the system eclipses; it will be close to 90°. The detailed shape of the light curve of an eclipsing variable provides information about the size, temperature, and luminosity of the two stars.

Eclipsing binaries are also of interest because they normally consist of stars in close orbit. If the stars were far away from each other, eclipses could be seen only if the observer was almost exactly in the orbit plane, which would be unlikely; if the stars were close to each other, the observer could be further from the orbit plane and eclipses could still be seen. If the two stars are close to each other, they will exert tidal forces, distorting each other's shape, and possibly pulling material from one star to another. This is especially true if one of the stars begins to swell as a result of its evolution. The evolution of close binary stars is quite different from the evolution of single stars; in many ways, it is

Table 5.1. *Bright and/or interesting eclipsing variable stars*

Name	HD	V Range	Period (d)	Spectrum	Comments
U Cep	5679	6.75–9.24	2.4930475	B7Ve + G8III–IV	EA/SD; deep eclipse
ζ Phe	6882	3.91–4.42	1.6697671	B6II + B9V	EA/DM
RZ Cas	17138	6.18–7.72	1.195247	A2.8V	EA/SD; deep eclipse
β Per	19356	2.12–3.39	2.8673043	B8V + G7III	EA/SD; prototype
λ Tau	25204	3.37–3.91	3.9529478	B3V + A4IV	EA/DM
ϵ Aur	31964	2.92–3.83	9892	A8Ia–F2epIa + BV	EA/GS; long period
ζ Aur	32068	3.70–3.97	972.16	K5II + B7V	EA/GS; long period
δ Ori	36486	2.14–2.26	5.732476	O9.5II–III + B0III	EA/DM
BM Ori	37021	7.90–8.65	6.470525	B2V + A7IV	EA; in Trapezium; θ^1 Ori B
V1016 Ori	37020	6.72–7.65	65.43233	O7 + B0.5Vp	in Trapezium θ^1 Ori A
β Aur	40183	1.89–1.98	3.9600421	A2IV + A2IV–V	EA/DM
UW CMa	57060	4.84–5.33	4.393407	O7Ia:fp + OB	EB/KE
YY Gem	60179C	8.91–9.60	0.81428254	M1Ve + M1Ve	Castor C
V Pup	65818	4.35–4.92	1.4544859	B + B	EB/SD
W UMa	83950	7.75–8.48	0.33363749	F8Vp + F8Vp	EW/KW; prototype
RS CVn	114519	7.93–9.14	4.797887	F4IV–V + K0IVe	EA/AR/RS; prototype
δ Lib	132742	4.91–5.90	2.3273543	A0IV–V +	EA/SD
α CrB	139006	2.21–2.32	17.359907	A0V + G5V	EA/DM
V1010 Oph	151676	6.1–7.0	0.66142613	A5V	EB/KE
u Her	156633	4.60–5.28	2.0510	B1.5V + B5III	–
W Ser	166126	8.42–10.20	14.15486	F5eIb(shell)	EA/GS
β Lyr	174639	3.25–4.36	12.913834	B8II–III +	EB; prototype
VV Cep	208816	4.80–5.36	7430	M2epIa + B8:Ve	EA/GS + SRc; prototype

more exciting. The radiation from one component, usually the hotter one, may also irradiate the other component, producing an observable effect in the light curve.

Nevertheless, professional astronomers' interest in eclipsing variables waxes and wanes. In North America, interest in stars and their fundamental properties has ebbed in recent years. With the advent of new optical interferometers in the near future, with the development of higher-precision radial velocities and photometry, and even with the need to distinguish exoplanet transits from shallow stellar eclipses, eclipsing variables may once again become 'popular'.

Thousands of eclipsing binaries are known, some little-studied, some well-studied, and some quite bizarre. As a result of sky surveys, even more are being discovered. Amateur astronomers could usefully learn to analyze and interpret these, as well as to observe them. Table 5.1 lists several typical or interesting eclipsing variables.

Research on eclipsing binaries is coordinated by IAU Commissions, 26 (double stars), 27 (variable stars), and 42 (close binary stars), which publish triennial reports and the *IAU Information Bulletin on Variable Stars*. There is a *Bibliography and Program Notes on Close Binaries* on the website of IAU Commission 42; see

http://www.konkoly.hu/IAU42/

and a good source of ephemerides at:

http://www.as.ap.krakow.pl/ephem/

Eclipsing variables are included in the GCVS. Annual predictions of minima are contained in the *Rocznik Astronomiczny Obserwatorium Krakowskiego, International Supplement* which is available on-line at:

http://www.oa.uj.edu.pl/ktt/rocznik/rcz_wstp.html

Amateur work on eclipsing variables is valuable, and is coordinated by the AAVSO Eclipsing Binary Committee and other variable star observing organizations.

Solar eclipses are, of course, a type of eclipsing variability, as are transits of Mercury and Venus. But we shall not discuss them here.

5.2 Ellipsoidal variable stars

Ellipsoidal variable stars are components of close binary systems. They vary, but not due to eclipses. The shape of each star is distorted by the gravity of the companion. As explained in section 5.4.6, this will cause the brightness of the star to be non-uniform ('gravity darkening'). It will also distort the shape of the star into an egg shape. The stars will therefore change their combined apparent brightness, with a period equal to the orbital period, because of changes in the brightness and projected emitting area turned to the observer. There will, however, be two minima and two maxima during each orbital cycle.

The periods of binaries which are close enough to produce distortion are usually less than five days (because of the relation between the period and the size of the orbit), and, in some cases, less than a day. The total range is only a few hundredths of a magnitude, so careful photoelectric photometry is required to analyze them. The orbital period can be determined from radial velocity observations; the light curve has two maxima and two minima per orbital period: this is a test for identifying ellipsoidal variables. Otherwise, they might be confused with small-amplitude pulsating or rotating variables.

Because ellipsoidal variability is caused by gravitational tidal distortion, which in turn depends on the masses, radii, and separation of the components, it follows that a careful analysis of the ellipsoidal light curve can provide information on the masses and radii of the component stars. One interesting application of this technique was to the binary system HDE 226868 (V1357 Cyg). This system

Table 5.2. *Bright and/or interesting ellipsoidal variables*

Name	Period (d)	ΔV	Spectrum	V
ψ Ori	2.526	0.03	B0+B0	4.58
α Vir	4.014	0.10	B2V+B3Vi	0.96
π^5 Ori	3.700	0.05	B3+?	3.72
o Per	4.419	0.03	B1III+B	3.82
b Per	1.527	0.06	A2+?	4.54
V1357 Cyg	5.599824	0.21	O9.7Iab + black hole	8.95

contains one normal component, and one component which is a black hole. The ellipsoidal variability of the normal component gives some information about the mass of the black hole.

Table 5.2 lists the names, periods (in days), amplitudes, spectral types, and *V* magnitudes of five notable ellipsoidal variables.

Figure 5.2 shows the light curve of ψ Orionis. Surprisingly, the ellipsoidal variable α Virginis (Spica) is not listed as ELL in the GCVS4. It was once a Beta Cephei star, but its amplitude decreased to zero.

5.3 Classification of eclipsing variables

As with other types of variable stars, the traditional classification scheme for eclipsing variables is based on the outward appearance of the light curves, which reflects the presence or absence of the complications discussed below. At one time, any variable star with brief minima was suspected to be an eclipsing variable. The GCVS still recognizes the following classes:

- *Algol* (EA in the GCVS) have light curves with almost flat maxima; the stars are almost undistorted.
- *Beta Lyrae* (EB in the GCVS) variables have light curves which are slightly rounded; the stars are distorted into ellipsoids.
- *W Ursae Majoris* (EW in the GCVS) variables have light curves which vary continuously; the stars are essentially in contact. The periods are short, generally less than a day, the minima are approximately equal in depth (a few tenths of a magnitude), and the stars are usually F to G type or later.

This classification is almost obsolete. In fact, it is rather misleading. Beta Lyrae is so bizarre that it should not be a prototype for any class. The EA and EW stars do, however, represent two ends of a spectrum. Algol binaries are now recognized as semi-detached systems, whereas they used to be considered

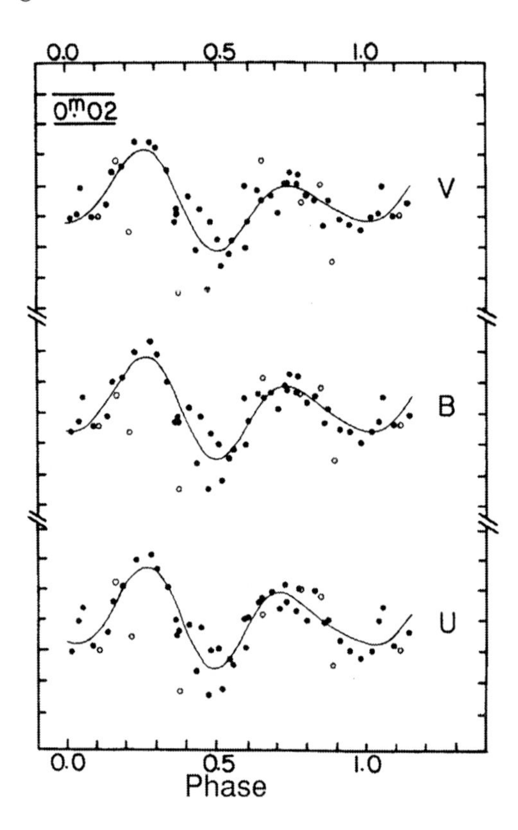

Figure 5.2 Light curve of the ellipsoidal variable ψ Orionis. The variability results from the gravitational distortion of one component by the other. (From Percy, 1969.)

together with detached binaries. EA is now taken to mean 'detached' and EB is associated with 'semi-detached'.

Modern classification of eclipsing variables (and binary stars in general) is based on the concept of *Lagrangian surfaces* and *Roche lobes*. These are shown in figure 5.3. Around a *single* point mass (or star) the gravitational potential energy is constant on spheres around the star, and is a function only of the distance from the centre of the star. Specifically, the potential energy is inversely proportional to the distance. In a system of *two* orbiting point masses (or stars), the gravitational potential energy is constant on the surfaces shown in figure 5.3. Close to each star, the surfaces are almost spherical. But there is a 'critical' surface – the one which is hour-glass shaped (or figure-eight shaped, for those not familiar with hour-glasses, though, as noted below, the surfaces are three-dimensional like an hour-glass, not two-dimensional like a figure eight). The two halves of the hour-glass are called *Roche lobes*. The point of intersection of the hour-glasses is the *inner Lagrangian point*. The surface of the ocean is

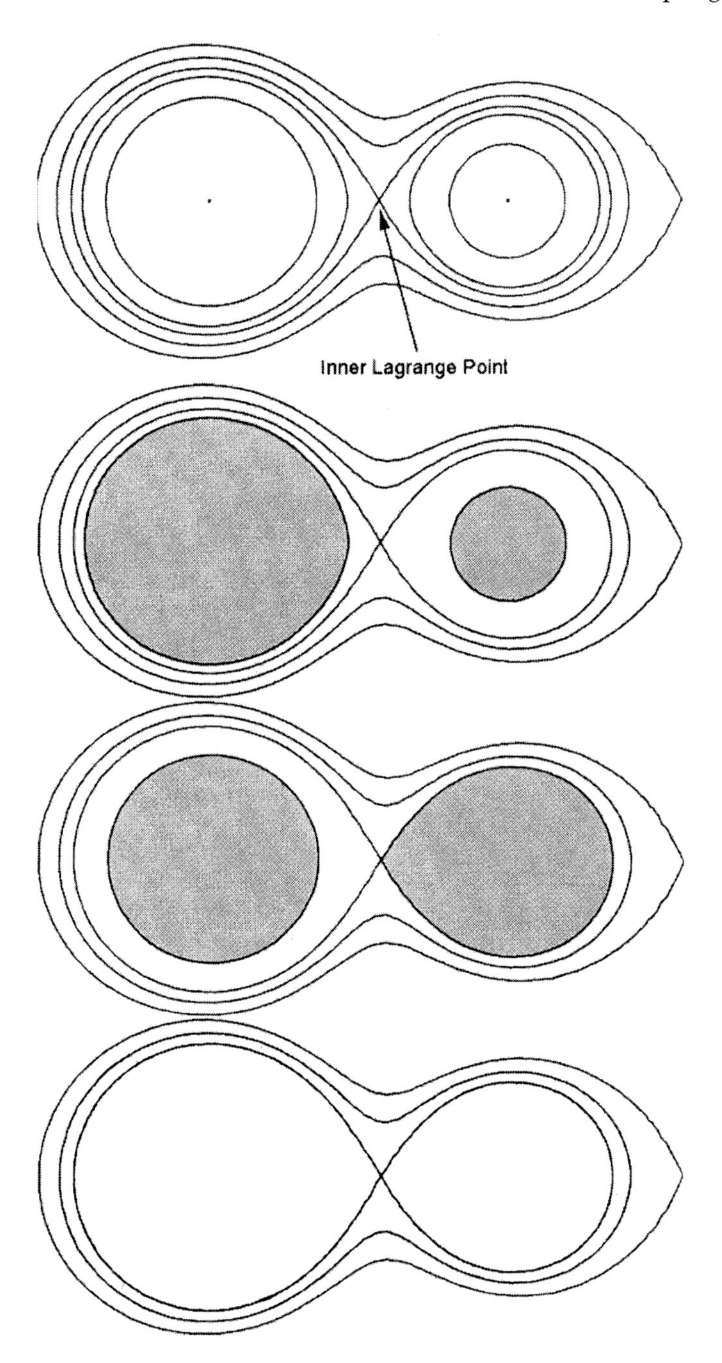

Figure 5.3 The Lagrangian surfaces around a pair of stars in mutual orbit. In order from the top: (i) the Lagrange surfaces in general; (ii) a detached binary, in which both stars are smaller than their Roche lobe; (iii) a semi-detached system, in which one star fills its Roche lobe but the other does not; (iv) in a contact system, both stars are contained within a common envelope of material. (From Terrell, 2001.)

an equi-potential surface, or surface of constant gravitational potential. To first approximation, it is a sphere. The rotation of the earth produces an equatorial bulge, and the tidal forces of the sun and moon produce additional deviations of the ocean surface from a sphere.

The first thing to remember is that the Lagrangian surfaces are three-dimensional; what is shown in the figure is a cut in the orbital plane of the stars. The gravitational potential is, in effect, a fourth dimension. And the stars exist within a gravitational potential 'well'. At each point on a Lagrangian surface, the same energy is required to remove a particle of a given mass (1 kg, for instance) to infinity. The next thing to remember is that the motion of a particle (of gas, for instance) in the system is determined not only by gravity, but also by the particle's *angular momentum*. The particle's angular momentum will be conserved (which means it will remain constant) as it moves.

If one of the stars slowly expands, due to evolution, it will eventually fill its Roche lobe, much like water rising up in a well from an underground spring. If the star expands further, the 'path of least resistance' for the expanding matter is through the *inner Lagrangian point* – called L1 – into the Roche lobe of the other star; the Roche lobe will overflow. The material falls toward the other star, but, since it is carrying angular momentum, it falls in a spiral pattern. The modelling of binary star evolution is even more challenging than the modelling of single stars, since the matter streams and accretion discs are very dynamic phenomena.

Binaries can thus be classified as follows:

In *detached binaries*, both components are well within their Roche lobes. Tidal distortion is minimal, and the stars are almost spherical. The masses, radii, and temperatures can be determined reliably from eclipse light curves and from radial velocities. The GCVS4 distinguishes between D, DM (detached main sequence systems), DS (detached systems with a subgiant), AR (detached systems of the AR Lacertae type, with subgiant components not filling their inner Lagrangian surfaces), and DW (systems similar to W Ursae Majoris systems, but not in contact).

In *semi-detached binaries* (SD in the GCVS4), one component fills its Roche lobe; the other is well within its Roche lobe. The former star is distorted, and is probably losing mass through the inner Lagrangian point; the latter star is almost spherical (figure 5.4).

In *contact binaries* (K in the GCVS4), both components fill their Roche lobes, and are essentially in contact. There may also be a common envelope of material surrounding the two stars, blurring their individual identities. In that case, they are referred to as an *over-contact binary*. The GCVS4 subclassifies them as hot (KE) or cool (KW).

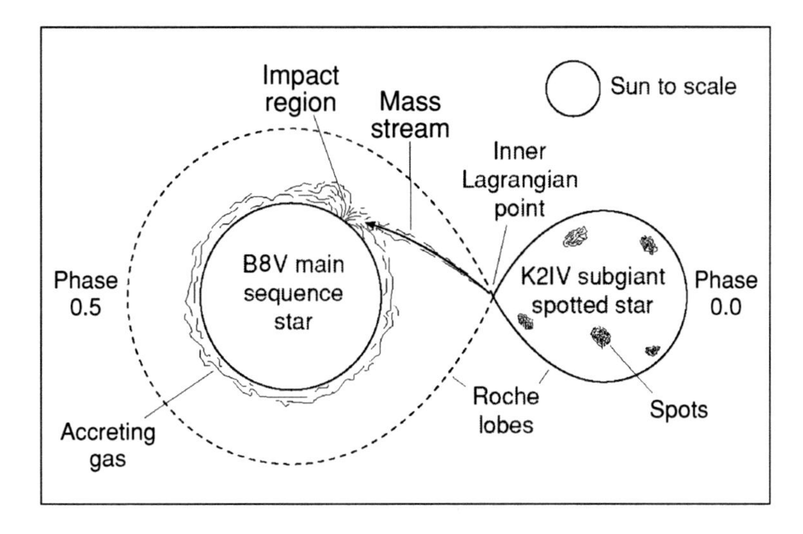

Figure 5.4 Schematic view of mass transfer from one component to the other. Specifically, this represents Algol, a K2IV star filling its Roche lobe, and a B8V star on to which the mass stream impacts. Note the sun, for scale. (Jeff Dixon Graphics.)

As is often the case, the GCVS classification system is somewhat archaic. Researchers in the field more commonly use the EA, EB, EW system, but modified as mentioned above.

Another means of classification, of course, is according to the types of stars in the system. Those containing white dwarfs, neutron stars, or black holes can be especially interesting. Those with giant or supergiant components will be interesting if they cause the system to be semi-detached, or contact. The GCVS4 includes this third system of classification: GS for systems with one or both giant components, PN for systems including a planetary nebula nucleus, RS for RS Canum Venaticorum systems, WD for systems containing a white dwarf, and WR for systems containing a Wolf–Rayet star. Since binaries with such exotic components are often eruptive variables, they will be discussed more in chapter 7.

The system of classification of binary variables is complex, and it is advantageous to focus on the physical *processes* which cause the variability in each case.

5.4 Analysis of eclipsing variables

The basic observational data for eclipsing variables consist of the light curve and, if possible, one or more colour curves (e.g. $(B - V)$ and $(U - B)$). The accuracy and reliability of these determines the accuracy of the stellar properties determined from them. The observations must be carefully made, and carefully reduced and transformed to a standard photometric system. Ideally, the magnitudes and colours should be free of both random and systematic errors at the millimagnitude level. Needless to say, spectroscopic observations are also important in the analysis of eclipsing variables. If radial velocity observations can be made, then the star can be analyzed as a spectroscopic binary, though rotation and mass transfer can sometimes complicate the measurement of the radial velocities. Recent advances in spectroscopic instrumentation and detectors, and analysis techniques (notably the use of cross-correlation to measure the velocities) have had a strong positive effect on our understanding of these stars. Radial velocities should now be measurable to much better than km s^{-1} accuracy.

Eclipses of a star may be *complete* if the eclipsing star covers all of the eclipsed star (a *total* eclipse) or leaves a ring of the eclipsed star uncovered (an *annular* eclipse) or it may be *partial* otherwise. The most conspicuous eclipses are total eclipses in which the eclipsed star is the brighter one.

A relatively simple form of analysis, and one to which amateur astronomers make an important contribution, is measuring the times of minimum of eclipsing binaries. This can be done using visual, photoelectric, or CCD techniques. Often, the times of minima are known approximately, so observers can measure the star around that time. The light curve is observed through the minimum and, using one of several possible techniques, the time of mid-eclipse is determined. Using the $(O - C)$ technique, these times can be used to monitor the period changes in the binary (Markham, 2004).

5.4.1 *Period*

The period is the interval between successive eclipses of one star by the other. The deeper eclipse is called the *primary eclipse*; the shallower one is called the *secondary eclipse*. The period is thus the interval between successive primary or secondary eclipses. The periods of eclipsing variables are generally in the range of 0.25 to ten days because, as mentioned above, most eclipsing variables are likely to be close binaries. But some eclipsing variables – especially those containing giant or supergiant components – have periods as long as several decades.

The study of period changes in eclipsing variables is particularly important, and is within the scope of amateur or student observers. If the eclipses are deep, they can be observed visually, and if they are relatively brief, the time of mid-eclipse can be determined with an accuracy of a few minutes. This kind of measurement must be continued over many years or decades, and organizations such as the AAVSO are able to co-ordinate and carry out such studies. Remember that the effects of period changes are *cumulative*! The period is studied using the standard $(O - C)$ or phase-shift method (section 3.3.4). The result for Algol is shown in figure 5.7. The curvature of the diagram shows that the period is not constant.

Period changes in binary stars are usually caused by *mass transfer* from one component to the other. Sometimes mass is lost from the system entirely. The rate of period change (denoted $\dot{P}$, which means dP/dt) is related to the rate of mass transfer or loss (denoted $\dot{M}$). In very close binaries, the period may also change due to the loss of energy through tidal interaction, or through gravitational radiation – an interesting and well-verified prediction of Einstein's general theory of relativity.

5.4.2 *Eccentricity*

If the orbit is circular, the velocities of the stars in their orbits will be constant, and the primary and secondary eclipses will be separated by exactly half a period. If the orbit is eccentric, this will not be the case unless the *line of apsides* (the major axis of the orbit) points to the observer. This is because the stars move more rapidly in their orbits when they are near *periastron* – their point of nearest approach. The eccentricity and the orientation of the line of apsides can also be determined from radial velocity observations, as well as from the relative timing of primary and secondary eclipses. In close binaries, the eccentricity is usually close to zero, because tidal interactions between the stars tend to circularize the orbit.

The line of apsides may slowly rotate in space – an important effect called *apsidal motion*. It arises because of the gravitational distortion of the stars by each other, and it provides important information about the interiors of the stars. If the system is particularly simple or 'clean', then it may even be used to test the general theory of relativity!

5.4.3 *Inclination*

The angle between the orbit and the plane of the sky is called the *inclination*, and is denoted by i. In an eclipsing variable, i is normally close to 90°. This information is particularly useful if the star is also a spectroscopic binary. In such binaries, the masses (multiplied by $\sin i$) can be determined; if i is close to 90°, then $\sin i$ is close to 1.0. If i is close to, but not equal to 90°, then the eclipses may be partial, but the inclination can still be determined, and the masses calculated.

5.4.4 *Size of the orbit and of the stars*

The determination of these quantities depends on a detailed analysis of the light curve. This is usually done automatically, using sophisticated computer codes developed for this purpose. However, consider the simple cases shown in figure 5.1. One of the stars is pictured as small, hot, bright, and more massive; the other is large, cool, faint, and less massive. This is the case in Algol, for instance. Then we may have: (i) a total and an annular eclipse, both central; (ii) a total and an annular eclipse, both non-central; (iii) two partial eclipses. The following discussion will be based on the schematic light curve and orbit in figure 5.1. The orbits are assumed to be circular and to be seen exactly edge-on, so that the eclipses are central.

On the light curve, eight times are marked; they are called *contacts*. The significance of the first four contacts is shown in the figure, which shows the small, bright star being eclipsed by the large, faint one. Now note that

$$\theta(rad) = 2\pi(t_4' - t_1')/P = D_2/a_2 \tag{5.1}$$

$$\phi(rad) = 2\pi(t_2' - t_1')/P = D_1/(a_1 + a_2) \tag{5.2}$$

where D_1 and D_2 are the diameters of the two stars, a_1 and a_2 are the radii of the orbits about the centre of gravity, and P is the period. To understand the above formulae, think of the times as fractions of an orbital period, and the angles as fractions of an orbital circle. These formulae show how the light curve can be used to deduce the sizes of the two stars, in units of the size of the orbit – which is obtained from the spectroscopic analysis of the star.

5.4.5 *Luminosity and temperature*

Now consider the brightness changes in the light curve, (Figure 5.1). The brightness is given in physical units, rather than in magnitudes. The brightness levels are x (maximum), y (secondary eclipse), and z (primary eclipse). The depths of primary and secondary eclipse are therefore $x - z$ and $x - y$, respectively.

Let the areas of the smaller and larger stars be A_1 and A_2, respectively, and let their surface brightnesses (a measure of the energy emitted from each unit of area) be B_1 and B_2. In each case, $A \sim D^2$ (geometry) and $B \sim T^4$ (Stefan's law), where T is the temperature of each star. Now note that

$$x = A_1.B_1 + A_2.B_2 \tag{5.3}$$

$$y = (A_2 - A_1).B_2 + A_1.B_1 \tag{5.4}$$

$$x - y = A_1.B_1 \tag{5.5}$$

$$Z = A_2.B_2 \tag{5.6}$$

$$x - z = A_1.B_1 \tag{5.7}$$

Thus

$$z/(x - y) = A_2/A_1 = (D_2/D_1)^2 \tag{5.8}$$

$$(x - y)/(x - z) = B_2/B_1 = (T_2/T_1)^4 \tag{5.9}$$

$$(x - z)/z = A_1.B_1/A_2.B_2 = L_1/L_2 \tag{5.10}$$

where L_1 and L_2 are the luminosities of the two stars. These formulas show how the physical properties of the stars can be deduced from the light curve.

5.4.6 *Complications*

The preceding discussion was based on a highly schematic picture of the light curve and of the orbit. Some of the complications of reality are:

Partial eclipses

If neither star is totally eclipsed, the analysis is obviously more complicated, though the principles remain the same.

Limb darkening

It has been assumed that all parts of the stellar discs have the same brightness. In fact, the disc becomes less bright towards the limb (edge) of the star. This effect is called *limb darkening*, and it occurs because, near the limb, we are seeing obliquely into the star's atmosphere. We see only to shallower, cooler levels, which are not as bright. This effect can be seen directly on the sun, even with a small telescope.

The standard formula for limb darkening is

$$B(\theta)/B(0) = (1 - k) + k \cos \theta \tag{5.11}$$

where θ is the angle between the observed point on the disc and the observer, as seen from the centre of the star, and k is a constant which has a value of approximately 0.6; it can be calculated from a model of the star's atmosphere.

The effect of limb darkening is to make the bottom of the eclipses bowl-shaped, as well as to change the shape of the decline from and rise to maximum light. These parts of the light curve therefore provide useful information about the value of k, which in turn can be used to test models of the atmosphere of the star.

Gravity darkening

If the two stars in the system are particularly close to each other, they will distort each other due to tidal effects, in much the same way as the moon and sun distort the ocean layer on the earth. Each star assumes an ellipsoidal shape, and the ends of the ellipsoid are cooler and therefore less bright. Even if an eclipse does not occur, the changing orientation of the ellipsoids, as seen by the observer, causes a slight change in the brightness of the system.

The effect of gravity darkening is therefore to cause changes in the brightness of the stars outside eclipse. The top of the light curve is no longer flat, but is rounded. It is brighter between the eclipses then near them.

Since the size of these gravity darkening effects will depend on the masses, radii, and separations of the stars, it follows that an analysis of the light curve *outside eclipse* will provide useful information about these properties of the stars, as is the case with ellipsoidal variables.

Reflection

If the two stars in an eclipsing variable are very close, then some of the flux from each star will strike the other star and be 'reflected' (or rather absorbed and re-emitted). This effect is most pronounced just before the eclipse of the cooler, fainter star; it produces 'shoulders' on the secondary eclipse. The reflection effect can be particularly interesting if one component of the star is a source of UV or other high-energy radiation, and the other is a cooler star.

Starspots

We have assumed that the surfaces of the stars are 'immaculate'. But cooler stars (like the sun) have starspots, which are produced by the star's magnetic field, which in turn is produced by the star's rotation. The rotations of the components of a close binary are often 'spun up' by tidal interactions between the two stars. The RS Canum Venaticorum stars, discussed in section 4.4, are an extreme example of this phenomenon. The starspots can affect any part of the light curve.

A third star

A somewhat different kind of complication is the presence of a third star in the system. This will complicate both the light curve analysis and the spectroscopic analysis. A substantial fraction of all binary stars are actually triple systems.

Several of these complications can occur in the same system. In principle, the more complicated a binary and its light curve are, the more interesting it is, and the more information it can provide. But the analysis becomes more and more complex, difficult, and challenging.

5.4.7 *Analysis: Modern approaches*

The foregoing discussion was based on an examination of individual features of a schematic light curve, using individual basic physical principles. For many decades, this approach was the one which was used in the actual analysis. Henry Norris Russell (of H–R diagram fame) was known for his research in this area. The complications, mentioned above, were allowed for in the parts of the light curve in which they were most prominent, and the simple geometrical principles, also mentioned above, were used to derive the stars' properties. With modern computers, however, it is possible to use all parts of the light curve, together with the full range of physical principles, including the complications.

Historically, one approach was the *Fourier analysis* of the light curve. This approach was developed by Zdenek Kopal and his students. In this approach, the shape of the light curve is expressed in terms of a set of periods or frequencies, in much the same way as a complex sound can be expressed as the sum of a set of component frequencies called *harmonics*. It is something like the method of Fourier decomposition discussed in section 3.3.3. It is mathematically elegant, but the connection between the harmonics and the physical properties of the stars is not directly obvious, and the method is no longer used.

The second approach involves *light curve synthesis*. Estimates of the physical and geometrical parameters of the stars are first made. An elegant computer code then synthesizes a theoretical light curve, using correct and complete laws of physics (as opposed to simplifications), compares it with the observed light curve, then modifies the estimates of the physical and geometrical parameters until the theoretical and observed light curves agree. The most widely used synthesis methods have been the LIGHT program developed by Graham Hill, Slavek Rucinski, and others, the WINK method developed by D. B. Wood, and the Wilson–Devinney method primarily developed by Bob Wilson. These methods have become more and more elegant, incorporating model atmospheres for the stars, and other technical and physical refinements. Powerful computers make these refinements possible. See Wilson (1994) and Kallrath and Milone (1999) for further details.

These methods are based on the classic Roche model (section 5.3) and its extensions. The stars are point masses, moving in circular orbits. The stars' rotations are synchronous with the revolution, and their structure is the same as if they were single objects. Note that, because of tidal interactions between the components, close binaries will normally have circularized orbits, and rotations synchronized with their revolution. Deviations from these assumptions can be introduced as extensions of the model. Unfortunately, computer modelling of eclipsing binary light curves is not totally automatic. It requires an experienced eye, both to guide the computer to plausible solutions, and to spot complications or peculiarities which should be followed up. Because the methods are simultaneously fitting many geometrical and physical parameters, there is a danger that they will find two or more sets of parameters which appear to fit the light curve. But only one set is 'right'. This is when the judgement of an expert human being is essential.

5.5 Detached eclipsing variables

Some of the most complex, unusual, and interesting eclipsing variables are those in which there is strong interaction between the stars, which distorts the stars and pulls mass streams from one to the other. On the other hand, the 'simple' eclipsing variables which are well-separated and non-interacting, are equally important. Precise photometry and precise radial velocities of these stars can reveal their physical properties – mass, radius, luminosity, and temperature – to an accuracy of better than 1 per cent. These data provide a powerful test of our understanding of stellar structure and evolution, particularly if one or both of the stars has evolved away from the main sequence. See Andersen (1991) for an excellent review.

The secret, of course, is to use radial velocity and photometry data which is free of both random and systematic errors. See section 2.16 for a discussion of modern radial velocity measurement techniques. Accuracies of up to a few m/sec is now possible. In measuring the radial velocities of the absorption lines of each component, any blending – either by other lines in the spectrum, or by lines in the spectrum of the second component – must be accurately known. Photometry in at least two wavelength bands is necessary. There are several methods and programs for light-curve modelling, as mentioned above.

In terms of clean geometry, and high-accuracy stellar parameters, perhaps the best example is DM Virginis (Latham *et al.*, 1996). Special attention was paid to radial velocity precision. The masses of the components were determined to an accuracy of 0.6 per cent or better. Comparison with evolutionary models gave an age of 1.75 $\pm$ 0.2 billion years, with the main uncertainty being in the metal abundance in the stars. Even more interesting are TZ Fornacis (Andersen, 1991) and especially AI Phoenicis (Andersen *et al.*, 1988), because these two systems are significantly evolved. For the latter system, the metal abundance was determined spectroscopically, so it was possible to derive the masses to better than 0.3 per cent, and the age (4.65 billion years) with very little uncertainty. Such is the power of binary stars in determining stellar properties! (figure 5.5).

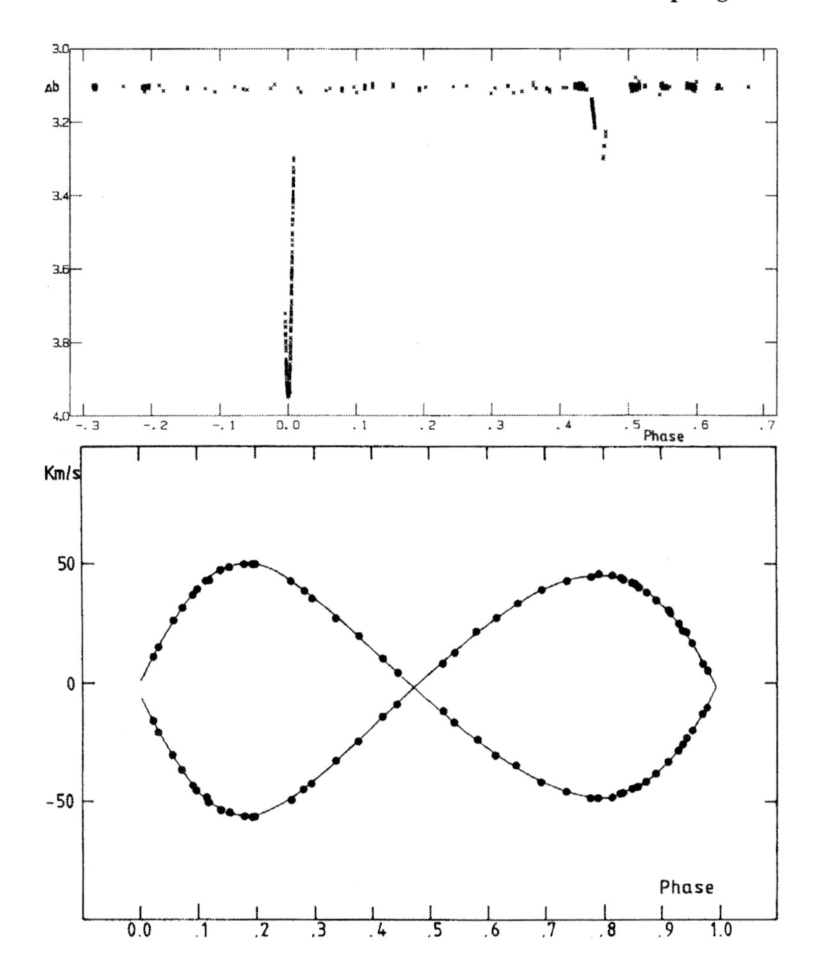

Figure 5.5 Light and velocity curves for AI Phoenicis, a 'clean' eclipsing binary. The velocity amplitudes of the two components (bottom panel) are about the same, so their masses are comparable. (From Andersen *et al.*, 1988.)

One of the problems with 'precision astrophysics', however, is that it requires astronomers to improve the most fundamental aspects of their science – the calibration of their photometric systems, and their knowledge of such atomic properties as opacity, for instance. And allowance must be made for the fact that many so-called binary stars are actually triple stars.

An interesting class of eclipsing binaries are those containing a pulsating component. If the binary is detached, and relatively wide, then the pulsation is only slightly affected by the presence of the binary companion. If the binary components are close, then the pulsation will be more complicated. For instance, I am interested in the behaviour of pulsating red giants in *symbiotic binary stars*, discussed later in this book.

Box 5.1 Star sample – Algol (β Persei)

Algol is the brightest eclipsing binary with deep eclipses, so it is easy for anyone to observe. Algol means 'Demon Star' in Arabic, which suggests that its variability might have been known in antiquity, though there is no concrete evidence to support this conjecture. The English astronomer John Goodricke is credited with the discovery of the periodicity of Algol in 1782–83, though it was apparently discovered independently by a German farmer named Palitzch. Goodricke and his colleague Edward Pigott also proposed that the variability might be caused by eclipses – not by a companion star, but by a planet revolving around the star. It was Edward Pickering who, in 1881, presented convincing evidence that Algol was an eclipsing binary star.

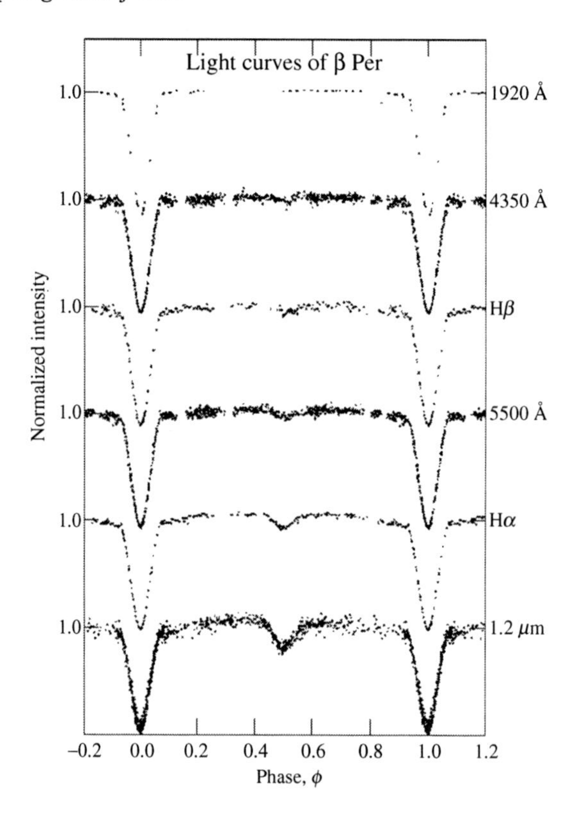

Figure 5.6 Light curves of Algol at different wavelengths. Note that the depth of the secondary minimum is negligible at UV wavelengths, because the eclipsed K2IV contributes negligibly at that wavelength. (From M. T. Richards, www.)

According to a study by Richards *et al.* (1988), the properties of the components of Algol are as follows: the primary is a B8V star with a mass of 3.7 ± 0.3 $M_{\odot}$ and a radius of 2.90 ± 0.04 $R_{\odot}$; the secondary is a K2IV star with a mass of 0.81 ± 0.05 $M_{\odot}$ and a radius of 3.5 ± 0.1 $R_{\odot}$; the latter is a rotating spotted star of the RS Canum Venaticorum type. There is a third component in a 1.86 year orbit around the close pair; it is a metallic-line A star with a mass of 1.6 ± 0.1 $M_{\odot}$. The system also contains circumstellar material which is lost from the K2IV star, and forms an accretion disc around the B8V star. For decades, it was a puzzle as to why the less massive star was more evolved, but mass transfer can now explain that puzzle.

Other consequences of mass transfer include radio emission including flaring, discovered by Hjellming in 1971, which arises from gas lost from the system, and X-ray emission, discovered by Harnden *et al.* in 1977, which arises from gas impacting on the accretion disc around the hot star.

Algol varies in *V* from 2.1 at maximum to 3.4 at primary minimum, with a period of 2.867315 days. This period, however, is slowly lengthening; long-term systematic visual or photoelectric observations by amateurs are useful for monitoring the period changes in this and other eclipsing stars. The primary eclipse occurs when the fainter K2IV star passes in front of the brighter B8V star, and lasts some ten hours in total. Because the eclipse is partial, the eclipse is not flat, but rounded. There is also a shallow eclipse when the B8V star passes in front of the K2IV star, but it can only be detected photoelectrically; see figure 5.6.

5.6 Semi-detached binaries

A semi-detached binary has one component which has filled its Roche lobe, and transfers mass to the other component, which is comfortably within its Roche lobe. The best-known example is Beta Persei – Algol; see the Algol 'star sample' for a good introduction.

As noted in the star sample, mass transfer has resulted in the more massive component being the less evolved. In fact, the less-massive component was initially the more massive, but it transferred a large fraction of its mass to its companion as evolution caused it to swell up.

A common feature of semi-detached binaries is an accretion disc around the accretor star. The accretion disc can be studied by taking the spectrum of the system, and subtracting the expected absorption-line spectra of the stars, suitably Doppler-shifted. The light from the accretion disc may also be visible in the light curve of the star, if it is carefully measured and analyzed. In some stars, the disc is so massive that its effects are clearly visible. In β Lyrae and ϵ Aurigae, it hides one of the stars completely.

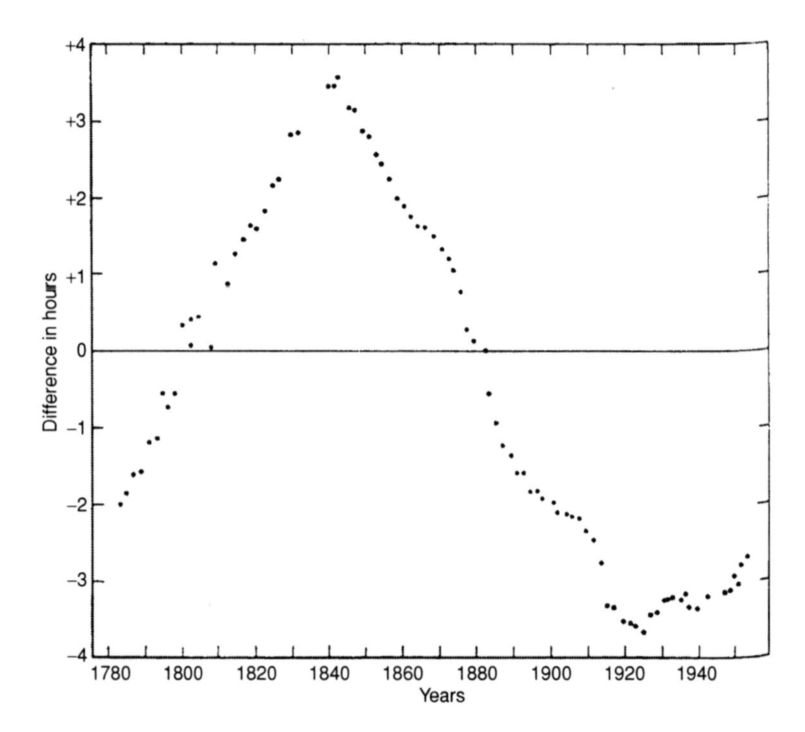

Figure 5.7 $(O - C)$ diagram for Algol, showing changes in period, due to mass transfer. The $(O - C)$s range up to several hours, and are easily measurable by visual observers. (From Herczeg and Frieboes-Conde, 1968.)

The accretion disc is not necessarily permanent, but may be transient if the mass transfer, or the emptying of the disc is episodic.

The gas in the accretion disc will eventually strike the accretor star, and produce a visible effect. If the accretor star is a white dwarf, neutron star, or black hole, the system may very well be an eruptive variable star; these are discussed in chapter 7.

The mass transfer may also lead to period changes in the system. These can be studied using the $(O - C)$ method. Long-term studies of this kind have been made, both by professional astronomers, and by organizations such as the AAVSO (figure 5.7).

5.7 W Ursae Majoris stars – contact binaries

W Ursae Majoris variables are 'eclipsing binaries with periods shorter than one day, consisting of ellipsoidal components that are in contact, and having light curves for which it is impossible to specify the moments of the beginning and the end of the eclipses; the depths of the primary and secondary minima are almost equal or differ quite insignificantly. Light amplitudes are

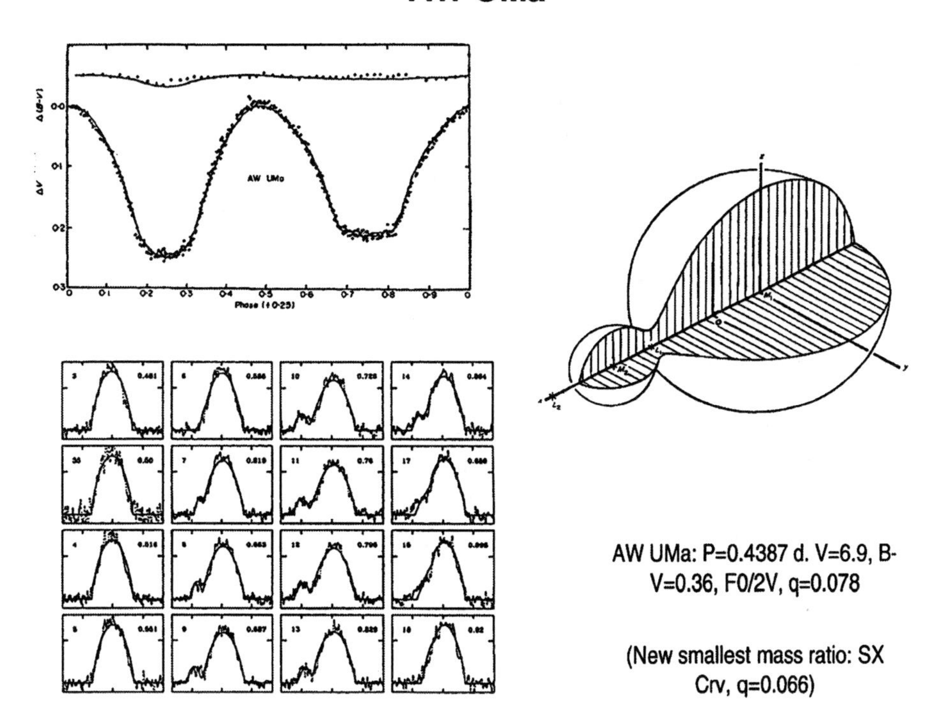

Figure 5.8 Light curve (upper left), line profiles showing two stellar components (lower left) and calculated model (right) for the W Ursae Majoris star AW Ursae Majoris, with the very low mass ratio 0.078 (the mass ratio is the ratio of the masses of the two components). The mass ratio of SX Corvi is slightly smaller. (From Mochnacki and Rucinski, 1999.)

usually 0.1 to one magnitude in V. The components usually belong to the spectral types F–G or later' (GCVS4). Unlike Algol and Beta Lyrae variables, the W UMa variables have a specific physical nature – they are *contact binaries*. The components fill their Roche lobes. Some over-fill their Roche lobes, and are connected via a *common envelope*. The two minima, per cycle, are of comparable depths, indicating that the two components have similar temperatures. This may be true even if the two components have unequal masses. That is because luminosity can actually be transferred between the components via the common envelope. So W Ursae Majoris stars are like 'Siamese twins'; they are physically connected, sometimes superficially, sometimes profoundly. Figure 5.8 shows a model of a W UMa star.

As dwarf stars in contact, they have small separations, and therefore short periods – 0.2 to one day. They range from A and F stars through G and K stars; the smaller, cooler systems have shorter periods. Most systems show period changes, some positive and some negative, suggesting that mass transfer can occur between the components in both directions. In some systems, the light curve shows evidence of starspots. These can be *Doppler imaged* by simultaneous radial velocity, light, and colour measurements. Furthermore, the light curve can change from week to week, indicating that the spots themselves undergo change. The study of W Ursae Majoris stars therefore illuminates not only the properties of the stars and their surfaces, but how they change with time. For instance, how do the spots, and their behaviour, differ in close binaries from what they would be like in single stars like the sun?

W Ursae Majoris stars are quite common; about one in 500 A to K dwarfs is a binary of this kind. Because of their brightness, large amplitudes, short periods, and changing light curves they can be usefully observed and analyzed by amateur astronomers.

Only a small fraction of known W Ursae Majoris stars have been well studied, and hundreds more of these systems are being discovered in sky surveys such as MACHO and OGLE. Good radial velocity curves are especially needed; at the University of Toronto's David Dunlap Observatory, Slavek Rucinski and his collaborators are using an elegant technique known as the *broadening function* to determine high-quality velocity curves for dozens of these systems; already, ten papers have appeared between Lu and Rucinski (1999) and Rucinski *et al.* (2005). Because it is difficult to disentangle the spectra of two interacting stars, this represents a significant contribution to the field. Rucinski and his collaborators have also shown that W Ursae Majoris stars obey a period–luminosity–colour relation, which can predict their absolute magnitude to within ± 0.25. When applied to large samples of W Ursae Majoris stars, this relation becomes a useful distance-determination tool.

These and other types of eclipsing binaries are potential 'yardsticks' within our galaxy, and within the local group – especially as massive sky surveys continue to expand. Millions of eclipsing binaries will be discovered by missions such as *Gaia*, the successor to *Hipparcos*. As with other types of variables, this raises the need for effective methods of *automatic classification* of variables from their light curves.

5.8 Symbiotic binary systems

Symbiotic binary systems are discussed in section 7.3.4. They are all binaries; many of them eclipse. We have included them under cataclysmic variables because they share many of the properties of those stars, including the close binary nature and – very often – outbursts. The binary nature is apparent from the spectrum of the star; it is a composite of the spectra of two stars, one hot and one cool (figure 5.10).

Box 5.2 Star sample – β Lyrae

Beta Lyrae (HD 174638, B8II-IIIep + (?), V = 3.25–4.36) is the brightest of eclipsing binaries; its variability can easily be seen with the unaided eye – just compare it with γ Lyrae. It is also the prototype of eclipsers with a continuous variation in brightness. Yet it has been so poorly understood that it is doubtful that it should be a prototype for anything! Harmanec (2002) has recently summarized the history, present knowledge, and unsolved mysteries of this star. Figure 5.9 is an artist's conception of the star system, taken from Harmanec's paper.

Its variability was discovered by 18-year-old John Goodricke in 1784, though Goodricke remarks that William Herschel suspected that either β or γ Lyrae was variable. Goodricke also proposed that the variability could be due to eclipses – or the rotation of a spotted star. The binary nature was later confirmed by spectroscopy, though the spectrum was (and still is) very complex. One star – the more massive – is not visible; it is a B1V star, hidden within a thick disc of gas and dust. The orbit is circular; it has a period of 12.94 days, and is increasing by 19 seconds per year; the period change was first noted by Argelander. At one time, it was believed that the total mass of the system was almost 100 solar masses. The best estimate of the masses is actually 3 solar masses for the visible

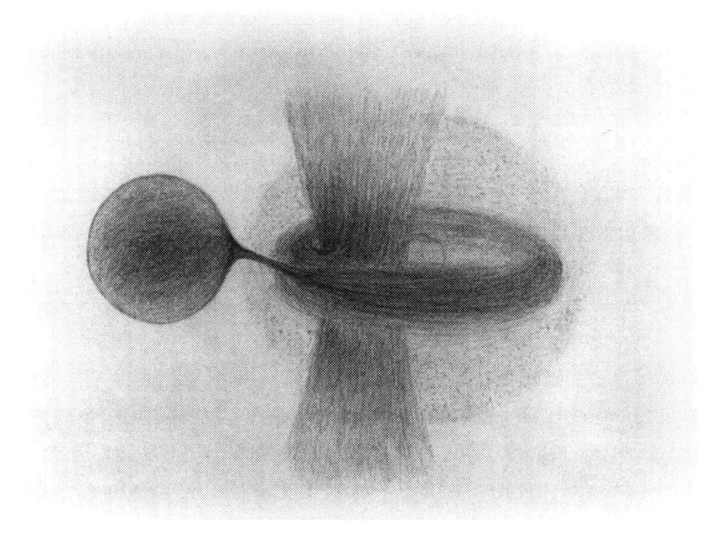

Figure 5.9 Artist's conception of Beta Lyrae, by Alexandra Kalasova, based on Harmanec (2002). (Figure provided by P. Harmanec.)

component, and 13 solar masses for the invisible one. Most of the mass of the invisible star has been transferred from the visible one!

The first evidence of the disc was the *visual* observation of H Beta emission by Secchi in 1866. Optical, IR, and UV spectroscopy since then has revealed its nature and motion. The spectrum, polarization, and radio observations have revealed jet-like structures flowing at right angles to the orbit.

The key to understanding the masses of the components, the presence of the disc, and the rapid change in period, is the fact that the visible component fills its Roche lobe, and is transferring mass and angular momentum through the inner Lagrangian point to the disc at a rapid rate – several earth-masses per year – as compared with other mass-transferring binaries. Modelling shows that this can explain many of the complexities of the system, the disc, and the radio jets.

With the next generation of optical interferometers, we should at last be able to 'see' this remarkable binary system.

The cool component is a red giant star, often pulsating. The hot component is usually a white dwarf. There is usually a gas stream and accretion disc around the white dwarf, and sometimes bipolar jets which are observed as expanding radio sources. The variability is extremely complex, with eruptions and flickering of the accretion disc, orbital variations due to ellipticity and illumination of the cool component, and pulsation of that component. The eclipses are often lost in the complexity!

5.9 VV Cephei stars

VV Cephei is the prototype of a small group of eclipsing variables with at least one of the components a supergiant. Such binaries have long periods, typically years to decades. This is because a binary must have a large separation in order to accommodate a supergiant, and larger separations mean longer periods.

VV Cephei consists of an M supergiant with $V \sim 5.25$ (but slightly variable as all M supergiants are) and a B9 companion with $V \sim 6.97$. The M star dominates in the IR part of the spectrum, the B star in the UV. Both components appear to be massive. Their separation is about 25 AU – greater than the distance of Uranus from the sun. The period is 20.3 years, which means that anyone who wishes to study the variability around the orbital cycle in detail must be patient. For about 600 days each period, the M star is in front of the B star. At this time, the

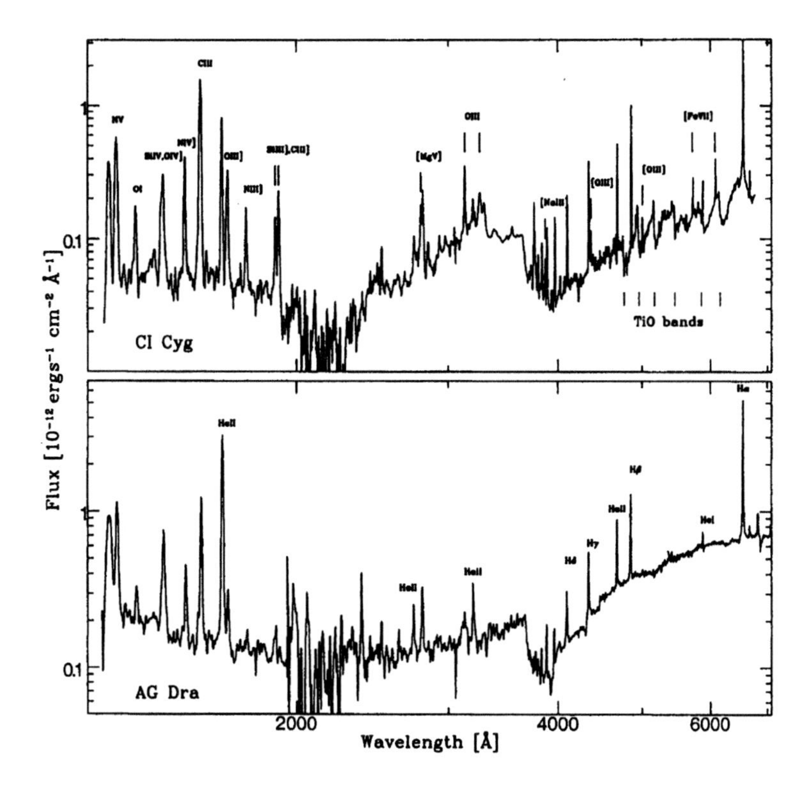

Figure 5.10 Spectrum of two symbiotic binary stars, CI Cygni and AG Draconis. In each case, the spectrum of the cool component can be seen at longer wavelengths (right). The spectrum of the hotter component, primarily the accretion disc, is visible at shorter wavelengths (left). Emission lines from the accretion disc are also prominent. (From Mikolajewska, 2003.)

blue magnitude of the system decreases by about 0.6, but the visual magnitude decreases only slightly; this is because, although the brightness of the B star is comparable with that of the M star in the blue, it is two magnitudes fainter in the visual.

As the B star enters and emerges from eclipse behind the M star, it shines through different layers of the M star's tenuous atmosphere. Here is the interest and importance of the VV Cephei stars: spectroscopic analysis of the light of the B star therefore tells us about the structure and properties of the M star's atmosphere (figure 5.11).

One star that is included in the VV Cephei class is ϵ Aurigae, even though it is a bizarre object; it is highlighted below. It is an excellent subject for observation by amateur astronomers using photoelectric techniques, but the interesting phases recur only every 27 years! Table 5.1 includes some bright VV Cephei stars, including ϵ Aurigae.

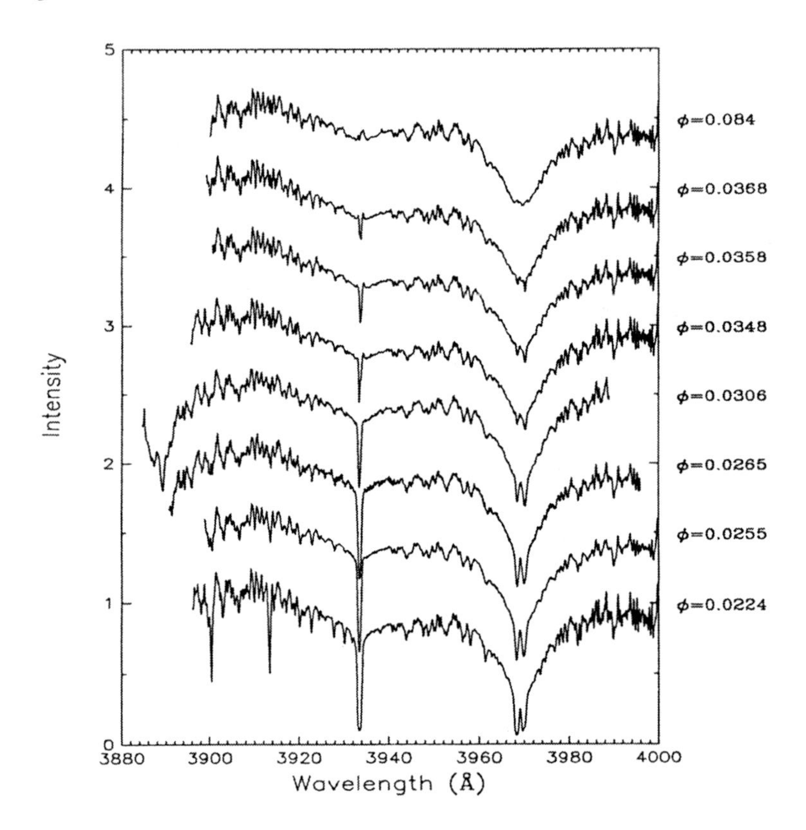

Figure 5.11 Spectrum variations of ζ Aurigae, a VV Cephei star, as the hot star emerges from behind the cool star. Time increases upward, and is expressed in terms of the phase in the binary orbit (scale at right). The lines shown are Ti II (3900 and 3915), Ca II (3934 and 3968) and hydrogen (3888 and 3969) Note the changing appearance of the absorption lines as the hot star emerges. (From Eaton, 1993.)

5.10 X-ray binaries

X-rays are high-energy photons; they are second only to gamma-rays in energy. They are therefore produced by high-energy processes in the universe, most commonly by the collisions of atoms in gases with temperatures of 10^6 K or more. X-rays are therefore ubiquitous in the hot interiors of stars, but they are not emitted by normal stars. The hottest normal stars (the O stars) have surface temperatures of 50 000 K; they emit UV radiation profusely. Freshly formed white dwarfs have surface temperatures of 100 000 K. Only freshly formed neutron stars have surface temperatures of 10^6 K, and a few of these have been detected by satellite-borne X-ray telescopes.

The most common X-ray stars are *X-ray binaries*. In a binary system in which there is mass transfer and flow from one star (the donor) to the other (the gainer),

Box 5.3 Star sample – ϵ Aurigae

Epsilon Aurigae [HD31964, A8Ia-F2epIa + (?), V = 2.92–3.83] is a remarkable binary system which, because it contains a supergiant component, can be classified as a VV Cephei system. On the basis of its light curve, it is considered an Algol binary. Studies of this star extend back to 1824, when its variability was discovered, but it remains an enigma in many ways.

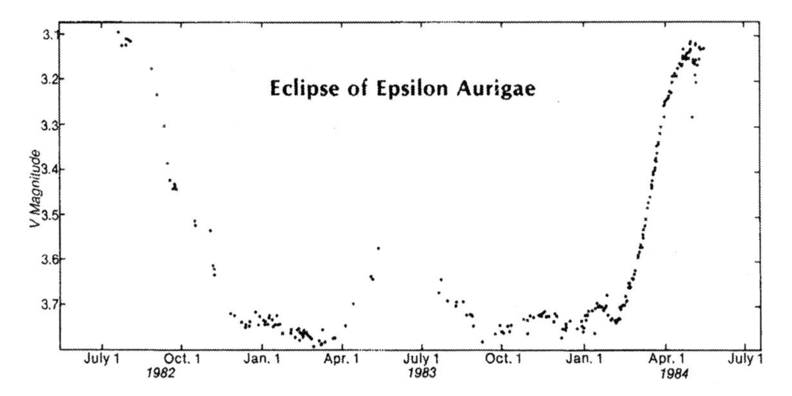

Figure 5.12 Light curve of ϵ Aurigae during the 1982–84 eclipse. Note that the eclipse lasted for almost two years. Note the slight brightening at mid-eclipse. The other variations, during eclipse, are due to the pulsation of the supergiant star. (From MacRobert, 1985.)

The orbital period is 27.1 years, so each eclipse is observed by a new generation of astronomers. The eclipse lasts for two years; it is a leisurely affair. The most recent eclipse, in 1982–84 (figure 5.12), was the first which could be observed with space telescopes such as *International Ultraviolet Explorer*. It also coincided with the time when photoelectric photometry could be done routinely by amateur astronomers, so there was a great deal of interest and participation by the amateur community. See Kemp *et al.* (1986) for a summary of the 1982–84 eclipse results.

The supergiant component is also of interest because it pulsates irregularly on a time scale of about 100 days, as many yellow supergiants do. But the greatest mystery is the nature of the invisible (so far) secondary component. It is generally agreed that the secondary star (or stars) is not visible, partly because it is surrounded by a huge doughnut-shaped disc of dust, seen edge-on. At mid-eclipse, the disc obscures about half of the photosphere of the supergiant, dimming it by 0.8 magnitude. But there is a brightening of the system at mid-eclipse, which is what suggests that there is a central hole in the doughnut. The masses of the two components are similar – about 15–20 $M_{\odot}$.

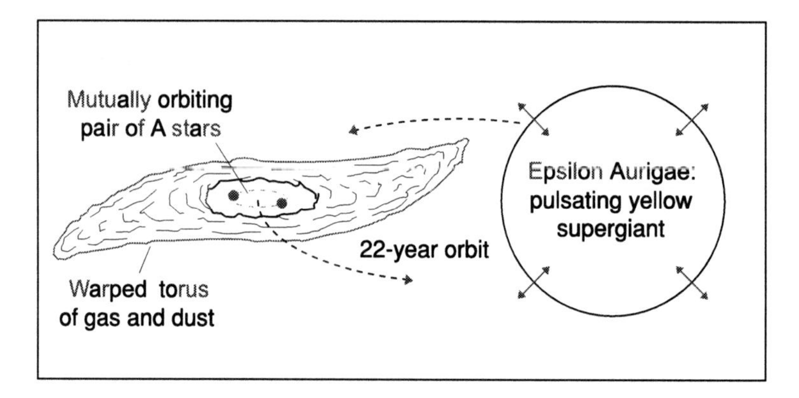

Figure 5.13 Model of ϵ Aurigae, based on the light curve in figure 5.12; see text. The 'hole' in the torus explains the brightening at mid-eclipse. (Jeff Dixon Graphics, after MacRobert, 1985.)

Their separation is about 25 AU – comparable with the distance from the sun to Uranus or Neptune. The supergiant would normally be expected to be massive, but it may have lost much of its mass to the disc, and to the secondary star(s). The secondary component could be massive, but not very luminous, if it was actually two stars with masses about 2 $M_{\odot}$. (Two stars of this mass are much less luminous than one star with a mass of 4 $M_{\odot}$.) Two orbiting stars might also cause a puffing out or 'fattening' of the disc, sufficient to eclipse half the photosphere of the primary, and hide the central stars from view. See figure 5.13 for a current model of this star system.

Perhaps by the time of the next eclipse in 2009–11, there will be a new generation of space and ground-based telescopes to clear up its mysteries once and for all.

energy can be released when the mass stream strikes an accretion disc around the accreting star, or strikes the surface of the accreting star itself. The source of that energy is ultimately gravitational energy which is gained as the mass stream falls toward the accreting star. If the accreting star is a neutron star or black hole, then the mass stream can fall deep into its gravitational field, and the energy gained, per unit mass, is large enough to heat the gas to millions of degrees. When the stream strikes the disc or star, X-ray photons can be produced and emitted.

X-rays (and UV and gamma rays) must be observed from above the earth's atmosphere, of course, because they do not penetrate the atmosphere. The first X-ray binaries were detected in rocket flights of X-ray detectors. X-ray satellites such as *Uhuru* and *Einstein* opened up the study of these energetic systems.

The brightest X-ray source in the sky is Scorpius X-1, which includes a neutron star with an accretion disc; it was discovered in 1967. An even more famous source is Cygnus X-1; its 'normal' component is a blue supergiant; its compact component is a black hole with an accretion disc. (The designation X-1 indicates that each is the brightest X-ray source in its constellation.) Several hundred X-ray binaries are now known in our galaxy, including many in globular clusters, and in nearby galaxies such as the LMC and SMC.

The development of X-ray astronomy has resulted in the addition of a whole new set of classes in the GCVS4: binary X-ray sources in general (X), X-ray bursters (XB), rapidly fluctuating X-ray systems (XF), irregular X-ray systems (XI), X-ray systems with relativistic jets (XJ), nova-like X-ray systems with a cool normal star (XND) or a hot normal star (XNG), X-ray pulsar systems with reflection effect and a hot primary (XPR) or cool primary (XPRM) or XP if no reflection effect present.

The optical and X-ray variability of X-ray binaries can be extremely complicated. There can be eclipses and other orbital phenomena. If the gainer is a neutron star, then its rotation may produce X-ray pulses, and it will be an X-ray pulsar. There may be eruptions or flickering in the accretion disc. As a result, many X-ray binaries are also eruptive variable stars; these are discussed in chapter 7. The accretion disc may grow or decay, or may wobble or rotate (precess) in space.

Despite this complexity, X-ray binaries can be divided into two relatively distinct groups. *High-mass X-ray binaries (HMXBs)* have OB stars as the normal components; they are losing mass to the gainer through a stellar wind. *Low-mass X-ray binaries (LMXBs)* have GKM stars as the normal components; they are losing mass to the gainer through Roche lobe overflow.

Many HMXBs are X-ray pulsars. The accreted material spirals slowly through the accretion disc, then is channelled to the pulsar's surface by its magnetic field. This may cause the spin of the pulsar to increase or decrease, depending on the angular momentum which is transferred to the pulsar by the accreted gas. If such a system is eclipsing, it is possible to determine the mass of the neutron star, as well as the mass of the OB star; the former turns out to be 1–2 solar masses.

Another subclass of high-mass X-ray binaries is the Be X-ray binaries, of which there is a handful. In this case, the donor is a Be star (chapter 9) and the gainer is a neutron star. The periods range from a few days to a year or more in the case of X Persei, the brightest example at magnitude 6. The spin periods of the neutron stars range from a fraction of a second to several hundred seconds.

The Be star loses mass, not only through a wind, but especially through an expanding equatorial disc. The formation and dispersal of the disc, on a time scale of weeks to years, adds another dimension to the variability.

In *low-mass X-ray binaries*, the principal mass transfer mechanism is Roche lobe overflow from the GKM star to the accretion disc, and thence to the gainer. In some systems, the radiation from the system may be completely dominated by that from the disc.

Globular cluster X-ray sources are an interesting and significant subset of LMXBs. Many of these are ultra-short-period binary systems, undergoing X-ray bursts. It is believed that interactions between stars in the dense cores of globular clusters are implicated in the formation of these extreme systems.

Unlike the HMXBs, the LMXBs are all faint; with the exception of the 12th magnitude Sco X-1, the rest are 15th magnitude or fainter. The orbital periods are mostly between 0.2 and 9 hours, though a few have periods of a day or more. All show complex light curves, with outbursts. A few show eclipses or other orbital phenomena. Several are pulsars. In particular, some are millisecond X-ray pulsars; the mass transfer has been extensive enough to spin up the neutron stars to very short periods. The most extreme examples are the *anomalous X-ray pulsars*, which also emit bursts of X-rays and gamma-rays, in which the neutron star is a *magnetar* with an exceptionally high magnetic field.

See chapter 4 for a more complete discussion of pulsars and magnetars.

5.11 The evolution of binary systems

These different kinds of binary systems can be linked together, and better understood, by summarizing the evolution of binary systems.

If the pair is well separated, as is the case with α Centauri A and B, the evolution of each star has little or no effect on the other. The stars will evolve as single stars do. At the other end of the spectrum are the contact binaries which, like conjoined twins, live out their lives in mutual embrace.

In the case of binaries with intermediate separation, the evolution and swelling of the more massive primary component results in mass transfer to the less-massive star. Depending on the separation, the primary may become a red giant with a strong pulsation-driven wind as well. As the primary's mass decreases, its Roche lobe shrinks, and the mass transfer accelerates, and the less-massive star soon becomes the more massive. Beta Lyrae may be in this phase of rapid mass transfer; W Serpentis may just have completed it.

The mass transfer then slows, and we have a system like Algol. The less massive star is now the most highly evolved; for many years, this was the 'Algol paradox'.

At some point, this more highly evolved component ceases mass transfer. It may have exposed its helium-rich core, in which case it is a *helium star*. If it is massive, it may be a *Wolf-Rayet star*. Eventually it will become a stellar corpse – a white dwarf, neutron star, or black hole, depending on its initial and final mass.

Eventually, the secondary star (now the more massive) begins to evolve and expand. It may transfer mass, through an accretion disc, to its dead companion. If the companion is a white dwarf, the result can be a dwarf nova, a nova, or even a supernova if enough mass accumulates on the white dwarf. If the companion is a neutron star, the result may be an X-ray pulsar, or something equally exotic.

Eventually the secondary, too, will complete its evolution, leaving a pair of stellar corpses in mutual orbit, perhaps surrounded by a slowly expanding nebula of material from the final stages of mass transfer and mass loss. Most commonly, the result would be a pair of white dwarfs. If one or both of the corpses was a neutron star, then it could be a binary pulsar, or even a very rare double pulsar – at least until the neutron star(s) rotation spun down.

5.12 Transiting exoplanets

In 1995, an exciting new class of potential variable stars was discovered: those which could be dimmed by a transiting exoplanet. An *exoplanet* (or extra-solar planet) is a planet around a star other than the sun. Astronomers have suspected for over half a century that many other stars have planets, but the evidence was tenuous at best. In the 1980s, astronomers developed a powerful technique for discovering and studying exoplanets: using the *Doppler effect* (see section 2.16) to look for the gravitational effect of the exoplanet on its star. As the exoplanet orbits, it causes the star to move in a small orbit, with the same period as the planet. The star's orbital speed depends on the mass of the planet; the sun moves in a 12-year orbit with a speed of 12 m/s as a result of the gravitational pull of Jupiter. By measuring the radial velocity of the star with an accuracy of a few metres a second – an order of magnitude better than was possible before – astronomers were able to discover the first exoplanet in 1995, orbiting the sunlike star 51 Pegasi. Less than a decade later, over a hundred exoplanets were known. Unfortunately, only exoplanets with masses comparable to those of Jupiter and Saturn could initially be discovered in this way, though, by 2004, exoplanets with masses as low as those of Uranus and Neptune had been discovered. Earth-mass planets would have too small a gravitational effect on their star; the motion of the star would be much less than the error of 1–3 km s^{-1}, which is set by the observational error, and the natural velocity 'jitter' in the spectrum of a sunlike star.

If an exoplanet were to *transit* or pass in front of its star, the star would dim slightly, confirming the exoplanet's existence. It would also reveal the size of the exoplanet, since the amount of the star's dimming is the area of the exoplanet (πR^2_{planet}) divided by the area of the star (πR^2_{star}); R_{star} would be known from the

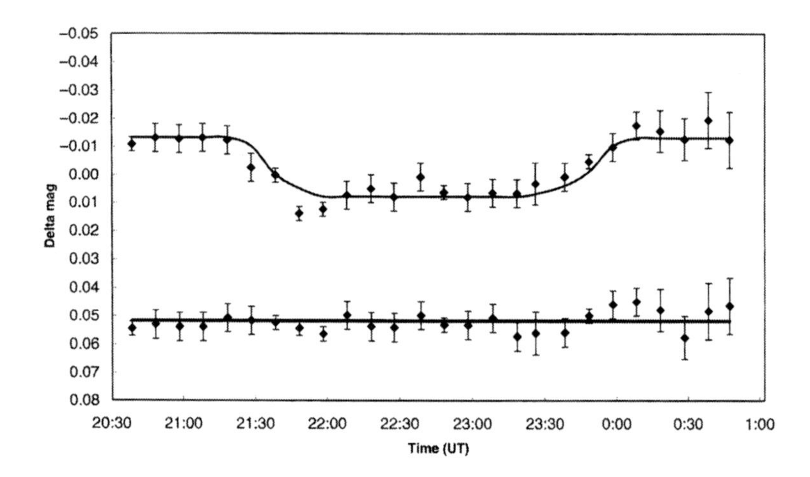

Figure 5.14 Light curve of HD 209458, showing the effect of the transit of an exoplanet, similar in mass and size to Jupiter, but much closer to its star. The depth of the transit light curve is proportional to the area of the exoplanet, hence the radius of the exoplanet can be determined. This light curve was obtained by amateur astronomer Arto Oksanen, at Nyrölä Observatory, Finland. (From A. Oksanen – private communication.)

spectral type of the star. But if the Jupiter-sized exoplanets were as far from their star as Jupiter is from the sun, the probability of an alignment and a transit would be close to zero. The remarkable result of the first decade of exoplanet hunting, however, was that many of the exoplanets orbit very *close* to their star – up to ten times closer than Mercury is to the sun. This result is partly an observational bias; close-in planets have a much shorter period, and a larger pull on their star, and are therefore easier to discover. But close-in or 'hot' Jupiters were not expected, prior to their discovery.

For close-in Jupiters, geometry makes the probability of a transit much higher. HD 209458b, the first transiting exoplanet, was discovered in 1999 by Henry *et al.* (2000), and quickly confirmed by Charbonneau *et al.* (2000). The depth of the minimum was 0.017 magnitude, which is typical of close-in Jupiters around sunlike stars; the transits last a few hours. The planet's radius is about 1.3 times Jupiter's, and its density is only about a quarter that of water – much less than the density of Jupiter, or Saturn. Imagine this puffed-up Jupiter-mass planet with a 1000-degree atmosphere! What would the weather be like?

Recently, an interesting project (www.transitsearch.org) has been developed to photometrically monitor stars that are known, from radial velocity surveys, to have planets, using a network of widely spaced observers with small telescopes. These observers monitor candidate stars during the time intervals when the radial velocity observations predict that a transit *might* occur, if the orbit was seen nearly edge-on. Photometric accuracy of better than 0.01 magnitude is required. This is well within the capability of experienced amateur photoelectric photometrists. Figure 5.14 shows a light curve of a transit of HD 209458b, observed by a skilled amateur astronomer with a small telescope.

In 2003, the first exoplanet *discovered* (rather than confirmed) by the transit method was announced, and about 20 groups are presently using the transit approach to search for exoplanets. It requires monitoring thousands of stars, looking for the characteristic dip in brightness. A significant fraction of these stars turn out to be variable, which is frustrating for the transit-searchers, but useful for the rest of us who study variable stars. We can use their 'chaff' as our 'wheat'! Transits can be confused with shallow eclipsing variables, or with variables with spots. But it opens up the possibility of finding thousands of new exoplanets, even among the fainter, more distant stars in our galaxy. Typically, these searches take about 30 days of observing. In principle, a discovery could be made with a small telescope, if sufficient stars were imaged, and if millimagnitude accuracy could be achieved. In 2004, a 12th magnitude transiting exoplanet was discovered with a 10 cm telescope. Within a week, transits of this planet had been observed by amateur astronomers; shortly after, the exoplanet was confirmed by the radial-velocity technique.

The transit provides direct information about the radius, mass, and density of the planet. As of 2004, six exoplanets had been observed by this technique – HD209458b which had previously been discovered by the radial-velocity technique, and five which had been discovered through their transits. Their periods range from 1.2 to 4.0 days, their masses from 0.5 to 1.5 $M_{\odot}$, and their densities from 0.4 to 1.2 times water.

With sufficiently accurate photometry, even earth-mass planets could be found with this technique. The space mission *Kepler* is being designed specifically for this purpose. The MOST satellite (chapter 3) presently has *almost* enough photometric precision needed to detect earth-sized planets. And it is already beginning to provide information about the 'hot Jupiters'.

It is also possible to detect exoplanets by gravitational microlensing (section 3.2); the first such discovery was made in 2005. The background star and planet would have to be aligned almost perfectly, so a discovery of this kind would be a rare event.

The most obvious way to discover an exoplanet would be by direct visibility. The usual problem is that the exoplanet is a billion times fainter than its parent star, and only a few arc seconds away from it. Recently, with adaptive optics, a 7.5-earth-mass exoplanet has been imaged around the red dwarf Gliese 876.

For up-to-date information about exoplanets, see:
www.exoplanets.org or www.obspm.fr/planets

6

Pulsating variable stars

Pulsation is the astronomer's word for *vibration* or *oscillation*. Every physical object has natural patterns or *modes* of vibration, each with a corresponding *period* – the time required for one vibration. We commonly observe this in musical instruments. A bugler can make the air column in a bugle vibrate in different modes or *harmonics*, with different periods which we perceive as different pitches. A violin string can vibrate in several pitches at once, producing a complex sound which may be pleasant or not, depending on the quality of the violin and the skill of the musician. It is tempting to call pulsation 'the music of the spheres', especially as most stars are spherical; the sun is spherical to 1 part in 100 000. This concept comes from an excellent recent review article on stellar pulsation, part of a conference in July 2005, and published in *Communications in Asteroseismology*, volume 147:
http://www.univie.ac.at/tops/CoAst/archive/cia147.pdf

6.1 Pulsation modes

Stars are, to a first approximation, spherical. The simplest form of pulsation is *radial* pulsation – a simple, spherically symmetric in-and-out expansion and contraction. A star has an infinite number of modes of radial pulsation. The simplest is called the *fundamental* mode. In this mode, all parts of the star expand together and contract together, in unison. The next-simplest mode is the *first overtone*. In this mode, there is a *nodal sphere* in the star, where the material remains at rest. When the part of the star outside this sphere is expanding, the part inside is contracting, and *vice versa*. In the *second overtone* mode, there are two nodal spheres, where the material remains at rest. The large-amplitude

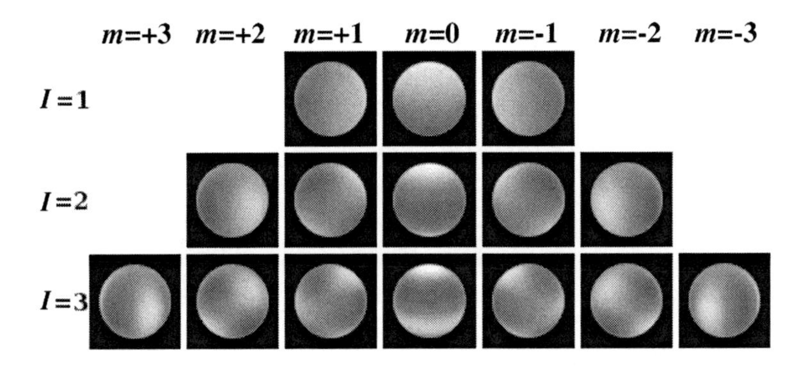

Figure 6.1 Illustration of non-radial pulsation modes. When the light-coloured regions are moving downward, the dark-coloured regions are moving upward. In radial pulsation, all parts of the star move in and out in unison. (*Source*: Whole Earth Telescope)

pulsating stars – Cepheids, RR Lyrae stars, and Mira stars – pulsate primarily in radial modes. Radial pulsation produces substantial changes in luminosity (and therefore brightness), temperature (and therefore colour), and radial velocity; these are observed, and provide direct evidence for the radial pulsation. Through techniques such as the *Baade–Wesselink method* (section 6.2), they can be used to determine the average radius of the star.

It is less easy to tell *which* radial mode a star is pulsating in. Most, but not all large-amplitude variables pulsate in the fundamental mode. If the radius and mass of the star can be estimated, the observed period can be compared with the theoretical or expected period for each mode. If the star pulsates in two or more radial modes, then the period *ratio* can be used to deduce the modes, as explained below.

Unlike the harmonics in a musical instrument, whose frequencies are related to the fundamental frequency f by f/2, f/3, f/4 etc., the periods of the radial modes of a star are not simply related to each other. A star is much more complicated than a stretched string.

In *non-radial* pulsation, the star changes shape, not volume (figure 6.1). There is a triply-infinite set of modes, corresponding to the three different co-ordinate axes on the star. These are chosen to be: the distance from the centre, the angular distance ('latitude') above or below the star's equator, and the angular distance ('longitude') around the star's equator. Non-radial modes can be divided into (i) *p (pressure)* modes, in which the motion is primarily radial and the restoring force is pressure (as it is in radial modes), and (ii) *g (gravity)* modes, in which the motion is primarily horizontal, and the restoring force is buoyancy or gravity (as it is in water waves). A wide assortment of stars pulsate non-radially: most of the white dwarf pulsators, the slowly pulsating B stars and the Gamma

Doradus stars, and many Beta Cephei and Delta Scuti stars which pulsate in both radial and non-radial modes. Non-radial pulsation tends to produce smaller amplitudes than radial modes in the brightness and colour variation, but, if they can be observed, their relative amplitudes and phases (along with the periods themselves) can be used to identify the precise non-radial mode. Non-radial pulsation also produces characteristic *absorption line profile variations* (lpv). With modern high signal-to-noise spectrographs, these can and have been observed and studied over time. Also, there are often very close non-radial periods, whose differences are a clue to the mode identification. In non-rotating stars, these close periods are actually identical or *degenerate*, but in a rotating star, they become different; the difference is proportional to the amount of rotation. The situation is complicated by the fact that many of these non-radially pulsating stars rotate *rapidly*, and the theory of rapidly rotating pulsating stars is not well understood.

6.2 Pulsation mechanisms

In 1879, August Ritter proposed that stars might pulsate. Prior to that time, it was assumed that stellar variability was caused by rotation or eclipse. In 1914, Harlow Shapley showed that pulsation was the most likely explanation for the variability of δ Cephei. Pulsating stars have since been proven to be a powerful tool for determining distances in astronomy, because the period of pulsation is correlated with the luminosity of the star, and this relation can be calibrated (section 6.3). Pulsation is also useful for probing the interiors of stars, and testing models of stellar structure and evolution.

The process of radial pulsation allows astronomers to use a technique called the *Baade–Wesselink method* (named after Walter Baade and Adriaan Wesselink) to measure the radius of the star. It is based on the assumption that the radiant flux (per unit surface area) of the star is a function of temperature only, and that the photometric colour of the star ($(B - V)$, for instance) is also a function of temperature only. Therefore, at two times t_1 and t_2 when the colour of the star is the same, $L_1/L_2 = (R_1/R_2)^2$. This is then combined with the radial velocity curve of the star. The radial velocity is the rate of change of the radius, so the integral of the velocity curve is the radius curve. The integral of the velocity curve between t_1 and t_2 is $(R_2 - R_1)$, so R_1 and R_2 can be obtained. By repeating this procedure around the pulsation cycle, the radius curve, and the mean radius can be obtained.

The *Barnes–Evans method*, named after Tom Barnes and David Evans (1976) is another useful tool. They point out that, with the sun as a calibrating point, there is a relation

$$\log T_e + 0.1C = 4.2207 - 0.1V_0 - 0.5 \log \phi \tag{6.1}$$

where T_e is the effective temperature, C is the bolometric correction, V_0 is the unreddened apparent V magnitude, and ϕ is the angular diameter in milliarc sec; the right-hand side of this equation is defined as the surface brightness F_V. The surface brightness is a function of colour index. Simultaneous measurements of the colour index and V magnitude, around the pulsation cycle, give the value of log ϕ around the pulsation cycle. As noted above, the integral of the velocity curve then gives the diameter or radius of the star around the pulsation cycle, in linear units. Comparison of the angular (apparent) and absolute diameter curves gives the distance.

With modern optical interferometers, such as that on the European Southern Observatory's Very Large Telescope, it is now possible to observe the expansion and contraction of pulsating stars, such as Cepheids, directly. This has led to improved calibrations of the surface brightess parameter in the Barnes–Evans method, and to improved distance to Cepheids using this technique (Kervella *et al.*, 2004).

6.3 Modelling stellar pulsation

The theory of stellar pulsation (Cox, 1980 is the classic text) begins with a model of the star (section 2.8), which gives the physical properties of the star as a function of distance from the centre. The first models of stars were primitive (and so were the deductions from stellar pulsation theory). Now, with improved understanding of the physics of stars, and powerful computers to implement this knowledge, the models are much more detailed and accurate.

The simplest level of modelling is the *linear, adiabatic theory* (LAT) of radial pulsation of a spherically symmetric, non-rotating, non-magnetic star. This was formulated by Arthur Eddington, early in the twentieth century. It assumes that the pulsations are infinitesimally small, and that there is no transfer of energy between parts of the star. It provides reasonable estimates of the periods of the radial modes, and their dependence on the physical properties of the star. For instance, it shows that *the period of any mode is approximately inversely proportional to the square root of the mean density of the star*. Alternatively, the period times the square root of the density is approximately constant; we call it the *Q-value*, or *pulsation constant*. It is this relationship which leads to the period–luminosity relation; since both period and luminosity are related to the radius of the star, and since the temperatures of pulsating stars are about the same, period and luminosity will be related to each other. The LAT also shows that the relative amplitude of pulsation tends to be largest in the outer layers, where the density is lowest. Corresponding calculations can be done for non-radial pulsation. The LAT assumes that the pulsation takes the form of a standing wave in the star, i.e. a wave which is perfectly reflected from the centre and the surface.

The next level of complexity is the *linear, non-adiabatic theory* (LNAT) (e.g. Castor, 1971). This also assumes that the pulsations are infinitesimally small, but it allows for the possibility that any layer of the star may gain or lose energy. In particular, the amplitude of the infinitesimal pulsations can increase or decrease, and we can tell what modes may be unstable, and grow. We cannot tell what the eventual amplitudes may be, i.e. which mode(s) may grow to become observable. The LNAT gives periods which are slightly more realistic than the LAT, though they are about the same in most cases. In the sun and some other stars, for instance, more than one pulsation mode and period is observed. The observed period ratios can be measured accurately, and compared with those determined from the LNAT models.

The next level of complexity is the *non-linear, non-adiabatic theory*, which amounts to a full, time-dependent, hydrodynamical model of the star (e.g. Christy, 1964). This includes the equation which relates the motion of a layer in the star to the forces acting on it. The calculation begins with the static, 'equilibrium' model of the star. Infinitesimal random motions are imposed on the model (such as from the round-off errors in the calculations), to see if they grow or decay. After several hundred cycles, the model (usually) reaches some constant, stable amplitude, in one or more modes. The non-linear theory accurately predicts the amplitudes of Cepheids and RR Lyrae stars, which pulsate in one or two radial modes. It has not been able to explain stars like Delta Scuti stars, which can and do pulsate in many radial and non-radial modes. An ongoing problem in stellar pulsation is: what determines what mode(s) a star will pulsate in, and with what amplitude?

These models are non-rotating and non-magnetic. Non-linear analysis of rotating, magnetic models is at the frontier of present research. But we know that all stars rotate – some of them very rapidly – and that many of them have significant magnetic fields. In reality, there is no such thing as a truly spherically symmetric star; for most of them, however, that is a reasonable approximation.

6.4 Non-linear effects

The pulsation waves in a star are not reflected perfectly from the surface, but move outward in an atmosphere of decreasing density. In many variables, such as Mira stars, this creates *shock waves*, as the atmospheric layers pile up against each other. In the Mira stars, this creates layers of higher density, where gases condense into dust. The dust absorbs the star's radiation, and is pushed outward, carrying gas along with it. The result is *mass loss*, which has a profound effect on the star, and the space around it. In the largest, most luminous supergiants, the outward 'radiation pressure' almost balances or perhaps even exceeds gravity; the star borders on instability; and non-linear effects may be significant.

6.5 The instability strip(s)

Normally, stars are stable. The inward pull of gravity is balanced by the outward pressure of the hot gas in the star. If the star begins to contract, its internal pressure and temperature increase, and reverse the contraction. If the star begins to expand, its internal pressure and temperature decrease, and gravity restores the initial balance. Furthermore, if a star expands or contracts, the energy of motion is dissipated by friction. Eventually, the star returns to rest. Imagine a child's swing in a playground. If the child sits still in the swing then, even if the swing is set in motion, it will eventually return to rest. But if the child (or someone pushing the swing) applies the correct periodic force to the swing, at the right moment, it will build up and maintain its vibration.

For a star to begin and continue to pulsate, there must be some mechanism which converts radiant energy into energy of motion in a timely way, overcoming friction and building up the motion. Such a mechanism can exist in a star: it is the thermodynamic effect of an *ionization zone* in a star. This conclusion was reached independently by S. A. Zhevakin in the USSR in the 1950s, and Norman Baker (1966) and others in the US, a few years later.

In a star of moderate surface temperature, the layers below the surface have temperatures up to a few tens of thousands of degrees. The helium tends to be once-ionized; it has lost one of its two electrons, but not the second. If the star contracts, it heats slightly, and this causes the helium to lose its second electron. But this requires energy. This energy comes from the radiation flowing outward through the gas; it is absorbed by the *opacity* of the gas. The energy then becomes stored in the form of *ionization energy*. When the star subsequently expands and cools, the electrons recombine with the helium atoms, and release the stored-up ionization energy. This is the 'push' which maintains the pulsation in all of the pulsating stars in the Cepheid instability strip (figure 6.2). If there is sufficient mass in the ionization zone to store up significant radiant energy during the pulsation cycle, then the star will pulsate. This mechanism is referred to as *self-exciting*, in contrast to the pulsations of the sun, and other sunlike stars, which are excited by the turbulence in their convection zones.

But what about the other types of pulsating stars? The Beta Cephei stars were a real mystery, because they are much hotter than the Cepheid instability strip.

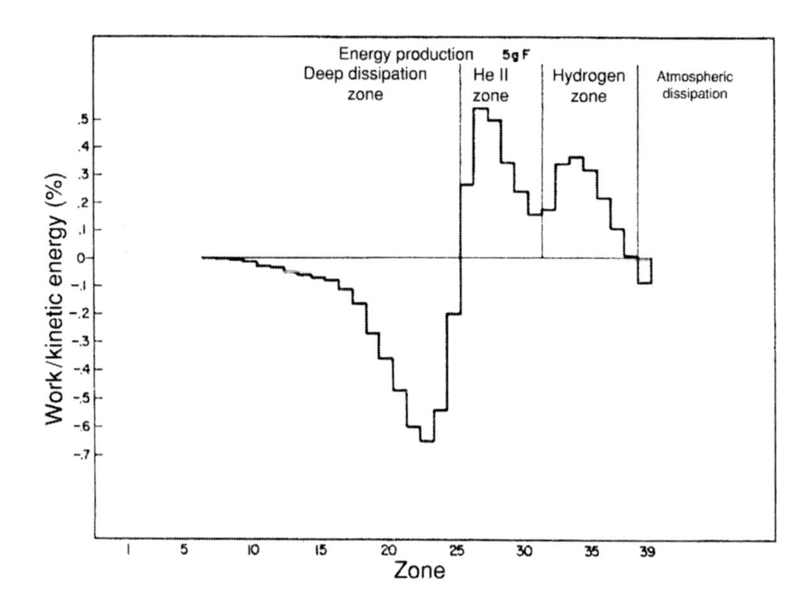

Figure 6.2 Diagram showing the zones in an unstable star which convert radiant energy into kinetic energy, and *vice versa*, in a model of a pulsating RR Lyrae star, expressed as the fractional energy production (positive) or dissipation (negative) per period. The centre of the star is at the left, the atmosphere at the right. The two energizing zones – the hydrogen zone and the ionized helium (He II) zone – are shown, along with the principal dissipation zone, deeper in the star. (From Christy, 1966.)

Only in the 1980s was it realized that the ionization of *iron*, at temperatures near 150 000K, deep within the star, had sufficient effect to maintain the pulsation. This realization came about as a result of physicists' recalculation of the opacity of iron at these temperatures (e.g. Iglesias *et al.*, 1987). This solved the long-standing mystery of the Beta Cephei stars, and some other nagging problems in pulsation theory as well.

Figure 6.3 shows the location of various types of pulsating variable stars in the H–R diagram. There is a misconception, often included in astronomy textbooks, that pulsation is something that happens only in evolved stars such as giants and supergiants. In fact, the Cepheid instability strip extends from the top of the H–R diagram to the bottom, at approximately constant temperature. That is because the instability mechanism occurs at a particular temperature – the temperature at which there is a helium ionization zone in the outer layers of the star. So supergiant Cepheids pulsate, main-sequence Delta Scuti stars pulsate, and white dwarfs pulsate – all as a result of the same mechanism.

The Cepheid instability strip is not exactly vertical: its temperature is cooler at its high-luminosity end. That is because more luminous stars have outer regions

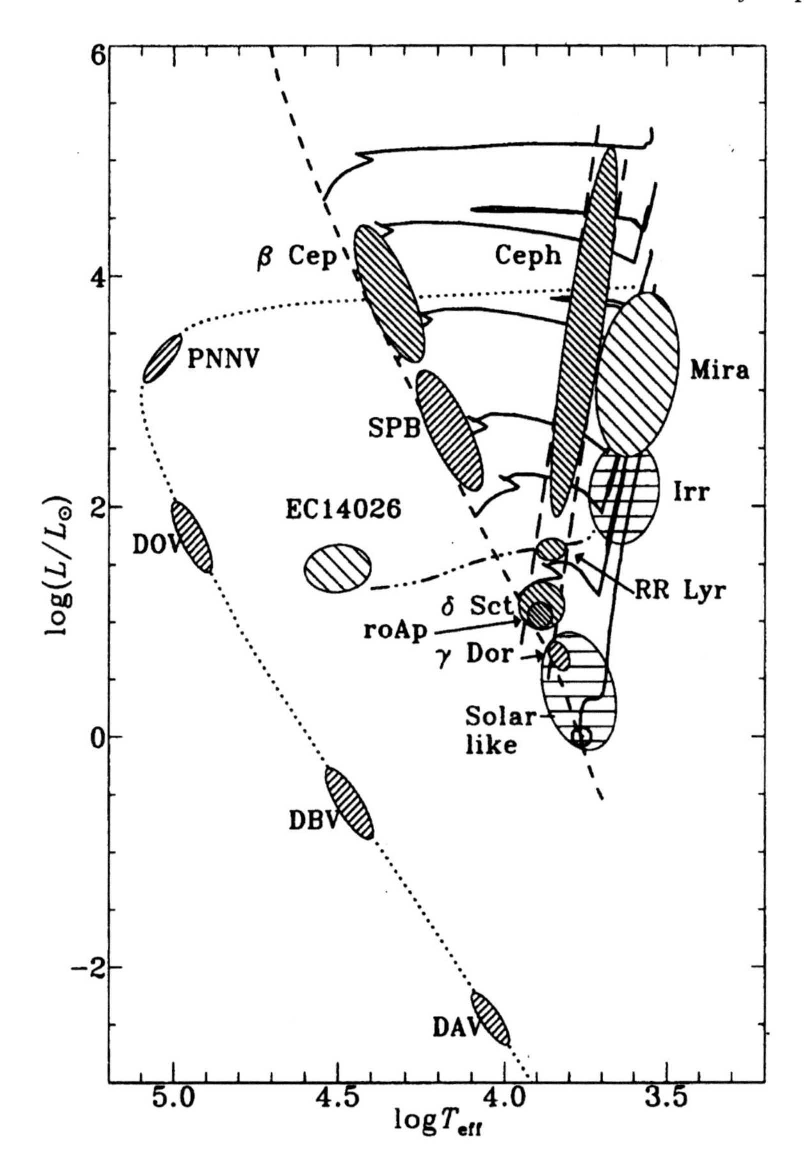

Figure 6.3 The location of various types of pulsating star on the H–R diagram. The main instability regions are the Cepheid strip, including its intersection with the main sequence (dashed line), the region of the coolest stars (Miras and their relatives), and the region of the Beta Cephei and SPB stars. There are also instability regions along the white dwarf cooling track (dotted line) at the bottom of the diagram, including the nuclei of planetary nebulae (PNNV) and the DOV, DBV, and DAV pulsating white dwarfs. (From J. Christensen-Dalsgaard, private communication.)

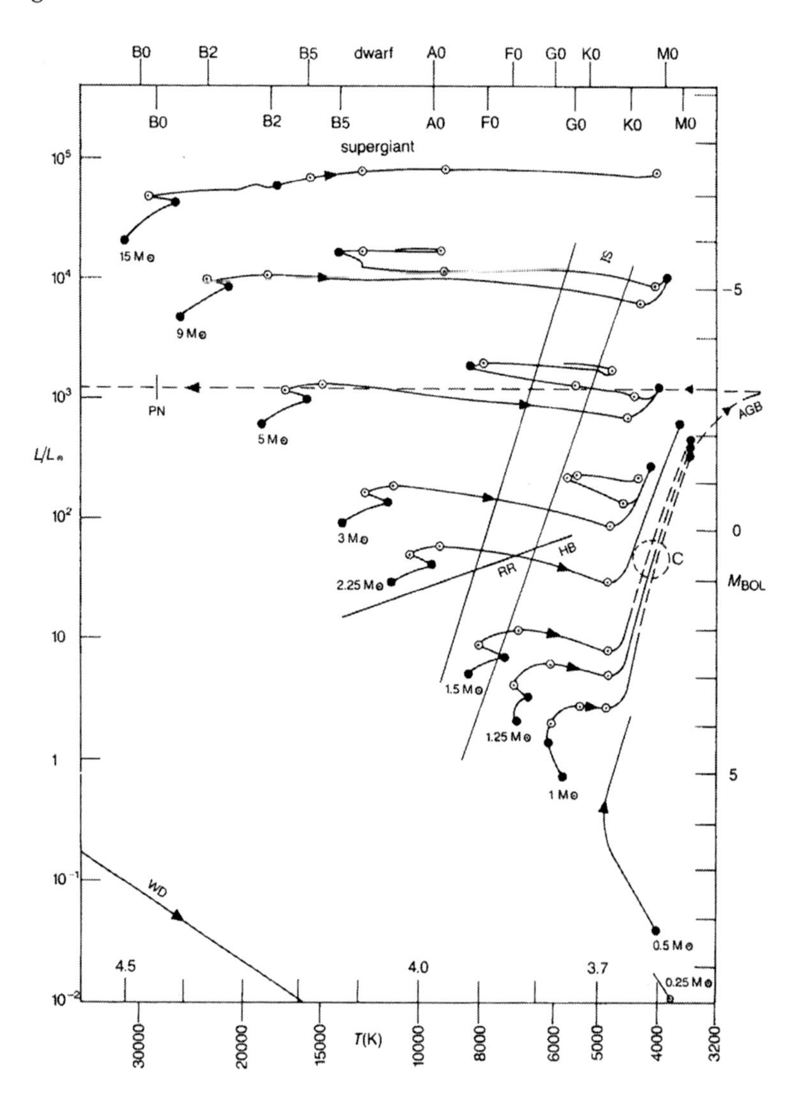

Figure 6.4 Evolution tracks on the H–R diagram, showing the changes which take place in a star as it evolves. The tracks are marked by the mass of the star in solar units. The Cepheid pulsation instability strip is marked IS, and the horizontal branch where RR Lyrae (RR) variables are found is marked HB. The time scales for evolution depend on the mass: the more massive the star, the faster the evolution. (From Kaler, 1997.)

of lower density; at lower density, ionization can occur at a lower temperature because there is lots of 'room' for the electrons that are released.

Stars can evolve into the instability strip in different ways (figure 6.4). Massive young stars can evolve quickly from left to right in the H–R diagram; old low-mass stars can evolve slowly to the same position. So there may be Population I and II stars in the same region of the instability strip.

6.6 Helioseismology: pulsations of the sun

The sun is the nearest star, and the only one that we can observe in detail. Its photosphere is a seething mass of convection currents, with occasional concentrations of strongly magnetic *active regions*. Especially around the active regions, there are streams of plasma, guided and heated by the magnetic fields, and eventually driven off the sun. At a distance, sunlike stars appear more benign, even though we know that they must be as active as the sun, or perhaps even more. But there is one kind of variability which is much less obvious on the sun, and at the limit of detectability on other sunlike stars – complex pulsations which are excited 'stochastically' by the turbulent convective motions below the stars' surface. These convective and pulsational motions are also important in heating the sun's chromosphere. Literally thousands of pulsation modes are excited by the convective turbulence, and have been detected and measured in the sun, by photometric and spectroscopic techniques (figure 6.5). Their time scale is typically a few minutes. They provide unique and remarkable information about the deep interior of the sun – a sort of CT (computed tomography) scan of our star. That is because different modes penetrate to different depths in the sun, and are therefore sensitive to conditions at different depths. They have been useful for probing 'second-order' effects in the sun such as rotation, convection, and composition gradients which are not well understood through basic stellar structure theory. They can even be used to probe the active regions near the sun's surface – even on the far side of the sun! The study of these pulsations is called *helioseismology*, by analogy with the technique that geophysicists use to probe the interior of our planet. This has turned out to have wide implications. For about three decades, astronomers faced the 'solar neutrino problem' – the number of these elementary particles observed by neutrino observatories was only a third of what solar models predicted. Were the models in error? Helioseismology supported the models. The solution was later found to be a consequence of the nature of the neutrino itself; about two-thirds of the emitted neutrinos changed, in flight, to a form which the first generation of neutrino observatories could not detect. For a good website on helioseismology, see:
http://soi.stanford.edu/results/heliowhat.html

6.7 Asteroseismology

Helioseismology has been outstandingly successful in probing the interior structure of the sun. Could the same techniques be used on other sunlike stars? The problem is that, whereas astronomers can resolve the disc of the sun and measure its brightness, temperature, and velocity at each point, other stars

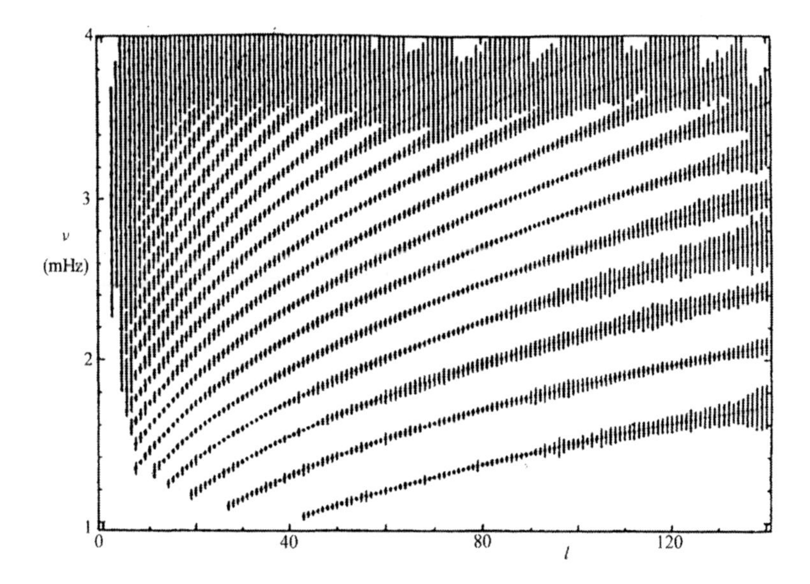

Figure 6.5 Frequency diagram for pulsational p-modes excited in the sun. The frequency in mHz is plotted against the degree l. The vertical lines over-estimate the probable errors. The distinct sequences of frequencies correspond to different values of n. The lowest sequence has $n = 1$. (From Libbrecht and Woodward, 1990.)

are only points of light – much fainter than the sun. What this means, in practice, is that astronomers can observe only low-order modes – ones which affect a significant fraction of the whole disc. The term *asteroseismology* is sometimes used to refer to measurements of a very few low-order modes, such as are found in Delta Scuti stars and pulsating degenerate stars, but the hope is to be able to observe dozens of modes in sunlike stars (it is not clear how many modes would have to be observed to qualify as 'seismology'!). Ground-based astronomers have claimed to have observed such modes, from radial velocities. Solar-type pulsations have been confirmed in α Centauri A, β Hydrae, η Bootis, and a star with at least three planets – μ Arae.

Ultra-precise photometric observations of Procyon by Canada's MOST satellite in 2004 did not detect the oscillations which were predicted by theory, and which had apparently been observed by less-precise ground-based observations. These spurious ground-based detections may be a result of a combination of gaps in the data and of noise produced by the granulation of the star. The non-detection by MOST suggests that either the amplitude of the oscillations is unexpectedly low, or the oscillations in Procyon have short lifetimes, and are therefore not coherent. This result emphasizes how important and productive it will be to have dedicated asteroseismology satellites such as MOST and France's COROT.

Nevertheless, important ground-based observations of solar-like oscillations *have* been made, of stars such as α Centauri and η Bootis.

6.8 Classical Cepheid variable stars

Cepheids are yellow supergiant pulsating variable stars which lie in a narrow instability strip in the middle of the H–R diagram. Their temperatures are 6000–8000K, and their periods are 1–100 days or more. We now know that there are two distinct classes of Cepheids – the *classical Cepheids* which are young stars, more massive than the sun, and the *Population II Cepheids* or *W Virginis* stars, which are old stars, less massive than the sun. The light curves of these two classes are rather similar, but can be distinguished by Fourier decomposition. The GCVS4 sub-classifies them as DCEPS if they have symmetrical, small-amplitude light curves, CEP(B) or DCEP(B) if they display two or more pulsation modes, DCEP otherwise.

Classical Cepheids are probably the best-known and most important of all pulsating variables. They are bright, numerous, and generally have large amplitudes. There is an important relation between their period and their luminosity. This allows their luminosity to be determined from their observed period. Then, using the inverse-square law of brightness, their distance can be determined from their observed apparent brightness. Thus, they can be used to survey and map our galaxy's spiral arms, where they are found, and to determine the distances to star clusters and nearby galaxies. They are the primary tool for establishing the size scale of the universe. Also, because they are bright, young, and often located in nearby star clusters of known distance and age, they can be used to test theories of stellar evolution, and models of stellar pulsation.

The first Cepheids to be discovered were η Aquilae (by Edward Pigott on 10 September 1784) and the prototype δ Cephei (by John Goodricke a month later). At that time, the cause of the variability was not clear. A rotational mechanism was first suggested, and in the late nineteenth century, the variability was thought to be caused by some form of eclipses. A. Belopolsky discovered in 1894 that δ Cephei showed radial velocity variations with the same period as the brightness variations, and in 1899 Karl Schwarzschild discovered that variations in temperature also occurred. Following theoretical studies of stellar pulsation by August Ritter, Harlow Shapley in 1914 showed that the brightness, temperature, and velocity variations in Cepheids were consistent with radial pulsation (figure 6.6).

At about the same time, the relation between period and luminosity was discovered by Henrietta Leavitt, from observations of 25 Cepheids in the Small

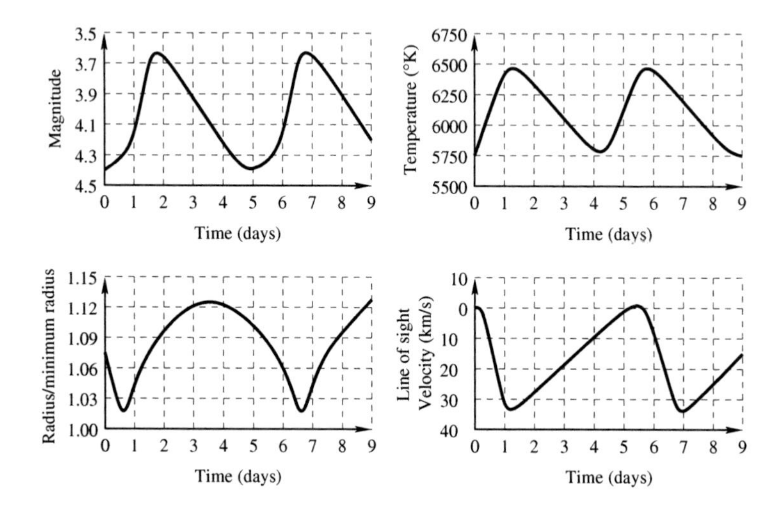

Figure 6.6 The magnitude, temperature (deduced from the colour), radius (deduced from the magnitude and temperature), and radial velocity of the prototype Cepheid δ Cephei. The radial velocity variations confirm that the star is pulsating, and are consistent with the variations in radius. (Based on a diagram produced by the author for *Scientific American*.)

Magellanic Cloud (part of a study of over 1700 variables in the Clouds); she noted a relation between the period and the apparent brightness, which, because all of the stars were at the same distance, reflected the true brightness or luminosity of the stars. Thereafter, much effort went into calibrating this relation, using Cepheids of known distance. For decades, the situation was complicated by the fact that, unknown to astronomers, there were two sub-classes of Cepheids, which had different period–luminosity relations. When Walter Baade realized this in the 1940s, the resulting revision of the period–luminosity relation immediately changed the size and age scale of the universe by a factor of two! With the discovery (by John B. Irwin) in the 1950s of Cepheids in galactic clusters, the calibration of the period–luminosity relation has become much more secure, because the distances of star clusters can be determined from observations of their main sequence stars.

6.8.1 *Variability*

Most Cepheids have ranges of 0.5 to 1 magnitude in *V*, and can be discovered efficiently in photographic surveys, especially in the Milky Way where they are concentrated, and in nearby galaxies such as the Magellanic Clouds. Thousands more have been discovered as a by-product of surveys such as MACHO and OGLE. These surveys have been especially useful in discovering thousands of Cepheids in nearby galaxies such as the Magellanic Clouds, M31, and M33.

These Cepheids help to refine the period–luminosity relation, and also provide information about the number and type of Cepheids which pulsate in different modes.

A key project of the Hubble Space Telescope was to find and study Cepheids in galaxies beyond the Local Group, in order to measure the value of the Hubble constant (Freedman *et al.*, 2001). For instance, in 1994, HST observed and studied Cepheids in M100, 56 million light years away. These Cepheids have apparent magnitudes of +25 or fainter!

Some Cepheids, however, have ranges < 0.3, and must be discovered through photoelectric or CCD surveys. Polaris, the nearest Cepheid, now (2004) has a range of between 0.02 and 0.06 mag in V. Bright; small-amplitude Cepheids are still being discovered – by the *Hipparcos* satellite photometry, for instance.

Particular efforts have been made to discover Cepheids in galactic clusters, or to establish the membership of known Cepheids in known clusters, because cluster membership allows the distance, luminosity, reddening, and age of the Cepheid to be determined. See Turner and Burke (2002) for an excellent review.

The periods of Cepheids range from about 1.5 to over 100 days. In our galaxy, Cepheids with periods greater than about 60 days tend to be irregular, so it is difficult to define an upper limit. The Cepheid-like star ρ Cassiopeiae, for instance, has a 'period' of several hundred days, but is usually classified as an SRd or hypergiant variable. In the Magellanic Clouds and some other galaxies, Cepheids with stable periods of over a hundred days are known to exist. At the low-period end, there is a star AC And, which, though long thought to be an RR Lyrae star, is more likely to be a classical Cepheid. It has three periods: 0.71, 0.53, and 0.42 day. Cepheids with periods as short as 0.8 day (fundamental mode) and 0.5 day (first overtone) have recently been found in the dwarf irregular galaxies of the Local Group – including the Magellanic Clouds – thanks to large-scale photometric surveys.

From systematic, long-term observation, the periods of Cepheids can be quite reliably determined, and period changes have been observed in many of them, using the (*O–C*) technique. Among the Cepheids with pronounced period changes are δ Cephei, ζ Geminorum, η Aquilae, and Polaris. The period of δ Cephei is decreasing; the radius of the star is decreasing. The period of η Aquilae is increasing; the radius of the star is increasing. For large-amplitude Cepheids, careful and systematic visual or photographic observations are useful for determining period changes, but photoelectric photometry is always preferred. In most cases, these period changes are probably a result of evolution, though random cycle-to-cycle changes have been noted in some stars. Comparison between observed period changes and the prediction of evolutionary models is hampered, to some extent, by limitations of the models, for instance, our lack of knowledge of

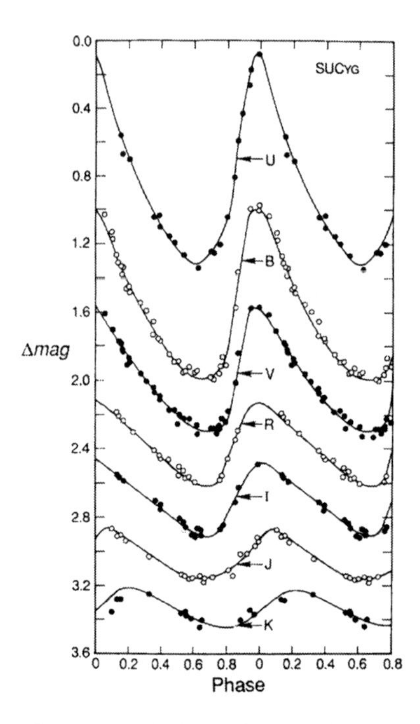

Figure 6.7 Variations of amplitude and phase of maximum seen in the light curve of a typical Cepheid as a function of increasing wavelength (UBVRIJK filters in the Johnson system). Note the monotonic drop in amplitude, the progression toward more symmetrical light variation, and the phase shift of maximum toward later phases, all with increasing wavelength. At longer wavelengths, the brightness changes are due primarily to changes in the radius of the star; at shorter wavelengths, they are due primarily to changes in the temperature of the star (Madore and Freedman, 1991).

how convection affects the structure and evolution of stars. The observed period changes may eventually be able to shed some light on this problem – and others. The evolution of long-period Cepheids is of special interest, since these stars may also lose mass, and contribute to the chemical enrichment of our galaxy in an important way.

Cepheid amplitudes range from a few hundredths to over a magnitude. Large-amplitude Cepheids can be observed visually, but photoelectric observations are preferred. The AAVSO has recently analyzed a large body of visual measurements of Cepheids, obtained over many decades. It appears that, by averaging a large number of systematic, careful observations, valid information about light curve shapes, and period changes can be obtained. But these results have not yet been formally published.

The amplitude is larger at shorter wavelengths (figure 6.7). This is because of the temperature variations in Cepheids: a small increase in temperature causes

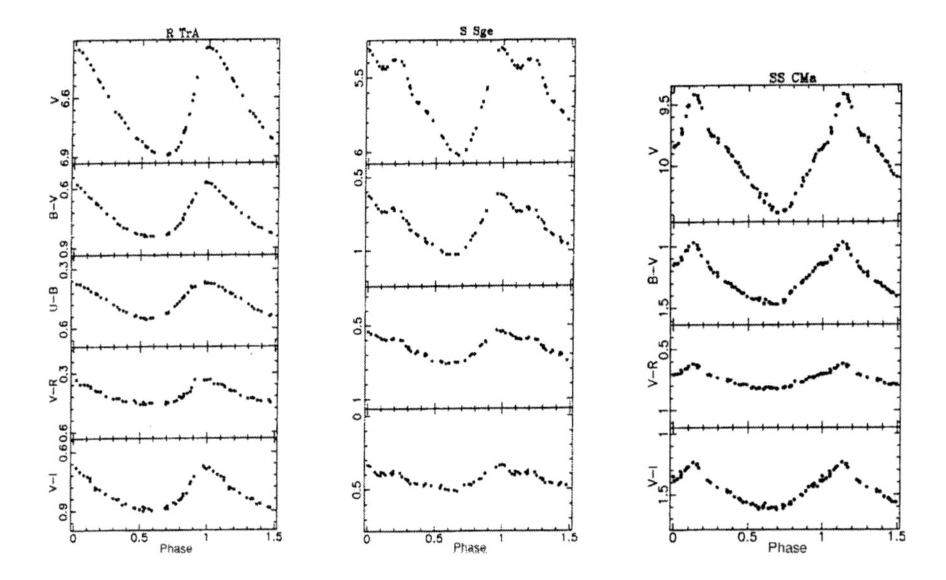

Figure 6.8 Light and colour curves of three Cepheids with different periods: R TrA (3.39 days), S Sge (8.38 days), and SS CMa (12.36 days), showing the characteristic shapes of Cepheids with different periods – the 'Hertzsprung progression'. (From Sterken and Jaschek, 1996.)

a much larger increase in the violet light than in the red. As a result, Cepheids show colour variations which are in phase with the brightness variations.

Are there non-variable stars in the Cepheid instability strip, as well as small-amplitude variables? Given the narrowness of the instability strip, and the uncertainties in determining the temperature and luminosity of supergiants, it is not possible to state, for certain, that such non-variables exist. There is no definitive explanation for the different amplitudes of Cepheids with similar luminosity and temperature. Figure 6.6 shows the brightness, temperature, and radial velocity variations in δ Cephei, and the radius variations deduced from Stefan's law.

The shape of the light curve of small-amplitude Cepheids is sinusoidal, but that of large-amplitude Cepheids varies in a systematic way with period (figure 6.8). This trend is called the *Hertzsprung progression*. It is caused by a shock wave which travels down into the star, and is reflected, and reappears at the surface as a bump in the light curve. Or equivalently, it is a result of interference between two radial pulsation modes.

The radial velocity of a Cepheid varies as the star expands and contracts. Figure 6.6 shows the radial velocity variations in Delta Cephei. The mean velocity (called the gamma velocity) represents the velocity of the star as a whole. The variations are a result of the motion of the star as it expands and contracts as it pulsates. The actual pulsational velocities are about 1.3–1.4 times the observed velocities, because, except at the centre of the disc of the star, we see only a projection of the pulsational velocity along the line of sight. This projection factor p requires a knowledge of the limb darkening of the star. This can be determined, with some uncertainty, from model atmospheres, and from observations of eclipsing variables. This knowledge is essential in this age of 'precision astrophysics', if the radial velocity variations are to be used for the Baade–Wesselink method, for instance.

Note the saw-toothed shape of the velocity curve, and the rapid change from contraction to expansion. This indicates a very strong outward acceleration of the star at minimum radius as the pulsation mechanism 'kicks' the photosphere upwards.

The observed velocity range 2K in km s^{-1} is generally about 70 times the range in magnitudes in blue light. The time of maximum light coincides approximately with that of minimum velocity, so that the light and velocity curves are approximately mirror images. Maximum light therefore occurs about a quarter of a cycle after minimum radius.

Other types of variability are seen in the spectra of Cepheids. The spectral type, which is a measure of temperature, varies throughout the cycle. Some Cepheids, particularly the low-mass ones, show emission lines before maximum light. These are caused by the passage of a shock wave through the stars' tenuous atmosphere; the emission lines arise from ions and electrons recombining after the shock wave has passed.

There is an interesting and important group of Cepheids known as *double-mode or beat Cepheids*, in which two radial modes are simultaneously excited. They are important because the two observed periods (usually the fundamental and first overtone radial periods) provide *two* accurate observable characteristics for comparison with theoretical models, leading to estimates of the masses and radii of the stars. This is normally done through the *Petersen diagram* (figure 6.9), named after Jorgen O. Petersen, which plots period against period ratio. The period ratio depends primarily on which modes are present, and secondarily on the radius (and therefore the period) of the star.

The periods of these are all 2 to 6 days, the amplitudes are moderate to large, and the period ratios are all about 0.70, indicating (from comparison with theory) that the periods are the radial fundamental and first overtone periods. Either the fundamental, or the first overtone, or neither can be dominant. The double-mode Cepheids seem to have normal Population I metal abundances and gravities. They tend to be hotter than average; the majority of short-period, blue Cepheids show double-mode behaviour, so their double-mode behaviour may be a consequence of their position within the instability strip.

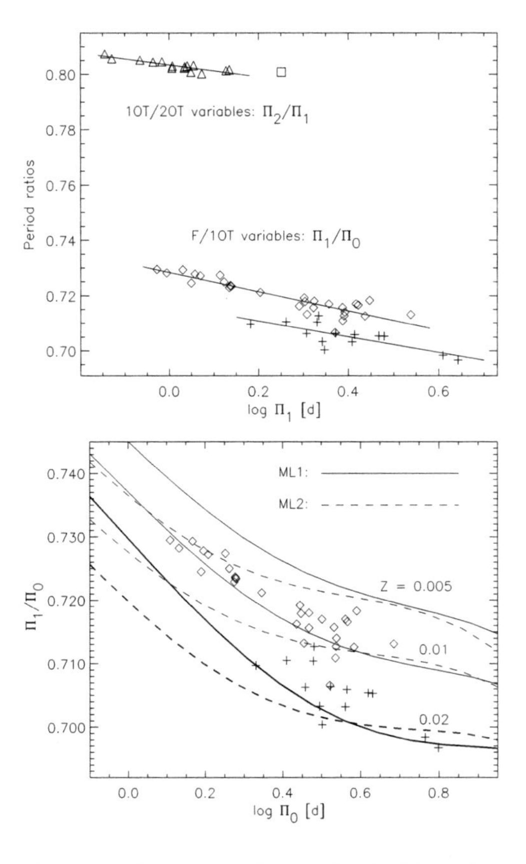

Figure 6.9 The Petersen diagram for double-mode Cepheids in the LMC (diamonds) and in our galaxy (crosses). The period ratio is plotted against the logarithm of the longer period. In the top panel, the stars clearly separate into two sequences. The one in the upper left corresponds to second and first overtone, the ones at lower right to the first overtone and the fundamental. In the bottom panel, the observed data points for the first overtone and fundamental are compared with pulsation models. (From Christensen-Dalsgaard and Petersen, 1995.)

A few triple-mode Cepheids are known, including two in the LMC, and these are even more useful for deriving stellar parameters (Moskalik and Dziembowski, 2005).

6.8.2 *Physical properties*

In 1908, Henrietta Leavitt noted that, in the Small Magellanic Cloud, the Cepheids which were brightest had the longest periods. Since the variables were nearly all at the same distance, the luminosities and the periods must be correlated. Since then, right up to the present era of the Hubble Space Telescope, astronomers have endeavoured to calibrate the period–luminosity

(P–L) relation, using Cepheids of known distance. Ejnar Hertzsprung carried out statistical studies of the motions of Cepheids in our galaxy; these indicated that, on the average, the luminosities of Cepheids were more than a thousand times that of the sun. Harlow Shapley noted that there were a few Cepheids in globular clusters, and, using the RR Lyrae stars in these clusters to determine their distance, established a P–L relation. Unfortunately, it was later discovered that the Cepheids in globular clusters – the W Virginis stars – are unlike those in the disc of our galaxy – the classical Cepheids – so Shapley's P–L relation was not applicable to the latter. However, it was good enough to enable Edwin Hubble to use 12 Cepheids in the Andromeda Nebula M31 to prove that the 'nebula' was actually a galaxy of stars, similar to our own. In the 1950s, John B. Irwin noted that several Cepheids were probable members of galactic star clusters, the distance to which could be determined reliably by comparing their main sequence stars to nearby main sequence stars whose distance and luminosity could be determined by parallax, or by the moving cluster method. According to the latest compilation by David Turner, there are 43 Cepheids known or strongly suspected to be members of galactic star clusters or associations. These provide the cornerstone for the Cepheid period–luminosity relation which, in turn, is a key method for determining the distance scale in our galaxy, and in the universe.

Identifying Cepheids in clusters is not trivial. Many clusters are ragged and sparse, and it is essential to confirm that the Cepheid is truly a member.

There are other methods for determining the distances of Cepheids: the Baade–Wesselink method and the Barnes–Evans method (section 6.2); these give the radius of the Cepheid which, if the temperature is known, gives the luminosity through Stefan's law. These methods provide a good independent check on the period–luminosity and period–radius relations.

There is one galaxy – NGC 4258 – whose distance (7.2 ± 0.5 MPc) has been determined very precisely from observation of stars which are *water masers*, orbiting the galaxy's central super-massive black hole. As of 2005, several hundred Cepheids have been discovered in this galaxy. It will be possible to use them to determine a period–luminosity relation.

Unfortunately, there are no Cepheids which are close enough for a parallax to be reliably measured. Polaris is the closest, but it is unusual in some respects, and believed to be a first-overtone pulsator.

The temperature of a Cepheid can be determined from its spectral type, its intrinsic colour, or from the distribution of energy in its continuous spectrum. In either case, the observed colour must first be corrected for reddening. This correction is usually large, since Cepheids are concentrated in the plane of our galaxy, where the dust is thickest.

In principle, the reddening can be determined by measuring the reddening of stars which are near the Cepheid in space, if the Cepheid and the other stars are seen behind the same screen of absorbing dust. This is not always the case. There may be differential or patchy reddening, extra dust in front of the Cepheid, for instance. Some improvement can be made by measuring the reddening of many stars near the Cepheid; this is particularly feasible for Cepheids in clusters. It enables the trends of the patchy reddening to be seen.

If the intrinsic colours of enough Cepheids are measured, it may be possible to find a relation between colour and period, and hence to determine the intrinsic colour of the Cepheid in that way. Unfortunately there is considerable scatter around that relation. It also assumes that any Cepheids to which it is applied are like the calibrating ones. The spectral type and luminosity class, and the strength of certain spectral features, such as the H line and the G-band, are unaffected by reddening, and could be calibrated in terms of intrinsic colour. However, the spectra are variable, and a bit peculiar, and difficult to obtain for very faint Cepheids. One of the most satisfactory approaches has been the use of multicolour photometric systems such as the Walraven VBLUW system. In this system, the Cepheids lie on a very narrow intrinsic locus in the $(V - B)$ versus $(B - L)$ diagram, and the reddening can be determined from the deviation from this locus. The VBLUW photometry provides a wealth of other physical data on these stars as well.

In order to obtain the temperature of the Cepheid, a relation between T and colour must be obtained. This is usually done by comparing the observed intrinsic colours, or the continuous spectrum, with theoretical models of the radiating layers of the stars. For instance, a relation derived by Robert Kraft in 1961 still appears to describe the temperatures satisfactorily

$$\log T_e = 3.886 - 0.175(B - V)_0, [0.3 < (B - V)_0 < 1.15] \quad (6.2)$$

The mass is the most fundamental property of a star, but few pulsating variable stars have a directly determined mass. This is particularly true of Cepheids. No Cepheid is a member of a visual or eclipsing binary system, with the exception of three faint stars in the Large Magellanic Cloud discovered by the MACHO project. Cepheids which are members of spectroscopic binary systems are extremely difficult to study: the periods are long; the velocities are complicated by the pulsational velocity variations; the companions are not easily visible (and may be binaries themselves). There are, however, many Cepheid binaries – up to a third of all Cepheids – and many ways of detecting them, in addition to the radial velocity technique. A few Cepheids (such as Polaris) have wide companions, which can be seen separately at the telescope, but too far away from the Cepheid for orbital motion to be observed. The companion to a Cepheid would likely be a B or A type star, which is still on the main sequence. If its spectral type can be determined, its luminosity – hence its distance, and that of the Cepheid – can be determined. If the companion cannot be resolved at the telescope, its slow orbital motion may cause periodic waves in the Cepheid's (*O–C*) diagram, which result from the finite light travel time across the orbit. Or its radiation may 'contaminate' the observed colours of the Cepheid, and its presence and nature can be inferred. For instance, the companion may cause the Cepheid to appear to describe loops in the $(B - V)$ versus $(U - B)$ diagram, or cause the colour curves to be out of phase with the light curve. So Cepheid binaries, though very useful, can often complicate their interpretation.

The spectrum of a bright, hot companion is easily detectable with a UV spectrograph in space. Nancy Remage Evans has gradually been accumulating high-quality observations of Cepheid binaries using the International Ultraviolet Explorer (IUE) satellite, and the Hubble Space Telescope, to observe the hot companions. She has determined the precise spectral type of the companions from their UV flux. From this, and the assumption that they are on the main sequence, she can then determine their absolute magnitudes, and hence their distances – which will, of course, be the same for the Cepheids.

Keep in mind that the stars which are progenitors of Cepheids – main sequence B stars – have a binary frequency of over half. The components are often close, and the periods short. What would happen if the more massive star were to evolve into a large yellow supergiant, given that it had a close companion to contend with? Mass transfer, and other complexities, would certainly occur.

One indirect approach to measuring the mass is to use the luminosity and temperature of the Cepheid, in conjunction with theoretical evolutionary tracks for different masses of stars, to estimate what mass of star would evolve into the observed Cepheid; this yields an *evolutionary mass*. This approach is slightly ambiguous in the sense that stars of different masses may end up at the same point on the H–R diagram at different stages of their lives. Also, stars cross the instability strip three or more times, with slightly different luminosities and periods. And, although most Cepheids pulsate in the fundamental mode, some pulsate in the first overtone. Ultimately, observations of period changes may help to identify the stage of evolution. Another indirect approach involves the use of the observed period, the theoretical pulsation constant *Q*, and some estimate of the radius to derive a *pulsation mass*. Double-mode Cepheids are especially useful for determining the mass and radius, by comparison with theoretical models, since two precise properties of the star are known; this method gives the *beat mass*. A more indirect method is to compare the shapes of the light and velocity curves, determined from non-linear pulsation models of stars of different masses and radii, with the observed light and velocity curves; this gives a *bump mass* (the reference being to the bumps which occur in the observed light curves of Cepheids).

For many years, the results of these mass determinations were rather disconcerting: the bump and beat masses were less than half the masses determined by other means. Eventually, it was realized that the models had been constructed with opacities which were incorrect, in that they underestimated the opacity contribution of iron at temperatures of 100–200 000K. The 'Cepheid mass problem' was resolved.

The chemical compositions of classical Cepheids are similar to those of other Population I stars, including those of non-variable stars of similar spectral type and class. There is, however, a slight variation in heavy element abundance in the sense that stars closer to the centre of our galaxy are richer in these elements; those in the outer regions of our galaxy are poorer. This is because there has been more starbirth, evolution, element creation, and recycling in the more populous inner regions of our galaxy. Because Cepheids are bright, and because their chemical composition and distance can both be accurately measured, they are useful for delineating the composition of different regions of our galaxy. In the Magellanic Clouds, the heavy element abundances are less: half of solar in the Large Cloud, and a quarter of solar in the Small Cloud.

6.8.3 *Interpretation*

In this section, we shall take a broad look at Cepheids: their position in the H–R diagram, their evolutionary history, and the cause of their period–luminosity (P–L) relation and of their pulsational instability.

Figures 6.3 and 6.4 show the H–R diagram, and the location of the classical Cepheids in it. There is a well-defined instability strip with a *blue edge* on the left, and a *red edge* on the right. The existence of these edges can be explained by the main pulsation mechanism: the effect of the ionization zones of hydrogen and helium. The position of the blue edge is sensitive to helium abundance in the star, and a comparison of the observed blue edge and the theoretical one confirms that the helium abundance Y must be about 0.25. The cause of the red edge seems to be related to the increasing importance of convection in cooler stars. Recall that the pulsation is caused by the interaction of radiation with the gas in the star; if convection replaces radiation as the dominant form of energy transport, then the pulsation mechanism will be quenched. But convection is a poorly understood process in stars, so the theoretical interpretation of the red edge is still incomplete.

We can now ask how Cepheids reached their present state, and their present position in the H–R diagram. This question can be answered by looking at theoretical evolutionary tracks, which model the changing luminosity and temperature of the star with time. Figures 2.14 and 6.4 shows such evolutionary tracks for stars with appropriate masses. These tracks bring the star through the instability strip up to five times: the so-called first to fifth crossings. The second is normally the slowest, so more Cepheids would be found in this stage. In principle, it should be possible to deduce the evolutionary state from observations of the period, the rate of period change, and the temperature, if possible. David Turner and his collaborators have done this with great success. One complicating factor is whether the star is pulsating in the fundamental mode, or in an overtone. Fourier decomposition of the light curve helps to resolve this issue.

The form of the evolutionary tracks is sensitive to the heavy element abundance Z. Low-Z models spend more time on the blue side of the instability strip (as is observed in the Magellanic Clouds), and would also cause stars of a given M_v to have a shorter-than-average period. Thus, the P–L relations, as well as the other properties of Cepheids in different galaxies, might be slightly different – something that needs to be kept in mind when using the P–L relation for distance determination. One way to investigate this question is to study Cepheids in the inner and outer parts of another spiral galaxy; they will be at approximately the same distance, but the inner parts of the galaxy are generally metal-richer than the outer parts.

The period of pulsation of a Cepheid depends mainly on its radius, so lines of constant period on the H–R diagram will be roughly coincident with lines of constant radius. These slant upwards and to the left, since R^2 is proportional to $1/T^4$ according to Stefan's law. In general, period increases with increasing luminosity and with decreasing temperature. Thus, there are period–luminosity and period–colour relations, respectively, of the form

$$\log P = a + b.M_v \tag{6.3}$$

$$\log P = c + d.(B - V)_0 \tag{6.4}$$

One recent version of the P–L relation is shown in figure 6.10. At a given M_v, however, there is a range of periods. The shorter periods occur for hotter stars, the longer periods for cooler stars. Thus, the scatter about the average P–L relation (above) is systematic. We can improve the situation by writing a period–luminosity–colour relation

$$\log P = e + f.M_v + g.(B - V)_0 \tag{6.5}$$

Recent applications of Cepheids to distance determination have generally used IR magnitudes. Whether at visual or IR wavelengths, the colour term adds

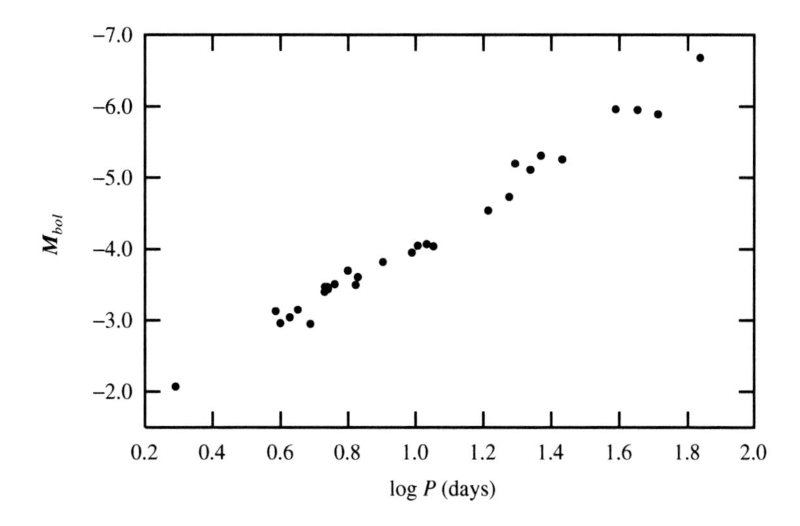

Figure 6.10 The period–luminosity relation for Cepheid pulsating variables. (Data from Turner, 2004.)

little strength to the process. Even the HST Key Project to determine the Hubble constant used only a P–L relation (Freedman *et al.*, 2001).

Interstellar reddening and absorption are major problems for Cepheids, particularly differential reddening. Reddening mimics the colour term in the period–luminosity–colour (P–L–C) relation, and it is sometimes difficult to separate the two effects. One solution is to use the so-called Wesenheit function

$$W = V - R.(B - V) = V_0 - R.(B - V)_0 \tag{6.6}$$

where R is the average ratio of absorption to reddening – about 3.1. The Wesenheit function is almost independent of reddening and absorption, and is a function of the intrinsic properties of the Cepheid only.

Another approach is to use IR or near-IR photometry. Reddening and absorption are much smaller at longer wavelengths such as J (1.25μm), H (1.65μm) and K (2.2μm). At these wavelengths, interstellar reddening and absorption, the effect of any blue companion, and the sensitivity to heavy-element abundance are all much smaller. Observations of the calibrating Cepheids in clusters and of Cepheids in several nearby galaxies were obtained in the 1980s. The approach has been successful in establishing the distance scale, and the value of the Hubble constant, to an accuracy of better than 10 per cent.

Doug Welch maintains an archive of Cepheid photometry and radial velocities at:
http://dogwood.physics.mcmaster.ca/Cepheid/

Table 6.1. *Bright and/or interesting Cepheid variable stars*

Star	HD	V Range	Period (d)	Spectrum	Comments
α UMi	8890	1.86 − 2.13	3.9696	F7–8Ib + hc[1]	amplitude decreasing!
β Dor	37350	3.46–4.08	9.8426	F4-G4Ia-II	–
T Mon	44990	5.58–6.62	27.024649	F7Iab-K1Iab + hc	Cl[2]Mon OB2
RT Aur	45412	5.00–5.82	3.728115	F4Ib-G1Ib	–
ζ Gem	52973	3.62–4.18	10.15073	F7Ib-G3Ib	Cl: ADS 5742AB
l Car	84810	3.28–4.18	35.53584	F6Ib-K0Ib	–
U Car	95109	5.72–7.02	38.7681	F6-G7Iab	Cl: Car OB1
V810 Cen	101947	4.95±	152.9	F5-G0Ia + hc	irregular, long-period
S Mus	106111	5.89–6.49	9.66007	F6-G0Ib + hc	–
R Mus	110311	5.93–6.73	7.510211	F7-G2Ib	–
S Nor	146323	6.12±	9.75411	F8-G0Ib	Cl: NGC 6087; first cluster Cepheid
X Sgr	161592	4.20–4.90	7.01283	F5-G2II	–
W Sgr	164975	4.29–5.14	7.59503	F4-G2Ib + hc	–
SU Cyg	186688	6.44–7.22	3.8455473	F2-G0I-II + B7V	spectroscopic binary
Y Sgr	168608	5.25–6.24	5.77335	F5-G0I-II	–
κ Pav	174694	3.91–4.78	9.09423	F5-G5I-II	brightest Pop. II Cepheid
FF Aql	176155	5.18–5.68	4.470916	F5-F8Ia +	–
S Vul	338867	8.69–9.42	68.464	G0-K2(M1)	Cl: Vol OB2
V473 Lyr	180583	5.99–6.35	1.49078	F6Ib-II	short period; amplitude varies
SV Vul	187921	6.72–7.79	45.0121	F7-K0Iab	Cl: Vul OB1; long period
η Aql	187929	3.48–4.39	7.176641	F6-G4Ib + hc	–
S Sge	188727 5.24–6.04	8.382086	F6-G5Ib +	–	
X Cyg	197572	5.85–6.91	16.386332	F7-G8Ib	Cl: Ruprecht 173/5
T Vul	198726 5.41–6.09	4.435462	F5-G0Ib +	–	
δ Cep	213306	3.48–4.37	5.366341	F5-G1Ib	prototype; Cl: Cep OB6

Cl = cluster or association. hc = hot companion.

6.9 Population II Cepheids (W Virginis stars)

Population II Cepheids – also called W Virginis stars – inhabit the same instability strip as the classical Cepheids, and pulsate for the same reason. Their evolutionary history is quite different, however; they are not massive, young stars like their cousins, but are older, low-mass stars – typically 0.5 to 0.6 solar masses. Not all Population II Cepheids belong to Population II; some are old disc stars with metal abundances not much different than the sun. So it might be preferable to use the W Virginis nomenclature, but to include the BL Herculis stars, mentioned below. The variability of W Virginis itself was discovered by Eduard Schönfeld, assistant to Argelander, in 1866.

The concept of Populations I and II was developed by Walter Baade, about 1950. Prior to that time, Cepheids of the two populations were included together. It is still difficult to distinguish them on the basis of their light curves alone. But their P–L relations are quite different, because their masses are quite different. Astronomers had previously calibrated this relation using RR Lyrae stars – Population II objects – and then applied it (erroneously) to Population I clusters and galaxies. When this error was corrected, the distance scale of the universe was increased by a factor of two!

It is therefore important to be able to distinguish between the two kinds of Cepheids. Population II Cepheids tend to be found further from the plane of the Milky Way galaxy, in orbits which are not circular and in the plane. Many (but not all) have lower metal abundance than the sun; this can be determined from the spectrum. These properties make them 'fossils' of the first generation of stars in our galaxy. They may also show emission lines in their spectrum, caused by shock waves passing through their low-density atmospheres. There are subtle differences in the shapes of the light curves, which can best be quantified through Fourier decomposition, and in the loops which they execute in the $(B - V)$–$(U - B)$ diagram. Observations in more astrophysically discriminating photometric systems, such as Walraven's VBLUW system, can now distinguish more easily between the two groups. Still, there are a few stars whose classification is uncertain.

As of 2002, there were 173 W Virginis stars found in the halo and disc of our galaxy, but the additional 60 or more which are members of globular clusters are especially important because their luminosity, temperature, chemical composition, and age can be independently determined. Such stars are listed in Helen Sawyer Hogg's *Catalogue of Variable Stars in Globular Clusters*, revised and placed on-line by Christine Clement. The cluster Omega Centauri contains 10 W Virginis stars – mostly short-period ones. Table 6.2 includes some of the brighter W Virginis stars in the field.

Box 6.1 Star sample – Polaris

Polaris (α UMi, HD 8890, F7:Ib-II, V = 2.005) is probably the best-known star in the sky for northern observers. It lies within a degree of the North Celestial Pole. Many people think it is the brightest star in the sky, but that is obviously not the case. There is some evidence that Polaris is 2.5 times brighter than it was in Ptolemy's time, 2000 years ago, though it is notoriously difficult to interpret observations from this early time. What few people know is that Polaris is variable in brightness; it is a Cepheid pulsating variable with a period of 3.9696 days.

Polaris is the closest Cepheid. The *Hipparcos* satellite measured its parallax to be 0.00756 arc sec. It also has a faint visual companion, and is a spectroscopic binary with a period of 30 years. The variability is not obvious; the range in V is presently only 0.02 magnitude. 'Presently' because, several decades ago, the amplitude was 0.2 magnitude. Armando Arellano Ferro (1983) noted that the amplitude had decreased significantly. At the rate it was decreasing, it would drop to zero by the year 2000. Did it? The most recent results (figure 6.11) show that it is still pulsating, but with a range of only 0.02. Will it now increase again?

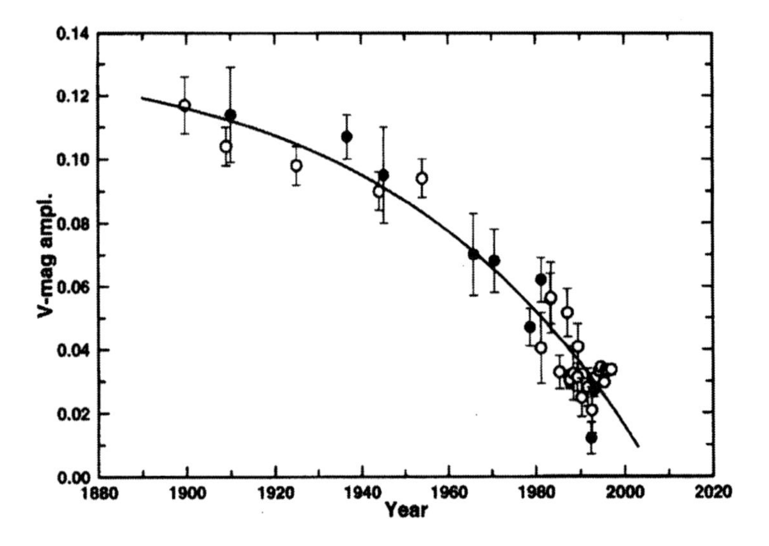

Figure 6.11 The V light amplitude of Polaris, as a function of time. The points show V ranges, either directly determined, or inferred from radial velocity variations. The star did not follow the trend (solid line); it is still pulsating. (From Turner, 2004.)

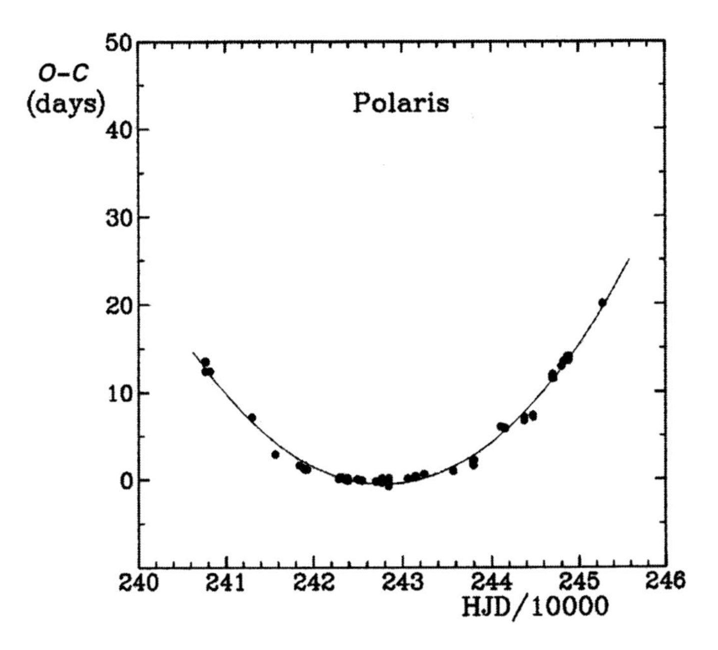

Figure 6.12 One interpretation of the (O–C) diagram of Polaris. Because of gaps in the observations, it is not possible to interpret the diagram definitively. But the simplest interpretation, represented by the parabolic fit (solid line), is that the period is increasing at a large and uniform rate. (From Turner, 2004.)

That will depend on what the cause is. A few pulsating stars have very long periodic variations in amplitude, caused by the presence of two close pulsation periods. In other cases, the variation can be a result of evolutionary changes in the star: it actually leaves the instability strip on the H–R diagram. There is no doubt that Polaris is evolving; figure 6.12 shows its $(O - C)$ diagram. The interpretation of the diagram is complicated by the gaps, but it is clear that the period is increasing by several seconds a year. This may indicate that the star is on its first pass from left to right on the H–R diagram; this passage is much faster than the second one.

Considering that Polaris is the nearest Cepheid, our understanding of its nature and evolution is still unclear. Its distance, according to *Hipparcos*, is 430 light years, but at least one astronomer thinks that it may only be 300. Depending on which of these distances is correct, Polaris may be a fundamental-mode pulsator or a first-overtone pulsator. There is even some evidence that Polaris is slowly brightening with time – by about 0.1 magnitude per century. But the interpretation of 100-year old photometry is difficult, and the interpretation of 1000- or 2000-year old photometry is even more so.

The periods of W Virginis stars range from 1 to 50 days, with the bulk of the 'standard' ones having periods of 10–20 days. The stars with periods of 1–7 (mostly 1–3) days are referred to as BL Herculis stars, and they are in a stage of evolution distinct from that of the longer-period stars. The GCVS4 classifies them as CWB, as opposed to CWA with periods longer than eight days. The upper limit to the period appears to be set by the maximum luminosity which a low-mass star can attain. The stars with periods greater than about 20 days often have RV Tauri characteristics, and it is difficult to distinguish W Virginis, RV Tauri, and SRd variables, without continuous, long-term observations.

Variables in globular clusters are particularly well suited to the study of period changes. Wehlau and Bohlender (1982) have investigated the periods of 12 of the 20 known BL Herculis stars known in globular clusters. The periods are either constant or slowly increasing at a rate which is consistent with the predictions of stellar evolution theory. The periods of the longer-period variables are sometimes constant, but are often erratically variable. This is consistent with their transitory phase of evolution. It may also be a consequence of random cycle-to-cycle fluctuations such as are found in RV Tauri and Mira stars.

Unlike the classical Cepheids, W Virginis stars seem all to have large brightness ranges (but see the note about RU Camelopardalis below). There are no small-amplitude variables, and no constant stars have yet been found in the instability strip. Of course, such stars are harder to find! Since the factor which determines the amplitude of classical Cepheids has not yet been identified, it is not possible to say why W Virginis stars are so homogeneous in this respect.

The $(B - V)$ colour variations are in phase with the brightness variations, but the $(U - B)$ variations are slightly out of phase, causing the star to describe a loop on the $(B - V)$-$(U - B)$ diagram. This behaviour is probably caused by violet emissions which result from shock waves; these affect the U brightness more than the B or V brightness.

The velocity curve also shows evidence of shock waves. The curve is discontinuous, and there is hydrogen emission around the phase of rising light. The shock waves are produced by pulsation waves which move up through the atmosphere, and collide with other layers of gas, which are moving downward.

The shapes of the light curves can be classified as either crested or flat-topped (rounded); the former appear to be on the blue side of the instability strip, the latter on the red side.

Figure 6.13 shows the location of some W Virginis stars in the H–R diagram. This information is based on stars in globular clusters; this enables the luminosity and intrinsic colour to be reliably determined. The intrinsic colour can be used to determine the temperature, using a $\log T_e - (B - V)$ relation appropriate to metal-poor stars.

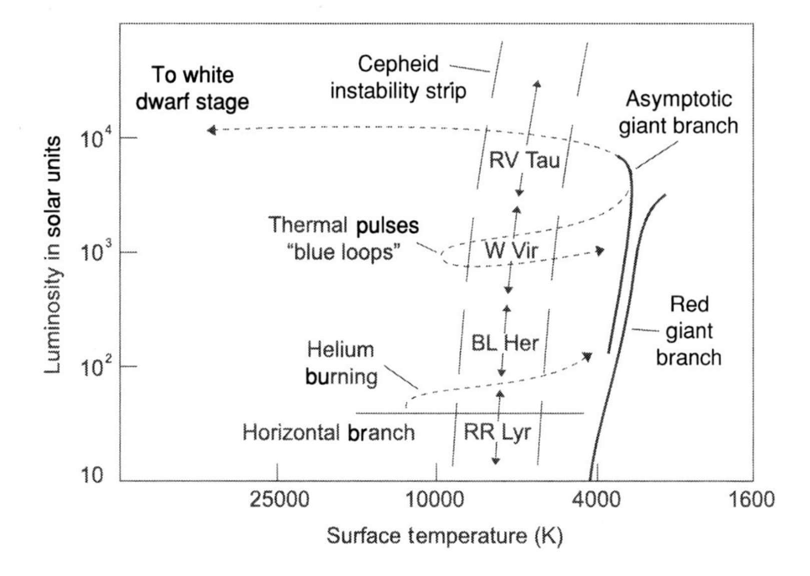

Figure 6.13 The position of Population II Cepheids, including BL Herculis, W Virginis, and RV Tauri variables, on the H–R diagram. As shown by the dashed evolution tracks, low-mass stars may reach these positions by helium-burning evolution from the horizontal branch, by thermal pulses (flashes) from the AGB, or by evolution from the AGB towards the white dwarf stage. (Jeff Dixon Graphics.)

The position of the observed blue edge is similar to that for classical Cepheids and, when compared with theoretical models, gives a helium abundance of ~ 0.25, as for the RR Lyrae stars. This is an important result. Since both these types of star are among the oldest in the universe, helium and hydrogen must *both* be primordial elements.

The evolutionary state of these stars has been deduced from general studies of the evolution of Population II stars, especially by Pierre Demarque and by Icko Iben. Old Population II stars evolve slowly from the main sequence to the tip of the giant branch (where they may become pulsating red giants), burning hydrogen in their core. Helium in the core then ignites, and the star moves to the horizontal branch (where it may become an RR Lyrae star), burning helium in its core, and hydrogen in a shell around the core. As the helium is consumed, the star reverses its evolution, moving across the supra-horizontal branch and upward along the asymptotic giant branch, burning helium in a shell around a carbon core, and hydrogen in a shell further out.

The BL Herculis stars are in the supra-horizontal branch stage of evolution, evolving away from the horizontal branch to larger luminosities and radii. Being slightly larger and more luminous than the RR Lyrae stars, they have slightly longer periods, though it its sometimes difficult to distinguish a short-period

BL Herculis star from a long-period RR Lyrae star. Since the stars are expanding, their periods should be increasing, and this is what is observed.

The evolutionary state of the longer-period (12–20 days) W Virginis stars was deduced by Martin Schwarzschild and Richard Härm (1970), who noted that a star on the upper asymptotic giant branch suffered *thermal instabilities* in its thin helium and hydrogen-burning shells. These cause the star to 'loop' to higher temperatures and smaller radii in the H–R diagram, and thus to enter the instability strip for a few thousand years, one or more times. This may explain the irregular period and amplitude changes in some W Virginis stars. The most remarkable example is V725 Sagittarii. Henrietta Swope discovered that, within a few years, this star increased its period from 12 to 21 days. As of 2000, the period was almost 100 days! The most likely explanation is that this star has undergone a blue loop, and is now returning to the AGB.

Thermal instabilities may also cause large-scale convection episodes, which 'dredge up' heavy elements such as carbon, from the interior, producing a few *carbon Population II Cepheids* such as RU Camelopardalis mentioned below.

Evolution beyond the tip of the asymptotic giant branch must, at some point, take the star across the H–R diagram to the white dwarf region, thereby passing through the instability strip one last time. The longest-period (20–50 days), most luminous W Virginis stars may be in this brief stage of evolution. Many of them are surrounded by dust – evidence of mass loss on the AGB – and many of them show RV Tauri characteristics. And most of the RV Tauri stars show the expected decreasing periods, though the presence of random cycle-to-cycle period fluctuations complicates this analysis.

Because W Virginis stars inhabit an instability strip, with a narrow range of temperature, and because the pulsation period depends primarily on the radius of the star, the W Virginis stars will obey a period–luminosity relation, or perhaps even a period–luminosity–colour relation. Alcock *et al.* (1998) recently published a study of W Virginis and RV Tauri stars in the LMC, using data from the MACHO gravitational lensing experiment. They derived the following P–L–C relation

$$M_v = -0.61(\pm 0.20) - 2.95(\pm 0.12)\log P + 5.49(\pm 0.35) < (V - R)_0 > \qquad (6.7)$$

assuming a distance modulus of 18.5 for the LMC.

It is not possible to derive the masses of W Virginis stars directly, since none of them is known to be in suitable eclipsing binary systems, and there are no double-mode W Virginis stars. Indirect methods suggest that they are 0.5 to 0.6 $M_\odot$. Four W Virginis stars are in binaries, however. The most interesting is ST Puppis, with an orbital period of 410 days, and a pulsation period of 19 days. ST Puppis has low abundance of the heavier elements, especially those that might have condensed to form dust grains. Presumably ST Puppis went through a red giant phase. How did the relatively close binary companion affect the evolution?

6.9.1 *Anomalous Cepheids*

There are very few W Virginis stars in the dozen dwarf spheroidal galaxies in our Local Group. There are, however, 47 'anomalous Cepheids', which, though they share some properties with W Virginis stars, differ in others. There is evidence that their masses are typically 1.5 $M_{\odot}$, so they must either be quite young (since stars of that mass have short lifetimes) or they must be binary stars which have coalesced. In either case, their evolutionary state would be quite different from that of the 'normal' W Virginis stars. The GCVS4 refers to these stars as BLBOO variables, after variable 19 (also known as BL Bootis) in NGC 5466.

6.10 RV Tauri variables

RV Tauri stars are pulsating yellow supergiant stars whose light curves are characterized by alternating deep and shallow minima. They have spectral types F to G at maximum light, and K to M at minimum. The period from one deep minimum to the next (the 'double' or 'formal' period) ranges from 30 to 150 days. The complete light amplitudes can reach 3-4 magnitudes in V. RV Tauri stars are executing blue loops from the asymptotic giant branch (AGB), or are in their final transition from the AGB to the white dwarf stage. Table 6.2 includes some of the brighter members of this class.

The RV Tauri stars are heterogeneous, and not well understood. The alternation of deep and shallow minima may not be strict, so, if the classification of the star is based on only a few cycles of the light curve, it may be misclassified as a Population II Cepheid, if the minima do not alternate, or a yellow semi-regular (SRd) variable, if the alternation appears irregular. The light curves are non-sinusoidal, and usually non-repeating. The range may vary considerably from cycle to cycle. The deep and shallow minima may occasionally (or not so occasionally) interchange. I am inclined to think that there is a smooth spectrum of behaviour from the periodic to the semi-regular variables.

There is, in fact, a problem with the definition of the period. Is it the interval between adjacent minima, or adjacent *deep* minima? One hypothesis for the alternating minima is that the stars are double-mode pulsators in which the periods are in the ratio of 2:1 (Takeuti and Petersen, 1983). The double period might be the fundamental radial mode, and the single period the first overtone radial mode. A second hypothesis, which is not totally independent of the first, is that they are exhibiting low-dimensional chaos (Buchler and Kovacs, 1987).

Table 6.2. *Bright and/or interesting RV Tauri and related stars*

Star	HD	Range	Type	Period (d)	Spectrum	Comments
RU Cep	–	8.2–9.8V	SRd	109.0	G6-M3.5III	–
RV Tau	283868	9.8–13.3P	RVB	78.731	G2eIa-M2Ia	prototype
SS Gem	41870	9.3–10.7P	RVA	89.31	F8-G5Ib	–
SU Gem	42806	9.8–14.1V	RVB	50	F5-M3	–
RU Cam	56167	8.10–9.79V	CWA	22	C0,1-C3,2e(K0-R0)	amplitude decreased
U Mon	59693	6.1–8.8P	RVB	91.32	F8eIb-K0pIb(M2)	brightest RVB star
W Vir	116802	9.46–10.75V	CWA	17.2736	F0-G0Ib	prototype
SX Her	144921	8.6–10.9P	SRd	102.9	G3ep-K0(M3)	–
UU Her	–	8.5–10.6P	SRd	80.1	F2Ib-G0	switches mode
TT Oph	–	9.45–10.84V	RVA	61.08	G2e-K0	maybe not RVT star
TX Oph	–	9.7–11.4V	RVA	135.0	F5-G6e	–
AI Sco	320921	9.3–12.9P	RVB	71.0	G0-K2	–
AC Her	170756	6.85–9.0V	RVA	75.01	F2pIb-K4e(C0,0)	bright RVA star
R Sct	173819	4.2–8.6V	RVA	146.5	G0Iae-K2p(M3)Ibe	brightest RVA star
κ Pav	174694	3.91–4.78V	CW?	9.09423	F5-G5I-II	brightest CW
AR Sgr	–	9.1–13.5P	RVA	87.87	F5e-G6	–
R Sge	192388	8.0–10.4V	RVB	70.77	G0-8Ib	–
V Vul	340667	8.05–9.53V	RVB	75.7	G4e-K3(M2)	–

Note: Range is given in visual (V) or photographic (P).

Box 6.2 Star sample RU Cam

From its discovery over a century ago, until the mid-1960s, RU Cam was an apparently well-behaved W Virginis star with an amplitude of about one magnitude, and a period of 22 days. In 1966, Serge Demers and Donald Fernie discovered that the star had apparently stopped pulsating! Astronomers have continued to monitor this star, especially at the Konkoly Observatory in Hungary (Szeidl *et al.*, 1992). They find that, although the star has made sporadic efforts, it has never regained its former range, though it still pulsates with an amplitude of about 0.1 magnitude (figure 6.14). The cause is not known. Since it is still pulsating, it must still be in the instability strip, but evolution may have changed some aspect of the pulsation mechanism of the star. This is not unreasonable; a W Virginis star may remain in the instability strip for only a few thousand years, so it is not surprising that, among the several dozen variables which have been studied by astronomers over several decades, one or two should be observed to enter or leave the instability strip. This is one more example of the need to monitor variable stars regularly – especially Population II Cepheids which are known to misbehave at times.

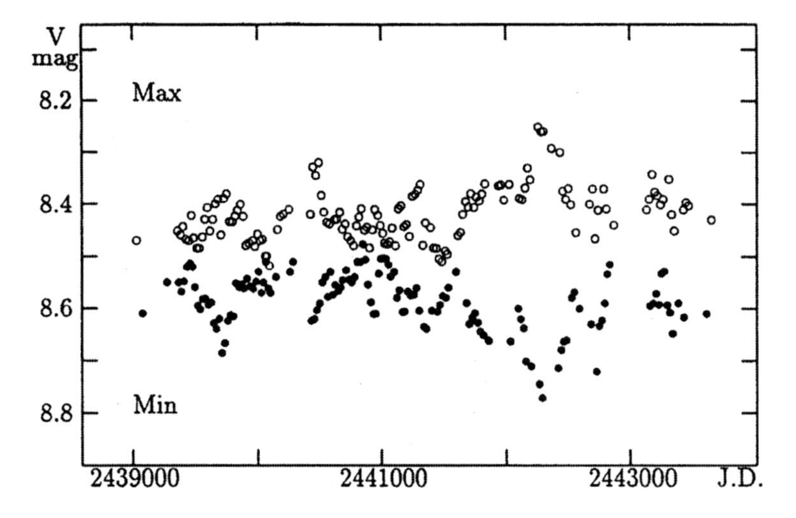

Figure 6.14 The observed photoelectric magnitudes of maxima (open circles) and minima (filled circles) of RU Camelopardalis after the sudden decrease in amplitude in 1966. The amplitude has varied from 0.1 to 0.5 magnitude on a time scale of years. (From Szeidl *et al.*, 1992.)

The $(O - C)$ diagram is often helpful in deducing the evolutionary state of pulsating stars. Yvonne Tang and Jonathan Hale, two high school students working with me, made some progress in understanding the star's current behaviour (see Percy and Hale, 1998). First of all, the light curve suggested that the star might be pulsating in two or more low-amplitude modes. The $(O - C)$ diagram is dominated by random cycle-to-cycle fluctuations but, beneath that, there is evidence for evolutionary period changes on a time scale of about 30 000 years – which is consistent with what would be expected for a star undergoing helium flashes from the asymptotic giant branch.

The magnitude of the period changes in these stars appears to support the AGB or post-AGB hypothesis for their evolutionary state, but the $(O - C)$ diagrams are dominated by random, cycle-to-cycle period fluctuations (Percy *et al.*, 1997).

The radial velocity variations are typically a few tens of km s^{-1}. The velocity curves are often discontinuous, showing periodic doubling of low-excitation absorption lines of metals. This is a result of the propagation of a shock wave upward through the low-density atmosphere. Other peculiarities include Balmer emission during rising light, and the superposition of TiO bands (indicative of a cool atmosphere) on an otherwise normal G or K spectrum around minimum light.

George Preston and his collaborators, in a pioneering study of these stars in 1963, showed that they could be sub-classified, spectroscopically, into:

- A: G and K type (though the spectral type varies over the pulsation cycle), probably old Population I. This class was later divided into **A1**: TiO bands present at minimum light; and **A2**: TiO bands not present at minimum light.
- B: Fp type; CH and CN bands of normal strength; probably old Population I.
- C: Fp type with weak CH and CN bands; Population II.

Do not confuse the spectroscopic A and B classes with the photometric a and b classes discussed below!

Several observers have attempted to derive temperatures and luminosities for these stars, but they were hampered by two problems. The spectra are peculiar, and difficult to compare with those of normal Population I stars. The apparent magnitudes and colours are affected by interstellar absorption and reddening, and also by absorption and reddening around the star itself. An interesting property of these stars is their infrared excess (figure 6.15), which results from gas and dust driven off the star – perhaps by pulsational shock waves during the RV Tauri phase or the AGB phase (when the stars would have been

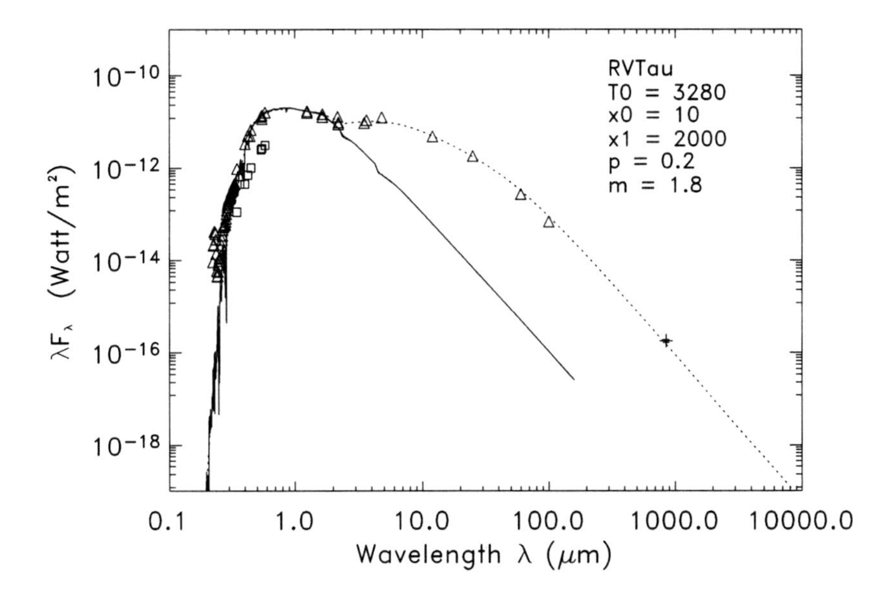

Figure 6.15 The continuous spectrum of RV Tauri in the visible and IR. The filled line, peaking at about 1.0 μm, is the model spectrum of the star; the dashed line is the model spectrum of the dust, which produces this IR excess. The symbols are the observed fluxes at different wavelengths. (From S. de Ruyter, *et al.*, 2005).

pulsating red giants). The stars with the larger heavier element abundances seem to have more infrared emission, which would make sense, since dust would form more easily in a heavier-element-richer environment. The dust may be related to an unusual aspect of the chemical abundances in these stars: elements with the highest condensation temperatures are the most deficient – apparently because they are the elements which have condensed on to dust grains.

The RV Tauri stars appear to be old stars, with masses similar to that of the sun. They find themselves in the yellow supergiant region of the H–R diagram near the end of their lifetimes. They would get there by the same process as Population II Cepheids do. One way is by describing 'loops' from the AGB in the H–R diagram. These loops occur in later stages of a star's lifetime, when it is burning helium and hydrogen in thin shells in its interior. Instability of the nuclear burning in these shells causes the changes which lead to these loops. When the star completes the AGB phase of its evolution, it makes one last trip across the H–R diagram, passing through the instability strip as it does. A few RV Tauri stars may be in this phase.

Until recently, the best estimates of luminosity were derived from membership in globular clusters, and from statistical studies of the motions ('statistical

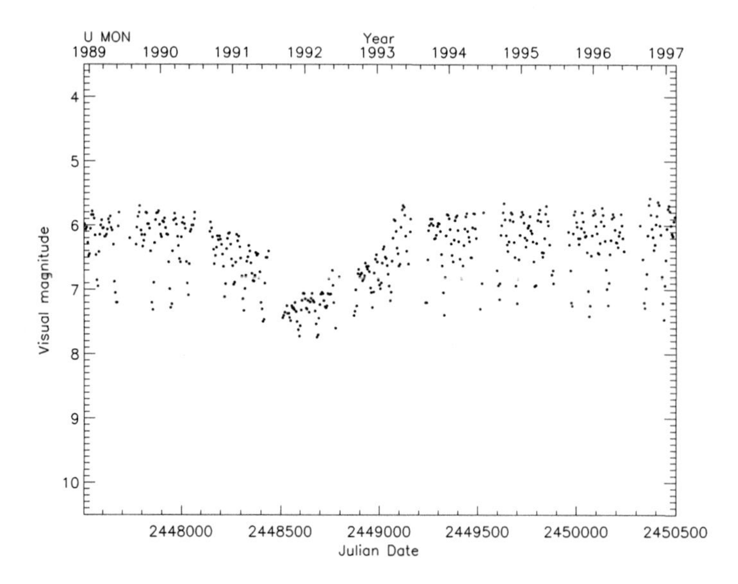

Figure 6.16 The eight-year light curve of the RV Tauri star U Monocerotis, based on visual observations from the AAVSO International Database. The pulsational variations with (generally) alternating deep and shallow minima can be seen, especially by looking at the light curve at an angle, from the bottom. The long-term eclipse-like variation can also be seen. (From AAVSO.)

parallax'). Recently, Alcock *et al.* (1998) have studied the RV Tauri stars in the Large Magellanic Cloud, and derived the period–luminosity relation given in the previous section.

A subset of the RV Tauri stars – the RVb stars – show regular long-term variations in mean brightness on a time scale of up to 2500 days, and are particularly enigmatic. The long secondary period is usually about ten times the pulsation period. The most extreme case is U Monocerotis, with a secondary period of almost 2500 days. When it first became possible to plot many decades of the AAVSO's archival data (figures 6.16 and 6.17), U Monocerotis long-term variations suddenly came into focus. The light curve is much like that of an eclipsing variable, but the shape of the minima is variable. Some form of binary model is, however, suggested. The RV Tauri star and its companion could be orbiting within a torus of dusty material, ejected from the pulsating star during its AGB phase. Most RV Tauri stars have circumstellar dust (figure 6.15). If the star was a binary, then it is not unreasonable for the dust to exist in a disc or torus.

RV Tauri stars which do not show long secondary periods are denoted RVa stars.

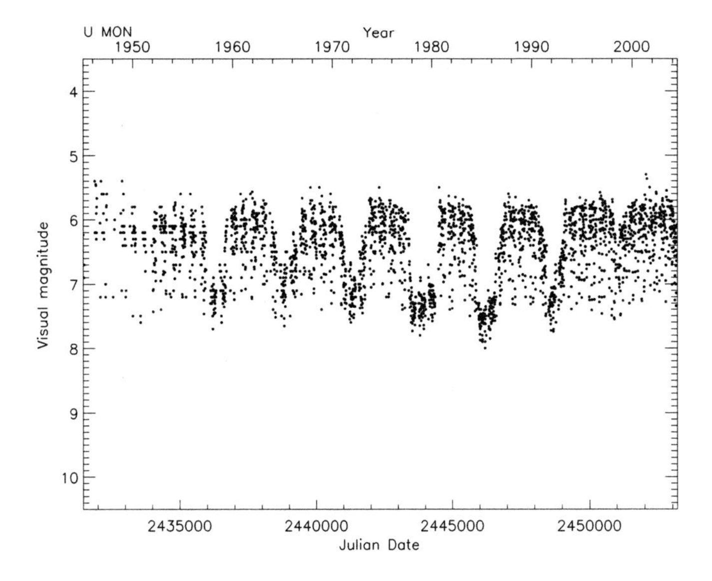

Figure 6.17 The 60-year light curve of the RV Tauri star U Monocerotis, based on visual measurements from the AAVSO International Database. Most of the vertical scatter is a result of the star's pulsation. The 2500-day secondary period is clearly visible. Its light curve is very much like that of an eclipsing variable. Note that the pulsational variations appear to be smaller during the 'eclipse' minima. (From AAVSO.)

6.11 RR Lyrae stars

RR Lyrae stars are pulsating variables with periods of 0.1 to 1 day, ranges in V of up to 1.5 magnitudes, spectral types A5 to F5, absolute visual magnitudes of about +0.5, masses about half a solar mass, and low metal abundances Z of 0.00001 to 0.01, depending upon their age. The variability of RR Lyrae itself was discovered by Williamina Fleming, around 1900. It has a period of about 0.567 day. The period shows small increases and decreases. By 1916, Richard Prager and Harlow Shapley had independently discovered that its light curve was modulated in amplitude and shape, in a period of 41 days. They had discovered the *Blazhko effect*, which remains a mystery, even today.

In practice, these stars are defined by their evolutionary state. They are stars which, having spent most of their lives burning hydrogen in their cores, have exhausted that fuel, and have begun to burn helium in their core. Stars in this stage have a wide range of temperatures, but a narrow range of luminosities (figure 6.18). They therefore occupy a 'horizontal branch' in the H–R diagram. If their temperature lies between about 6000 and 7500K, they are unstable against pulsation. Their narrow range of temperatures and luminosities expains their small range of periods.

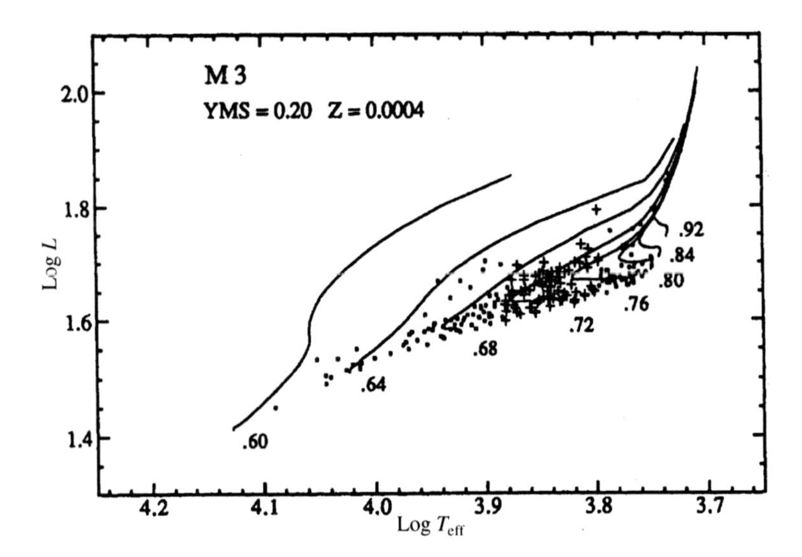

Figure 6.18 Synthetic horizontal branch model for the globular cluster M3 with some horizontal branch models ($Y = 0.20$, $Z = 0.0004$) shown. RR Lyrae variables are marked by crosses, and each evolutionary track is labelled by its total mass in solar units. (From Lee *et al.*, 1990.)

In order to lie on the horizontal branch, stars must be very old: either Population II or old Population I. Stars of this age must have masses less than the sun's, and will naturally have metal abundances lower than the sun's, because most of the heavy elements in the universe had not yet been produced by stars and supernovas when they were born.

Population II stars are found in the halo of our galaxy, old Population I stars in the disc. RR Lyrae stars are particularly numerous in some globular clusters, and were at one time called *cluster variables*. Variables in globular clusters are particularly convenient to study (because many can be recorded on a single image), and to interpret (because all the stars in the cluster have the same distance, reddening, composition, and age). RR Lyrae stars are also found outside globular clusters, in the 'field'. The nearest RR Lyrae star is RR Lyrae itself, $V = +7.5$ and 250 parsecs away. The nearest globular cluster known to contain a substantial number of RR Lyrae stars is M4, 14 000 light years away. Thus, RR Lyrae stars are rather faint for detailed study, compared with Cepheids, for instance. They have been found in the Magellanic Clouds, and with difficulty, in some other nearby galaxies. Helen Sawyer Hogg did important work in compiling a catalogue of variable stars (mostly RR Lyrae stars) in globular clusters. Christine Clement and her students have updated the catalogue, and made it available on-line.
www.astro.utoronto.ca/~cclement/read.html

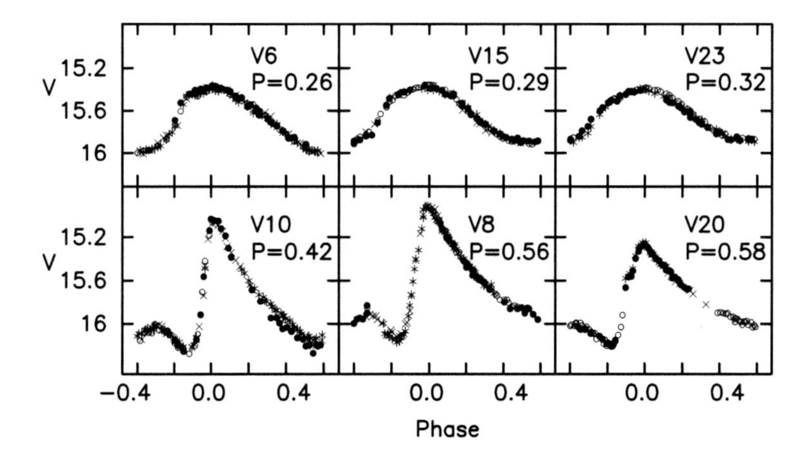

Figure 6.19 Light curves for a selection of RR Lyrae variables in the metal-rich globular cluster M107 (NGC 6171). Variables 6, 15, and 23 (top three curves) are RRc types, but note that the light curve shape changes with increasing period. Variables 10, 8, and 20 (bottom three curves) are RRab types. The different symbols represent observations on different nights: note that the light curve for V10 does not repeat; it shows the Blazhko effect. (From C. M. Clement, based on work by C. M. Clement and I. Shelton.)

It is also important to study RR Lyrae stars in the field. They provide important information on the distribution and motion of the oldest stars in our galaxy – the extreme Population II stars. The absolute magnitudes of the RR Lyrae stars, and how they vary with metal abundance, are increasingly well known, as are the intrinsic colours. These are necessary in order to measure the interstellar reddening of the star, which can then be used to determine the interstellar absorption. This is needed before the distance of the star can be determined.

There is a catalogue of field RR Lyrae stars at:
www.astro.uni-bonn.de/~gmaintz/

6.11.1 *Variability and classification*

RR Lyrae stars can be divided into several types, on the basis of their light curves (figure 6.19). *Type a* have moderately long periods, large ranges, and highly asymmetrical light curves. *Type b* have longer periods, somewhat smaller ranges, and less asymmetrical light curves. There is, in fact, a fairly smooth transition between types a and b, so there may not be any significant difference between them. *Type c* have much shorter periods, small ranges, and sinusoidal light curves. They are distinct in that they are pulsating in the first radial overtone, whereas the types a and b are pulsating in the radial fundamental period.

Traditionally, the V amplitudes of RR Lyrae stars were 0.3 to 1.5 magnitudes, but smaller-amplitude variables have been discovered through CCD photometry.

RR Lyrae stars also show variations in colour and radial velocity, in the effective gravity felt by their photosphere, and in their radius. Once each period, the photosphere feels a strong 'kick' from the pulsation mechanism in the star, and is driven outward. At this time, the photosphere feels a strong effective gravity, just as a jumper on a trampoline feels at the bottom of their bounce. The photosphere is launched upward, then is pulled back by the star's gravity, until it receives yet another kick. Because of the rapid acceleration of the photosphere, the radial velocity curve is quite non-sinusoidal. The spectrum and photometric colours show the effects of these rapid gravity changes, and of shock waves rising up through the atmosphere. The highly asymmetrical light curves frequently show a 'bump' near minimum light; this is also a result of shock waves, reflected back upwards from the deeper layers of the star.

6.11.2 *Period changes*

Because of the short periods and sharp maxima of RR Lyrae stars, it is relatively easy to determine period changes in these stars. This is particularly true of stars in globular clusters, where up to several hundred variables can be recorded on a single image. Some clusters have been studied for a century, so the periods (and the period changes) are known to a high degree of accuracy. Period changes have been studied more intensively in RR Lyrae stars than in any other kind of variable. Organizations such as the AAVSO monitor the period changes in field RR Lyrae stars, using visual and now CCD techniques.

Two representative $(O - C)$ diagrams are shown in figure 6.18. In general, a variety of forms is seen. A few diagrams appear parabolic, as might be expected from simple evolution theory. One or two appear sinusoidal; this could result from light–time effects in a binary system, or simply from random cycle-to-cycle fluctuations in period. Most appear abrupt, irregular, or (to the precision of the observations) constant. But, given the large amount of good data on period changes in RR Lyrae stars, it is disappointing that they have not told us more. Or perhaps they are telling us something, but we do not yet understand.

If the period changes were caused by evolution, they would be expected to be small, regular, and generally of the same sign, since, in a given cluster, most of the stars are evolving in the same sense. There is some tendency for this to be true in some clusters, but most of the large, abrupt changes cannot be explained by evolution. One possibility is that these abrupt changes are a result of *semi-convection* – an abrupt mixing process within the star, which changes its internal structure slightly.

6.11.3 *Light curve variations: the Blazhko effect*

Another interesting property of RR Lyrae stars is the *Blazhko effect*, a long-term modulation of the amplitude, shape, and phase of the light and radial velocity curve. The period of the Blazhko effect is 11 to 533 days, or 50 to 500 times the pulsation period. It occurs in about a third of the RRab stars (though not among long-period RRab stars), and has been observed in a few RRc stars as well. The incidence is lower in Large Magellanic Cloud stars. V10 in figure 6.19 shows this effect.

The Blazhko effect is not entirely regular, nor is its presence strictly correlated with any other property of the star. There is no well-accepted explanation for the Blazhko effect, but there are two promising theories: (i) resonance between the radial fundamental period of pulsation, and a non-radial period; or (ii) a deformation or splitting of the radial period by a magnetic field in the star. From time to time, claims of a detection of a magnetic field in an RR Lyrae star are made, but none has been confirmed. There is some interesting recent evidence, based on spectral line profile analysis, that favours the hypothesis that it is an effect of one or more non-radial periods which are slightly different from the basic radial period. This is the explanation for the 'beat effect' in Beta Cephei stars. As of 2005, there is a major 'Blazhko Project' underway to study a small sample of Blazhko-effect stars (and a few non-Blazhko-effect stars for comparison) over several weeks, using both photometric and spectroscopic techniques, to try to solve this mystery. See:
http://www.astro.univie.ac.at/~blazhko/

Some RR Lyrae stars pulsate simultaneously in the fundamental and first overtone periods; these *double-mode RR Lyrae stars* are called RRd stars, and they are very important because the two observed periods can be used, in conjunction with models, to determine the mass and radius of the star, using the Petersen diagram.

6.11.4 *Distribution and motions*

For the RR Lyrae stars, the distribution and motions are particularly significant, because they reflect the population type and age of the stars. This enables us to learn about both the RR Lyrae stars, and about their population in general.

Because all RR Lyrae stars have virtually the same absolute magnitude, measurement of their apparent magnitude (averaged over their cycle, of course) yields their distance. Correction must be made for interstellar reddening, but this can be done either statistically (using the average absorption per parsec in that direction in our galaxy) or individually for the brighter stars by comparing their observed colour with the intrinsic colour, and using the standard result that the absorption in V is about 3.1 times the reddening in $(B - V)$. Fortunately, it turns out that RR Lyrae stars have approximately the same intrinsic colours at minimum brightness. And they tend to be located away from the plane of our galaxy, in regions where the reddening and absorption are small. Surveys of RR Lyrae stars to magnitude +20 or fainter have been carried out; they constitute one of the most powerful tools for mapping the Population II stars in our galaxy.

The oldest stars formed when our galaxy was spherical, slowly rotating, and almost devoid of heavy elements. The younger stars formed when our galaxy was much flatter, more rapidly rotating, and enriched in heavy elements. Thus the age and metal abundance of RR Lyrae stars is related to their motion as seen from the sun. The oldest RR Lyrae stars have large apparent velocities, in a variety of directions. Younger stars move in circular orbits in the disc of our galaxy at a well-defined velocity.

6.11.5 *Physical properties*

RR Lyrae stars are a very homogeneous group in every respect but their metal abundance Z. The hydrogen and helium abundance are relatively uniform at about 0.75 and 0.25 by mass, respectively, but Z can be anywhere between 0.00001 and 0.01. A classical measure of the metal abundance in RR Lyrae stars is ΔS, which is defined as −10 times the difference in spectral subtypes between the star's spectral type as measured from the Ca II H and K lines, which are sensitive to metal abundance, and the spectral type as measured from the hydrogen lines, which are not. ΔS ranges from 0 for a normal Population I star to 10 for an extreme Population II star.

With modern large telescopes and sensitive spectroscopic detectors, it is increasingly possible to measure the abundance of individual elements. One specific measure is

$$[Fe/H] = \log((Fe/H(\text{star}))/(Fe/H(\text{sun})) \tag{6.8}$$

where Fe and H are the abundances of iron and hydrogen, respectively, by number. Note that this measure involves iron only. In many ways, it is a representative heavy element, though the abundances of the other heavy elements are not necessarily in proportion. According to one calibration

$$[Fe/H] = -0.16\Delta S - 0.23 \tag{6.9}$$

The metal abundance can also be determined from photometric measurements through selected filters. Walraven's photometric system is particularly useful for this purpose, and a large number of field RR Lyrae stars have been observed and analyzed in this system. Nevertheless, there is still some controversy about the absolute calibration of the scale of metal abundance in Population II stars.

Perhaps the most important property of an RR Lyrae star is its absolute magnitude, since precise knowledge of that quantity makes RR Lyrae stars excellent cosmic yardsticks. The absolute magnitude does, however, depend slightly on the metal abundance and other physical properties of the star, but the total range in M_V is small, and the relationships between M_V and other properties are reasonably tight. The following methods can be used to determine the absolute magnitude:

- Statistical study of the motions of field RR Lyrae stars – so-called *statistical parallax*. This gives values of M_V ranging from +0.9 for short-period, high-Z stars, to +0.5 for longer-period, lower-Z stars. Because of the statistical nature of this method, it must be applied to a large sample of stars, which might not be homogeneous.
- Fitting of the main sequence of globular clusters containing RR Lyrae stars to a standard main sequence determined for nearby Population II stars with known distances; there are, however, very few of these. This method gives a mean of about +0.4 for the RR Lyrae stars in several clusters.
- The Baade–Wesselink method has been applied to some of the brightest RR Lyrae stars; it gives an absolute magnitude of about +0.5.
- There is a relation between the mass and luminosity of an RR Lyrae star (discussed below) which enables the luminosity to be determined, if the mass is known. If the mass is between 0.5 and 1.0 suns, then M_V is between +0.1 and +0.5.

Unfortunately, there are very few RR Lyrae stars which are close enough for an accurate trigonometric parallax to be measured – even with *Hipparcos*; for RR Lyrae itself, the parallax is 4.38 ± 0.59 mas. Even more recently, the parallax has been measured by the HST to be 3.82 ± 0.2 mas, giving an absolute magnitude of +0.61 ± 0.11.

But the *Hipparcos* parallaxes are still useful for determining the distances of a large sample of RR Lyrae stars, statistically. The *Hipparcos* proper motions are also useful for statistical distance determination.

Although an M_V of about +0.6 seems to be well established, the relation between M_V and Z is still slightly uncertain, though a slope of $\sim$ 0.2, for this relation, is gradually receiving general acceptance.

For distance determination, it is also important to know the reddening and absorption. This, in turn, requires a knowledge of the intrinsic colours of the stars. Many observers have carried out large-scale surveys, primarily in the UBV system, but more recently in the Walraven VBLUW system, which provides considerably more information: temperature, reddening and absorption, gravity, and metal abundance.

6.11.6 *Correlations*

To a first approximation, all RR Lyrae stars have the same M_V, a value which is imposed by their similar age, mass, and evolutionary state. The properties of RR Lyrae stars *are* slightly sensitive to metal abundance, however; a star with lower metal abundance is hotter and brighter, though the effect is only a few tenths of a magnitude in M_V.

Within a given cluster, there is only a small range of M_V among the RR Lyrae stars, so there is no significant period–luminosity relation, as there is for Cepheids. There is a range of temperatures, however. The radii are given by Stefan's law, so the coolest stars are the largest, have the lowest density, and therefore longest period. This relation is found, both among the type ab stars, and the type c stars. The type c stars, being overtone pulsators, have consistently shorter periods than the type ab.

There is also a correlation between the amplitude and the period (and therefore the temperature). Stars of similar periods have similar (but not identical) amplitudes. This is unlike the situation with other types of pulsating variables, in which stars of similar temperature and luminosity can have quite different amplitudes.

6.11.7 *Theoretical interpretation*

The RR Lyrae stars lend themselves to theoretical analysis and interpretation, perhaps more than any other class of variable stars. This is partly for observational reasons: thousands are known, especially in globular clusters; the clusters allow the physical parameters of the stars to be determined, and these are not much affected by interstellar reddening and absorption. On the theoretical side: the stars are so homogeneous that it becomes feasible to construct a grid of models of different mass, luminosity, temperature, and composition, which will encompass all of the observed variables.

The evolutionary state of the RR Lyrae stars is also well known: they are burning helium in their cores, and are on the horizontal branch in the H–R diagram, evolving from hotter to cooler (figure 6.18). Above the horizontal branch are the *supra-horizontal branch stars*, which are in a slightly more advanced stage of evolution. If these stars are in the instability strip, they are called *BL Herculis stars*, and are included here with the Population II Cepheids.

If the periods of RR Lyrae stars in a particular cluster are plotted against the temperature or colour, the type ab and type c variables are observed to lie on two parallel sequences. For a given colour, the ratio of the periods of the two sequences – namely about 0.70 – is the ratio of the first overtone period to the fundamental period, which suggests that the type ab are fundamental mode pulsators, and the type c first overtone pulsators. Furthermore, the type c variables are the ones on the hotter side of the instability strip, the type ab on the cooler side. And recall that the type ab have larger amplitudes and more asymmetrical light curves; the type c have smaller amplitudes and sinusoidal light curves. (The hot boundary of the instability strip is called the *blue edge*; the cool boundary is called the *red edge*. There are separate edges for fundamental and first overtone pulsation.) This observation is confirmed by analyzing the pulsational stability of model RR Lyrae stars: the hotter models are unstable to higher-overtone pulsation.

There is a physical relationship between the masses (M) and luminosities of RR Lyrae stars. If we combine Stefan's law, the period–density relation for pulsating stars, and the definition of magnitude, we get

$$\log P_o + 3\log T_e = a - 0.3M_{bol} - 0.5\log \mathrm{M} \quad (6.10)$$

where a is a known constant involving the pulsation constant Q, which can be calculated theoretically; here, M is the mass. But from the period–temperature sequences, we can deduce that

$$\log P_o + 3\log T_e = b \quad (6.11)$$

assuming that we have a relationship between the observed colour and the effective temperature, and a correction for reddening. Combining the two preceding relations, we obtain

$$0.3M_{bol} + 0.5\log \mathrm{M} = c \quad (6.12)$$

where c is a known constant.

If M_{bol} is between +0.5 and +1.0, then the mass is between 0.5 and 1.0 $\mathrm{M}_\odot$.

The colour of the blue edge of the instability strip provides important information about the helium abundance in these stars. Recall that the driving mechanism for the pulsating variables in the Cepheid instability strip is the effect of the helium ionization zone. Calculations show that the temperature of the blue edge is a function of the helium abundance Y. Observations of the colour of the blue edge in several clusters, corrected for reddening and calibrated in terms of T_e, give $Y \sim 0.3$, with no large differences from one cluster to another. This tells us that even the oldest stars in our galaxy – the globular clusters – had almost as much helium as young stars like the sun. Unlike the heavy elements, which were synthesized in stars and supernovas, most of the helium in the universe was created in the Big Bang.

The cause of the red edge is less certain but, as with Cepheids, appears to be connected with the fact that, in cooler stars, most energy is transported out of the star by convection, as it is in the sun. Since the driving mechanism for pulsation is a radiative process, it will be less effective when most of the energy is being carried by convection.

Only a non-linear (hydrodynamic) theory of stellar pulsation will provide information about the amplitude and shape of the light curve. This was first done for the RR Lyrae stars, with great success, by Robert Christy (1966) in the 1960s. He constructed a grid of nearly 100 models of RR Lyrae stars whose mass, luminosity, temperature, and metal abundance covered the observed range. He tested each model for pulsational stability, and then determined the period, light and velocity amplitude, and other observable quantities which the models could predict. He found that:

- In a typical unstable RR Lyrae model, the energy of the pulsation increases by about 1 per cent per cycle, until the amplitude reaches a stable value. Thus, it takes 100–200 cycles for the star to reach its full amplitude.
- Hotter models were unstable to pulsation in the first overtone, cooler models to pulsation in the fundamental, as observed.
- For a helium abundance of $Y = 0.30$, the blue edge was between $T_e =$ 7250 and 7550K, in reasonable agreement with observations.
- The amplitude of pulsation is a function of period; the relation is similar to what is observed.
- The theoretical light and radius variations are in good agreement with observations, if $Y \sim 0.30$.
- The amplitude of pulsation increases to a point at which the pulsation driving mechanism saturates; beyond that amplitude, the driving no longer exceeds the damping by the deeper layers in the star.

More recently, Bono and others (e.g. Bono *et al.*, 1997) have generated state-of-the-art pulsation models for RR Lyrae stars. But Christy's models provided a rather complete and accurate simulation of real stars, and represented a triumph for pulsation theory (especially considering the modest power of computers at that time). There are still RR Lyrae mysteries to be solved: the cause of the Blazhko effect, and the nature of the period changes, for instance.

6.12 Delta Scuti stars

Delta Scuti stars are pulsating variables of type A and F, with short periods, and generally small amplitudes. They lie in the downward extension of the Cepheid instability strip, where it crosses the densely populated main sequence. As a result, they are the most numerous pulsating variables among the bright stars. See Breger and Montgomery (2000) and the Delta Scuti Network (www.deltascuti.net) for a review, and current information.

Delta Scuti stars were known as early as 1900. The large-amplitude ones were classified as RR Lyrae stars, sub-type RRs. The small-amplitude ones were often included with the Ap stars and the Beta Cephei stars, in a 'miscellaneous' class. In 1955, Harlan J. Smith introduced the term *Dwarf Cepheid*, which was later used to denote the large-amplitude variables or RRs stars. The small-amplitude variables were called *Delta Scuti stars*. Dwarf Cepheids are sometimes called *AI Velorum stars*, after the prototype. Nowadays, they are often called *high-amplitude Delta Scuti stars*. In the 1970s, Michel Breger showed that the majority of dwarf Cepheids were not fundamentally different from Delta Scuti stars, and he proposed that the term should be abandoned. There were, however, a few dwarf Cepheids and Delta Scuti stars which were much older and metal-poorer than the others, and were presumably in a different stage of evolution. The name *SX Phoenicis stars* is now applied to this group, after its most prominent member. The GCVS4 includes a subclass of low-amplitude variables (DSCTC), as well as SXPHE and normal DSCT variables. Perhaps this all sounds confusing, but it was included deliberately in order to illustrate the origins and problems of classification and nomenclature.

Because of the small amplitudes of the Delta Scuti stars, they are normally discovered by photoelectric or CCD photometry. They make up about a third of all A5-F2 III-V stars, so precision surveys of such stars are a fruitful source of new variables. CCD photometry has been especially fruitful in discovering Delta Scuti stars in clusters, since millimag photometry is now routinely possible with this technique. Over 600 have been discovered in this way, a good fraction of them among the naked-eye stars. They are the most common type of bright pulsating variables, and are therefore frequently found as small-amplitude variables in massive surveys of stellar variability. Needless to say, *careful* photometry is necessary, and every newly discovered Delta Scuti star should be confirmed by an independent observer. Radial velocity variations are doubly difficult to detect: not only are they small, but they are also rapid, so time resolution was a problem, especially in the era of photographic spectroscopy. Multiple low-amplitude pulsation modes were recently discovered in the bright star Altair, using the WIRE satellite (Buzasi *et al.*, 2005).

The periods of Delta Scuti stars are typically 0.02 to 0.3 day, this being the range of the natural periods of radial pulsation of A5-F2 III-V stars. Many of these stars are multiperiodic. This, along with the small amplitudes, makes period determination very difficult, especially since there is a danger of alias periods in single-site photometry. But multiperiodicity raises the possibility of *asteroseismology* – determining the interior structure of the star by comparing the many periods with theoretical calculations. This has been done effectively by multi-longitude campaigns, which are able to acquire two to three weeks of almost-continuous photometry for analysis. In a few Delta Scuti stars, a dozen or more periods have been determined (see Star Sample – FG Virginis; XX Pyxidis also shows dozens of periods), and this provides a goldmine of information for comparison with theoretical models.

Period changes can be measured, and – at least in the case of some large-amplitude variables – this could be a worthwhile activity for amateur or student observers, using photoelectric or even visual techniques. The nature and cause of the observed period changes are still not understood. The AAVSO observes a few such stars as part of its RR Lyrae program.

The amplitudes of Delta Scuti stars are typically a few hundredths of a magnitude, with smaller amplitudes being more common. Specifically, the probability of an amplitude Δm is found to be approximately proportional to $e^{-\Delta m}$. In this context, 'constant' stars in the instability strip may simply be ones with sub-millimagnitude amplitudes. Large amplitudes are rare; less than 1 per cent have amplitudes greater than 0.3, though such variables are easier to detect. The radial velocity amplitude 2K is related to the light amplitude, as in the case of the Cepheids, such that $2K/\Delta m = 70$ km s^{-1} mag^{-1}. Minimum velocity lags about 0.09 period behind maximum light. There are also small variations in colour, which are in phase with the variations in light.

Large-amplitude Delta Scuti stars have asymmetrical light curves, as RRab stars do, with a rapid rise to maximum and a slower decline. Several of them show the same 'beat effect' as in double-mode Cepheids, and for the same reason: there are two or more interfering periods. Small-amplitude Delta Scuti stars have sinusoidal light curves, but there are often complex variations in amplitude, resulting from the presence of several pulsation periods.

There has been a tendency to think of large-amplitude Delta Scuti stars as separate from the small-amplitude ones, but there is no reason to do so. But what determines the amplitude of the star? A likely candidate is rotation; large-amplitude stars tend to be more slowly rotating, whereas rapidly rotating stars have small amplitudes. The presence of many pulsation modes is also a factor: it is not clear how the pulsation energy is shared.

The location of the Delta Scuti stars in the H–R diagram is well established. They occur between spectral types A5 to F2. The lower luminosity boundary is the zero-age main sequence, and the upper luminosity boundary is set by the hydrogen-burning stage of stellar evolution; beyond that stage, stars move rapidly away from the region of the main sequence. When stars moved a significant way across the H–R diagram from the main sequence to the instability strip, they

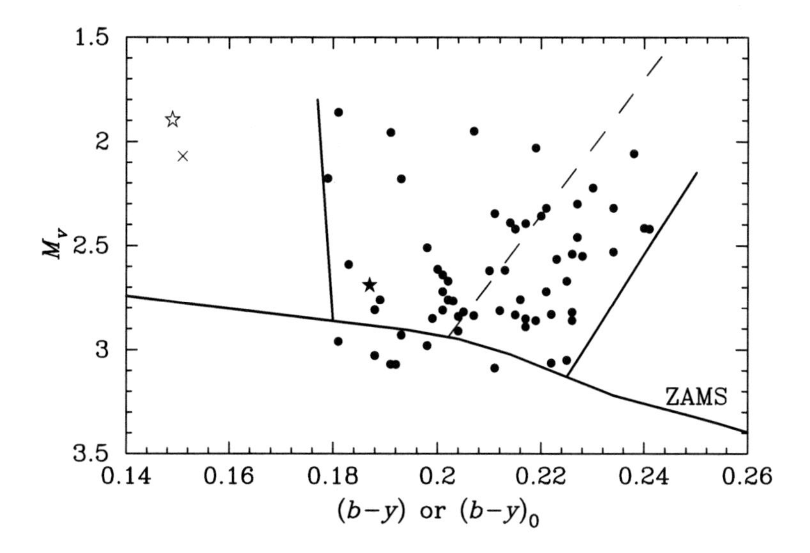

Figure 6.20 The position of Delta Scuti stars and Gamma Doradus stars in the H–R diagram. The dots represent Gamma Doradus stars. The stars and the cross are Gamma Doradus stars with special properties. The two thick lines, perpendicular to the main sequence are the boundaries of the Gamma Doradus instability strip. The dashed line is the cool edge of the Delta Scuti instability strip. (From Handler, 1999, updated to 2006.)

would be classified as Cepheids. The hot and cool boundaries of the instability strip are caused by the decreasing effectiveness of the pulsation mechanism at high temperatures, and the increasing role of convection at lower temperatures. Figure 6.20 shows the location of both the Delta Scuti and Gamma Doradus stars on the H–R diagram; see also Figure 6.29.

Virtually all Delta Scuti stars have normal masses and chemical composition, as determined from observations of their spectra; some exceptions are discussed below. Most are in the core hydrogen-burning stage of evolution, as deduced from their location in the H–R diagram, and from their frequent occurrence in young clusters like the Hyades. The exceptions are the high-amplitude variables, which are almost certainly in the shell hydrogen-burning phase, and the SX Phoenicis stars whose evolutionary state is discussed below.

Delta Scuti stars, like other pulsating variables, obey a period–luminosity or period–luminosity–colour relation, which reflects the way in which the radial pulsation period depends on the radius and mass of the star. One recent period–luminosity relation (Petersen and Christensen-Dalsgaard, 1999) is

$$M_v = -3.725 \log P(fundamental) - 1.969 \tag{6.13}$$

This reproduces the observed values of M_V to within $\pm$ 0.1. This satisfactory result is rather surprising. Aside from the uncertainty in measuring the distance, luminosity, and colour of the star, the periods of Delta Scuti stars are not always well known, because of the complexity of their light curves. And not all of them pulsate in the same radial or non-radial mode. There is a tendency for the hotter stars to pulsate in the first overtone, and the cooler stars to pulsate in the fundamental mode, as is the case with the RR Lyrae stars.

On the other hand, the amplitudes of the Delta Scuti stars behave quite differently from those of RR Lyrae stars. In the latter, stars of the same period have about the same amplitude. In the former, the amplitude is determined by some combination of factors, including temperature, luminosity, rotation, multiperiodicity, and perhaps other things. For stars above the main sequence, slow rotation seems to favour larger amplitude. (As mentioned earlier, these large-amplitude stars are sometimes called dwarf Cepheids, but they do not otherwise differ from other Delta Scuti stars.) For stars on the main sequence, the opposite seems to apply. Slowly rotating stars undergo *diffusion* (as in the Ap stars described earlier). Some elements 'float' to the surface, because of an imbalance between the downward force of gravity on the atom, and the upward force of radiation, causing the star to be classified as a *metallic-lined (Am)* star. Helium, however, sinks inward, removing the driving mechanism, and preventing the stars from pulsating. *Delta Delphini* stars appear to be a mild version of Am stars; they have slightly abnormal compositions (probably a result of diffusion), but they still manage to pulsate. *Lambda Bootis* stars are another group with mild chemical peculiarities; most of them pulsate, but they tend to do so in high-order radial pulsation modes. Current research suggests that these stars have peculiar chemical compositions because they have accreted interstellar gas which has been depleted of some chemical elements.

The pulsation mechanism of the Delta Scuti stars is well understood; it is basically the same as for Cepheids; the Delta Scuti stars simply lie in the downward extension of the Cepheid instability strip. Nevertheless, a few important questions remain: what determines the mode (or modes) in which the star pulsates, and what determines the amplitudes?

6.12.1 *SX Phoenicis stars*

There are a few large-amplitude Delta Scuti stars which do appear to differ fundamentally from the others: they have low metal abundances, and space motions typical of old Population II stars. They are called *SX Phoenicis stars*; they have short periods of 0.03 to 0.08 days, and amplitudes of a few tenths of a magnitude. These stars must be in an advanced evolutionary state, since a normal A5-F2 main sequence star has a lifetime of less than a billion years. As well as being found in the field, they also occur in dwarf galaxies and in globular clusters, among the *blue straggler* population. This is an upward extension of the main sequence, above the *turnoff point* where stars are beginning to leave the main sequence as they exhaust their hydrogen fuel. Blue stragglers are believed to be stars which have increased their mass and luminosity late in life, by gaining mass from a binary companion, or even by merging with one. This hypothesis is supported by observations of SX Phoenicis stars, which pulsate in two modes, with two different periods: the period ratio indicates that their mass is much greater than 'normal' old stars.

Period changes are easy to measure in these stars, because their periods are short, their amplitudes are large, and their maxima are usually sharp. And period changes are useful for comparison with evolutionary models of the stars. Unfortunately the $(O - C)$ diagrams are wave-like in appearance, perhaps caused by random cycle-to-cycle period fluctuations.

6.12.2 *Gamma Doradus stars*

Gamma Doradus stars are a homogeneous group of variables with spectral types of F0-F2 (they lie at or just beyond the red edge of the Delta Scuti instability strip) on or near the main sequence (figure 6.20), with one to five periods ranging from 0.4 to 3 days and with photometric amplitudes of up to 0.1 in *V*. Line profile variations, and radial velocity variations of 1–4 km s^{-1} are also observed. They pulsate in high-order (n), low-spherical-degree (l) non-radial gravity modes – g-modes; many have multiple periods, and are pulsating in multiple modes. The variability of γ Doradus, the bright F0V prototype star, was discovered in 1963. It had periods of 0.733 and 0.757 days, but the nature and cause were not understood; it was deemed 'a variable without a cause'. The rotation of a peculiarly spotted star was one hypothesis. The existence and nature of the Gamma Doradus stars as a group was established in the early 1990s. Variable stars with periods of about a day are difficult to discover with single-site photometry, because only part of a cycle of variability is observable in a night.

By 1999, when Kaye *et al.* (1999) defined this class of variables, there were 13 members known; there were about 30 as of 2002. Their absolute magnitudes ranged from +1.93 to +3.07, their luminosities from 5 to 15 $L_{\odot}$, their effective temperatures from 6950 to 7375 K, their radii from 1.43 to 2.36 $R_{\odot}$, and their masses from 1.51 to 1.84 $M_{\odot}$; they are thus a very homogeneous group. It least one star is both a Delta Scuti star and a Gamma Doradus star.

See www.astro.univie.ac.at/~gerald/gdor.html for an on-line list of Gamma Doradus stars.

Box 6.3 Star sample – FG Virginis

FG Virginis (HD 106384, A5, V = 6.560) is a Delta Scuti star. Most of the more famous pulsating variables, such as δ Cep, pulsate in a relatively simple fashion, in one or occasionally two pulsation modes. FG Virginis pulsates in 79 different modes, with periods ranging from 0.025 to 0.174 day, and amplitudes of 0.2 millimagnitude or higher (Breger *et al.*, 2005)! Other than the sun, it is the most complex 'normal' pulsating variable known (figure 6.21). The first challenge is to accumulate long and continuous datasets, and disentangle these 79 different modes in the data (figure 6.22). This would be impossible if the measurements were made from one site only; the daily gaps would not only restrict the number of

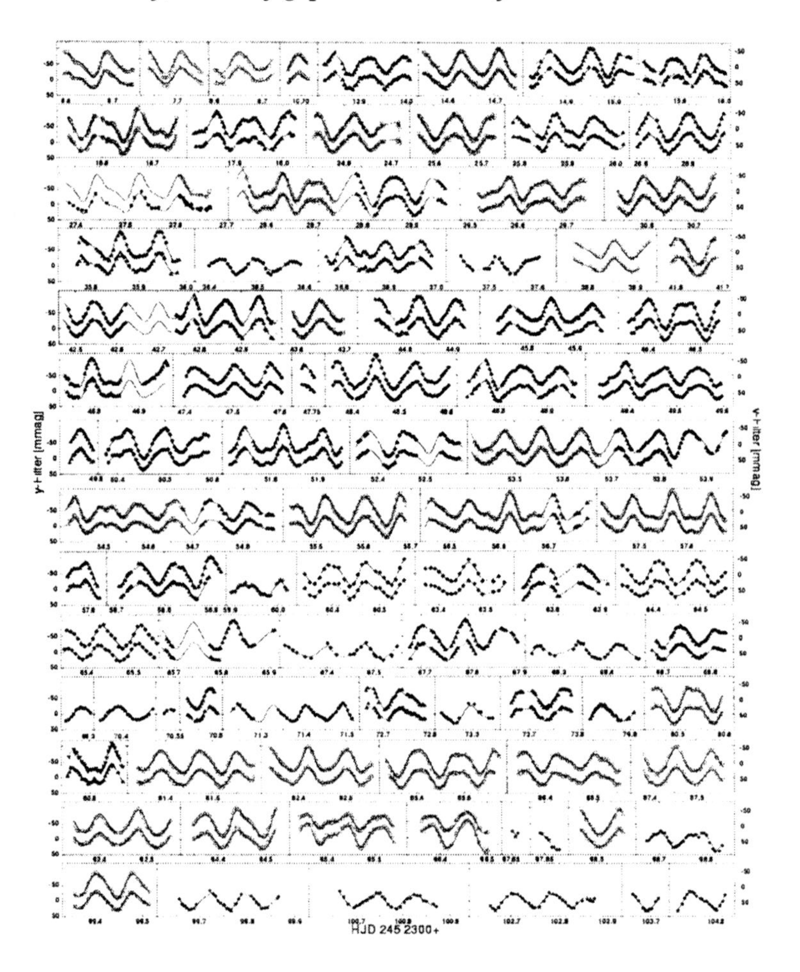

Figure 6.21 The complex light curve of FG Vir, obtained from the 2002 Delta Scuti Network campaign (Breger *et al.*, 2004.)

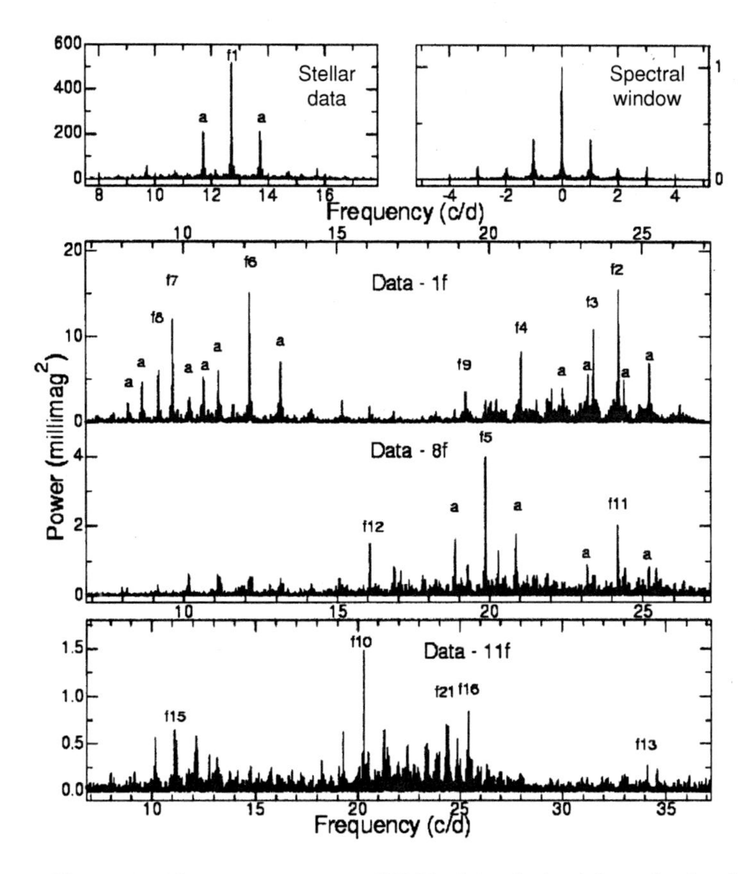

Figure 6.22 The power spectrum of FG Virginis, obtained from the data in the previous figure (Breger *et al.*, 2004). The top panels show the power spectra of the star data, and of the times of observation (the window function). The next three panels show the power spectrum with 1, 8, and 11 frequencies removed.

observations, but they would create alias peaks in the power spectrum. Continuous multi-longitude observations are essential. The *Delta Scuti Network* helps to organize such campaigns. The most recent published results from such a campaign are from January 2002, when 398 hours of data were gathered at five observatories.

The second challenge is to compare the observed periods with the periods predicted from stellar models. There is an infinity of possible periods; why are some excited, and with the amplitudes that we observe? One factor which leads to a multitude of possible periods is the presence of significant rotation, and this is true in many Delta Scuti stars. Their observed periods can be very helpful in making sure that our theories of stellar rotation – and other stellar processes – are correct.

6.13 Rapidly oscillating peculiar A (roAp) stars

Peculiar A (Ap) stars (section 4.6) are B8-F2 stars, temperatures about 6 500 to 8 500K, on or near the main sequence, which show unusual abundances of certain elements. These stars have strong global magnetic fields, and it appears that these stabilize the atmosphere of the star so that elements can diffuse upward or downward, according to the relative effect of the downward force of gravity, and the upward force of radiation on the atoms of each element. The magnetic field also causes certain elements to be more visible on different parts of the star's photosphere. The rotation axis of the star is generally not coincident with the magnetic axis so, as the star rotates, the spectrum, brightness, and colour vary slightly. The period of variation is the rotation period. The star is a rotating variable. This model of Ap star variability is called the 'oblique rotator model' (the rotation axis being oblique to the magnetic axis).

The Ap stars lie close to the Delta Scuti instability strip, so some of them were surveyed, from time to time, for Delta Scuti variability. In 1982, Donald Kurtz discovered several Ap stars which varied in brightness with the unusually short period of a few minutes. Thanks mainly to Kurtz and his collaborators, over 30 roAp stars are known today, with temperatures between 7000 and 8500 K, and periods ranging from about 5 to 15 minutes. Note that the GCVS4 refers to these as ACVO variables, ACV referring to Alpha Canum Venaticorum, and the O to oscillating. There are also dozens of 'noAps' or non-oscillating Ap stars, which help to define the instability strip. RoAp behaviour is more prevalent among cooler Ap stars – those showing enhancements of strontium, chromium, or europium.

The periods of the roAp stars are much shorter than the low-order radial pulsation periods. That is because the strong magnetic field controls the pulsation motion; it is in-and-out along the magnetic axis. Furthermore, the character of the pulsation changes according to the rotational phase of the star, as the observer sees a different aspect of the pulsation. This model, developed by Kurtz, is called the *oblique pulsator model* (figure 6.23). It predicts that the observer should see a complex light curve with $(2l+1)$ frequencies, each separated from the previous one by an amount equal to the rotation frequency. The pulsation can be even more complex than this, since the rotation may have an effect on the pulsation frequencies, as well as on the geometry by which the observer views the star.

Because the pulsation period is so short, it is difficult to measure the small radial velocity variations in these stars, because of the time and velocity resolution required. Because the pulsation periods are known, however, it is possible, in one or two stars, to construct a velocity curve by phasing spectra and radial

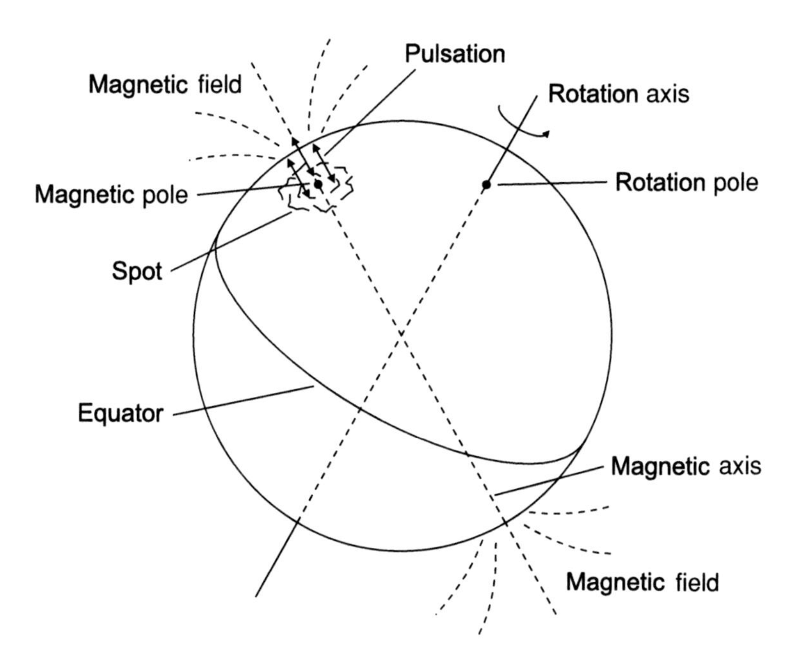

Figure 6.23 Schematic model of a roAp star. The star is rotating around an axis, which is inclined, at some angle, to the sky. The star has a strong approximately dipole magnetic field, which is inclined to the rotation axis (the rotation then produces the variations in brightness, colour, and spectrum, which were noted in chapter 4). The pulsation of the star is confined by the strong magnetic field of the star. (Jeff Dixon Graphics.)

velocities together with the known period (figure 6.24). This provides an even more direct proof of the pulsation.

Understanding these stars is not straightforward. In the interior, the effects of the strong magnetic field are less important. In the outer layers and atmosphere, the magnetic field is as important as the 'standard' physics. So current theoretical models must consider the magnetic field, the rotation, the detailed structure of the atmosphere, and convection if it occurs. In the best available current models, the observed modes do turn out to be unstable in models with temperatures and luminosities typical of real stars. The excitation occurs primarily in the ionization zone of hydrogen.

One interesting question about these stars is – why are some Ap stars roAp stars and others non-oscillating Ap stars (noAps)? In the case of the Delta Scuti stars, there was a similar question about what determined the incidence and amplitude of variability – a combination of temperature, luminosity, and rotation? In the case of the Ap stars, there is an even wider assortment of star properties, the additional ones being the strength and orientation of the magnetic field, and the nature and amount of diffusion which has taken place.

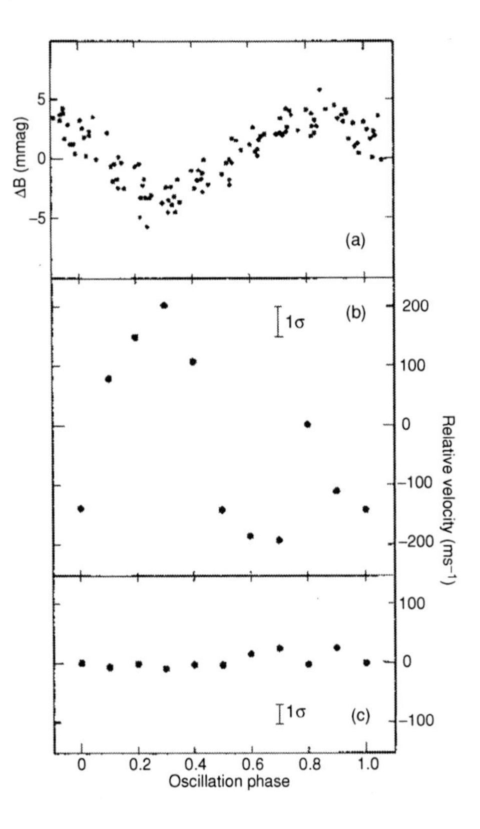

Figure 6.24 The B light curve (top) and velocity curve (middle) of the roAp star DO Eri. The bottom panel shows the non-variation of the velocity of a reference line in the spectrum. The known pulsation period was used to phase the radial velocity measurements. (From Matthews *et al.*, 1988.)

With their short periods, and complex variability, the roAps are a prime candidate for asteroseismology with the new generation of satellites, such as MOST and COROT. Ground-based studies have already revealed a rich spectrum of modes which contain a wealth of astrophysical information, from the luminosity of the star to the structure of its atmosphere and magnetic field (e.g. Kurtz *et al.*, 2003).

6.14 Pulsating degenerate stars

The majority of stars end their lives as red giant stars. As a result of the combined effect of radiation and pulsation, they lose their outer layers in a massive stellar wind. Their hot white dwarf cores, now devoid of nuclear energy, are exposed, and these cool like a cinder in a fireplace. Their radius and mass are now fixed. Their rate of cooling – their luminosity – depends on

Box 6.4 Star sample – DO Eridani

DO Eridani (HR 1217, HD 24712, A5p(Sr-Cr-Eu)) has been selected as the prototypical roAp star, but it could equally be considered as an 'average' α^2 Canum Venaticorum rotating Ap variable; it is very much 'average' within this class. Its period of rotational variability was determined to be 12.5 days by George Preston in 1972, on the basis of its spectrum variations. Its photometric period is the same, and its photometric amplitude is only 5.97 to 6.00. Minimum light occurs when the europium lines are strongest in the spectrum. Its rapid oscillations were discovered in 1983. There were several periods, the strongest being 6.14 minutes. Some of these periods could be explained as non-radial periods which have been 'split' by the 12.5-day rotation of the star. While it is relatively easy to measure photometric variations with six-minute periods, it is much more difficult to measure radial velocity variations with such periods. Many more photons are required, and this usually requires observation times of many minutes, even with the largest telescopes. And this is long enough to smear out the variations. But in a *tour de force*, a group using the Canada–France–Hawaii telescope were able to detect the radial velocity variations, by binning very rapid radial velocity measurements with the known photometric period (figure 6.24). This demonstrated conclusively that the star was actually pulsating. Further radial velocity observations showed that the pulsation was strongly concentrated at the magnetic poles.

Calculating the effect of a strong magnetic field on pulsation is very difficult but, by 2000, it had been done. The calculations predicted at least one pulsation mode which had not been observed, and what its period should be. A 44-person team then organized a *Whole Earth Telescope* run to obtain continuous, multi-longitude observations of the star. And the 'missing' mode was observed! See Kurtz *et al.* (2003) for the results of the most recent campaign on this star.

their temperature so, as their temperature decreases, so does their luminosity. In the H–R diagram, their 'cooling track' follows a line of constant radius. As they cool, they pass through several regions of pulsational instability. The last, but the best-studied, is the ZZ Ceti instability strip.

6.14.1 *ZZ Ceti stars*

ZZ Ceti stars are pulsating white dwarfs of DA spectral types, i.e. temperatures of about 10 000K. Although these stars are much smaller than Delta Scuti and Cepheid pulsating variables (and therefore have much shorter periods), they are still closely related. In the H–R diagram, they lie in a downward extension of the Cepheid instability strip, because their driving mechanism is the same as for Cepheids (figure 6.3). Unlike Cepheids, but like some Delta Scuti stars, they pulsate in non-radial modes.

ZZ Ceti stars are intrinsically difficult to study, both because of their faintness (the brightest ZZ Ceti star has $V = 12.24$!) and because of their small and rapid variability. Although the first ZZ Ceti star (HL Tau 76) was discovered (in 1964 by Arlo Landolt) by conventional photometry, further progress was mainly a consequence of the development of high-speed photometry by James Hesser, Barry Lasker, and a group of astronomers (John McGraw, Ed Nather, Ed Robinson, and Brian Warner) at the University of Texas at Austin.

The pulsation periods are in the range 100–1000 seconds, and the amplitudes are up to 0.3 mag. The stars can be meaningfully grouped into three sub-classes: (i) stable periods 100–200 seconds, ranges ~ 0.03 and sinusoidal light curves; (ii) less stable periods 300–800 seconds, ranges ~ 0.05 to 0.2 and approximately sinusoidal light curves; and (iii) even less stable periods 800–1000 seconds, ranges 0.2 to 0.3 and non-sinusoidal light curves.

In almost all of these stars, there are two or more modes which interfere to produce beats. In a few cases (including ZZ Ceti itself), the pulsation period has been tested for period changes, and found to be extremely stable. In fact, it may be possible to detect evolutionary (cooling) effects in white dwarfs by this technique.

White dwarfs lie in a band on the H–R diagram, roughly parallel to the main sequence, but several magnitudes below it. They cool along lines of constant radius. The radius of a white dwarf is about the same as the radius of the earth (about 0.01 $R_\odot$); the precise value of the radius is determined solely by its mass, which is usually about 0.5 $M_\odot$ (the upper limit to a white dwarf mass is 1.44 $M_\odot$). Their surface gravities are enormously high: about a thousand times that on the sun. The high density and pressure in the atmosphere produces characteristic broad absorption lines in the spectrum. The gravity also produces a *gravitational red shift* in the spectrum of about 50 km s^{-1}, which, among other things, can be used to estimate the mass of the white dwarf. This redshift is a consequence of the bending of space, around the white dwarf, by its strong gravity.

ZZ Ceti stars are confined to a narrow range of colour in the H–R diagram, namely about $+0.15$ to $+0.25$ in the BV system, and -0.38 to -0.45 in the Greenstein GR system. In this range, almost every white dwarf is a ZZ Ceti star. Outside this range, they are not variable.

The periods of ZZ Ceti stars are greater, by a factor of about 100, than the radial pulsation periods of white dwarfs. Furthermore, in ZZ Ceti stars with two periods, the periods are about equal. These results indicate that the pulsations are non-radial, specifically gravity (g) modes. The rotation of the white dwarf causes modes with different m-values to have slightly different periods. This is what produces the observed beats. Observations of the colour variations of ZZ Ceti stars confirm that the radii of the stars do not change significantly, i.e. the pulsation is not radial.

The fact that the ZZ Ceti instability strip is a downward extension of the Cepheid instability strip suggests that the two classes of variables are driven by the same pulsation mechanism. This is borne out by theoretical calculations.

Subsequently, two other groups of pulsating white dwarfs were discovered: the DBV and DOV variables. The ZZ Ceti stars are referred to as DAV, where the A, B, and O refer to the spectral type of the star. The DBV stars have temperatures of about 30 000K, the DOV about 100 000K. Each of the three types inhabits a narrow instability strip, and is driven by the ionization of different elements (figure 6.14). (See box 6.5).

6.14.2 *Central stars of planetary nebulae*

About a dozen central stars in planetary nebulae are known to be periodic variable stars. They exhibit semi-regular photometric and velocity variations on time scales of three to ten or more hours. They have effective temperatures between 25 000 and 50 000 K. Their luminosities are about 10 000 times that of the sun. Most likely, they are radially pulsating variables. Gerald Handler has proposed the name 'ZZ Leporis' stars for this group.

Figure 6.25 shows the location of the different types of degenerate pulsating variable stars on a version of the H–R diagram.

6.15 Beta Cephei (Beta Canis Majoris) stars

Beta Cephei stars are pulsating variables with periods of 0.1 to 0.3 days, and very small (visual) light amplitudes. They have spectral types B0.5–B2, and lie on or slightly above the main sequence in the H–R diagram. They are sometimes called Beta Canis Majoris stars, especially in Europe, because β Canis Majoris – like many of the class, but unlike β Cephei – is multiperiodic. Often, the periods are close, and produce a characteristic rise and fall in the amplitude called the *beat effect*. In the stars which are multiperiodic, there is usually one radial period, and one or more non-radial periods. In a spherical non-rotating star, these non-radial periods would be *degenerate* – they would have the same values. But the rotation removes the degeneracy, the periods become slightly different, and they combine and interfere to produce the beat effect. Although these multiple periods have been known for many decades, they are only now

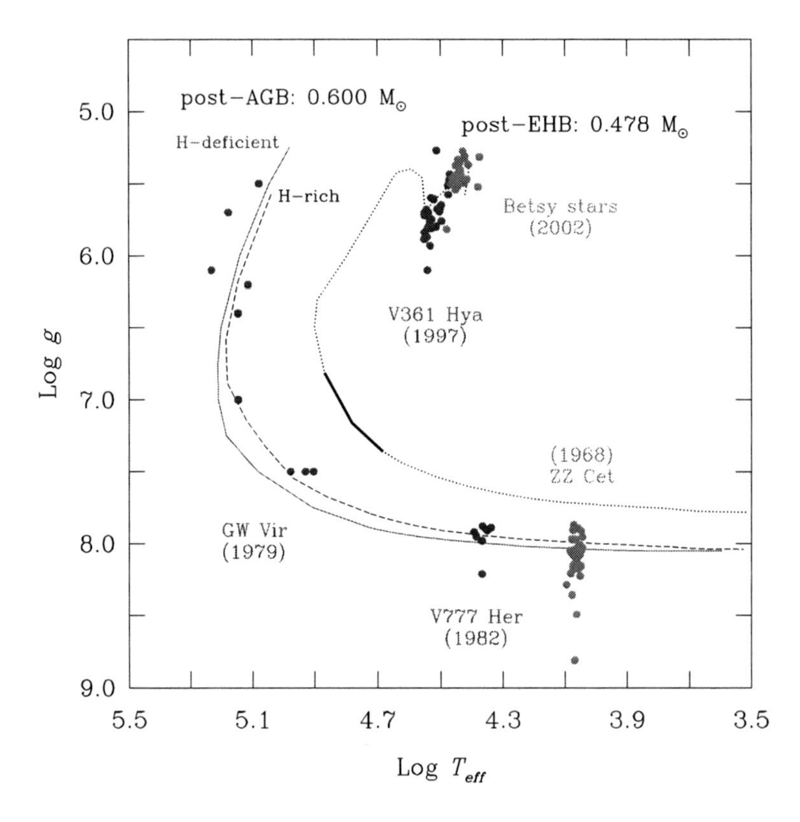

Figure 6.25 The position of ZZ Ceti stars and other degenerate pulsating stars in the H–R diagram. In this version of the diagram, the vertical axis is the logarithm of the surface gravity of the star, in CGS units. It shows the DOV (GW Virginis) stars, the DBV stars (V777 Herculis), and the DAV (ZZ Ceti) stars, as well as the sub-dwarf B (V361 Hydrae and 'Betsy') stars. It also shows models for post-AGB evolution, for two masses and compositions. (From Fontaine *et al.*, 2003, updated to 2006.)

starting to be used for asteroseismology of these stars. In the GCVS4, they are classified BCEP. There is a separate BCEPS class of short-period Beta Cephei stars, which may not have any physical significance.

The Beta Cephei variables were discovered early in the twentieth century. The cause of their pulsation was a great mystery, and about a dozen theories were put forward by various astronomers (including me). The breakthrough came when Norman Simon suggested that calculations of the opacity of the gas in stars might have been underestimated at temperatures of 100 000–200 000K. Detailed calculations confirmed this suspicion. The opacity of iron atoms at these temperatures is sufficient to drive the observed pulsations – and to solve other 'problems' in pulsation theory. Observation and theory now agree, with observation taking the lead (figure 6.16).

Box 6.5 Star sample – PG 1159–035 = GW Vir

This star, despite its obscure designation, is the prototype of the DOV stars, pulsating *hot* white dwarfs, or *PG 1159 stars*, and is a remarkable example of how comprehensive observations of a complex pulsating star (Winget *et al.*, 1991) can provide a wealth of astrophysical data (Kawaler and Bradley, 1994). From 264.1 hours of nearly continuous time-series photometry from nine sites around the world (figure 6.26), the 32 authors identify 125 individual periods between 385 and 1000 seconds (figure 6.27). These correspond to low-order non-radial pulsation modes. The total amplitude is only about 0.1 magnitude. (Note that, as long as the pulsation is strictly periodic, very small amplitudes can be detected and measured by Fourier analysis as long as the photometry is long and continuous.) From the observations, Kawaler and Bradley (1994) derive a mass of 0.59 ± 0.01 $M_\odot$, an effective temperature of 136 000K, a helium-rich layer of 0.004 $M_\odot$, a surface helium abundance of 0.27 by mass, a luminosity of 220 $L_\odot$, a log (gravity) of 7.4, and a rotational period of 1.38 days (derived from the splitting of close pairs of non-radial periods). These parameters are in good agreement with values (or estimates) derived in other ways – a remarkable illustration of the power of asteroseismology.

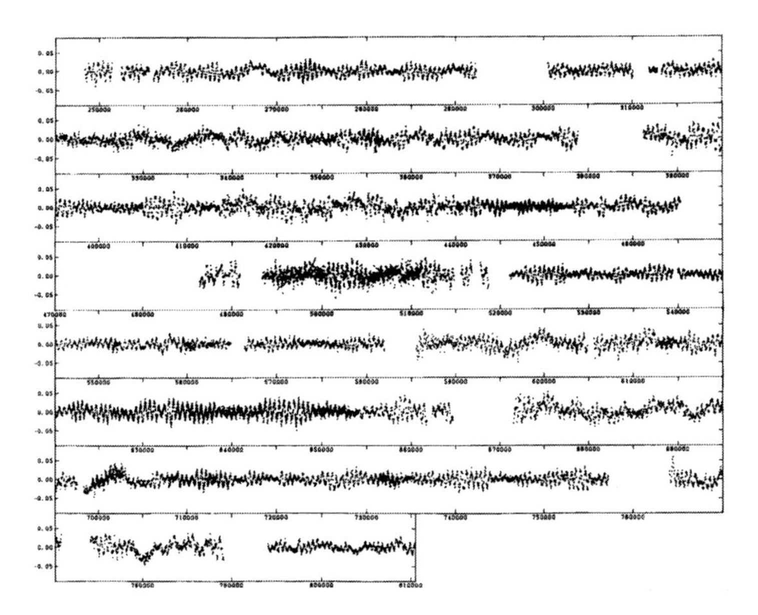

Figure 6.26 The complex light curve of PG1159-035 as determined from the central six days of a *Whole Earth Telescope* campaign. The horizontal axis shows the elapsed time in seconds, and the vertical axis shows the fractional intensity. (From Winget *et al.*, 1991.)

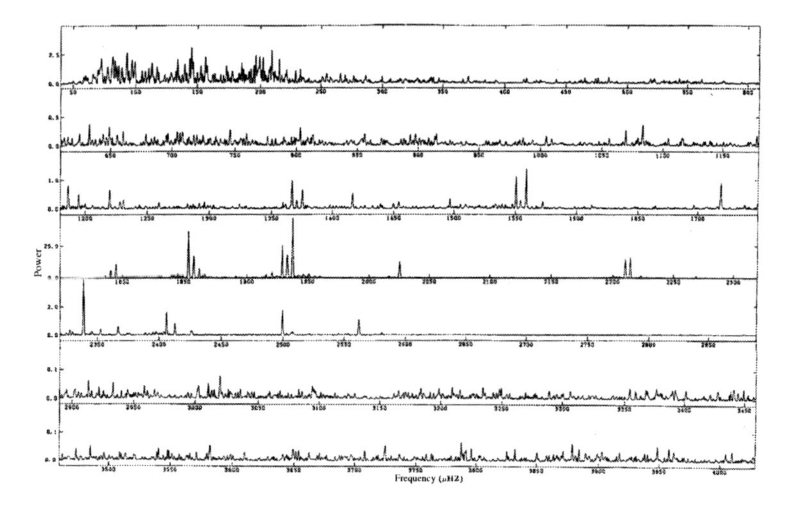

Figure 6.27 The power spectrum of the complete set of measurements of PG1159-035 from the WET campaign (figure 6.26). The power is shown in units of 10^{-6} and the frequency in μHz. The vertical scale is different for each panel, in an attempt to accommodate the dynamic range. (From Winget *et al.*, 1991.)

The light amplitudes of these stars are small, but the actual amplitude of pulsation is actually quite large: the star expands and contracts at up to 200 km s^{-1} – a speed which is difficult for us to imagine! The expansion and contraction was discovered in several Beta Cephei stars by radial velocity observations, starting early in the twentieth century. See figure 2.10 for a dramatic illustration of the radial velocity variations in the largest-amplitude Beta Cephei star. The small light variations were subsequently observed, once sensitive photoelectric photometers were available. In the 1970s, when some of these stars were observed at ultraviolet wavelengths with astronomical satellites, it was found that the ultraviolet variations were much larger than the light variations which had been observed from the ground. This was expected; for a hot star, a small change of temperature has a much larger effect in the UV than in the visual (figure 6.28). Here is one more example of the benefits of observing stars at a variety of wavelengths!

As with the Delta Scuti stars, there is the question of what determines the amplitude of pulsation. Again, the most likely candidate is rotation, though the position of the star in the instability strip, and the presence of multiple pulsation modes, may play a role.

There are a large number of Beta Cephei stars among the brightest stars in the sky – many α and β stars in their constellations. But there were surprisingly few fainter variables discovered up to 1980. At that time, there was

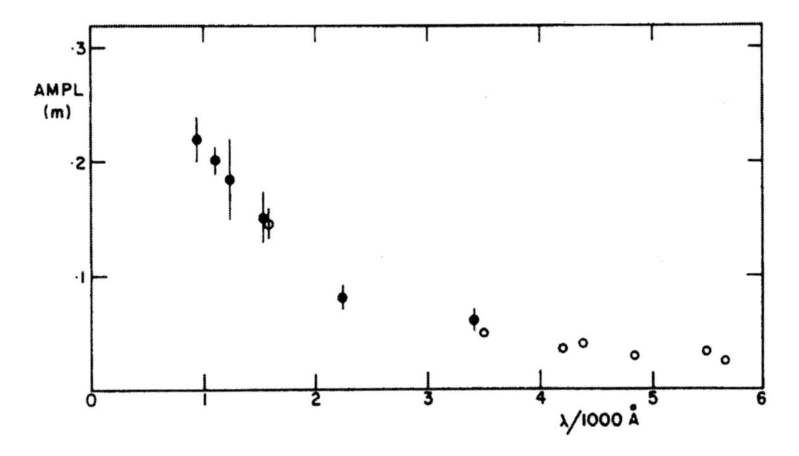

Figure 6.28 The relationship between the light amplitude of β Cephei and wavelength. Although the amplitude is only a few tenths of a magnitude in the visual, it is ten times greater in the UV. (From Hutchings and Hill, 1977.)

much interest in the newly discovered, slowly pulsating B stars, the 53 Persei stars, and pulsating Be stars. But by 1993, Sterken and Jerzykiewicz (1993) were able to list 59 certain members, and 79 suspects. Also, a dozen Beta Cephei stars were discovered in southern clusters, and another dozen or so in northern clusters, bringing the total number to almost a hundred.

There was some evidence that Beta Cephei stars were less common in the outer part of our Milky Way, where the stars had lower metal abundances. This might be expected, since the pulsation mechanism involves the opacity of one of the metals – iron. One test of this phenomenon would be to search for Beta Cephei stars in the Large and Small Magellanic Clouds, which have metal abundances of 0.01 and 0.005 as compared with 0.02 for stars like the sun. Initially, no Beta Cephei stars were found. Then three were found. Then the OGLE-II survey data were carefully reprocessed, and, from photometry of over 75 000 early B-type stars, 64 probable Beta Cephei stars were found – only one in a thousand, but not zero. These stars had some unusual properties. Their periods were almost twice as long as those of Beta Cephei stars in the Milky Way. That is partly because, when the metal abundance is low, Beta Cephei pulsational instability occurs only for cooler, larger stars. But it also appears that these stars may be undergoing low-order non-radial g modes, which are not observed in Beta Cephei stars in our galaxy. Or maybe they have not been looked for, carefully, in those stars.

Interest in Beta Cephei stars has been rekindled recently, thanks to the work of Conny Aerts and others, who have contributed to the organization of multi-longitude campaigns, and to the analysis of the results. Ultra-precise observations from the WIRE and MOST satellites have also been obtained. These studies have led to the detection of many radial and non-radial periods, which have then been used, through asteroseismology, to determine the properties and internal structure of the stars.

6.15.1 *Non-radially pulsating B stars*

Many years ago, the beat effect in the Beta Cephei stars was explained as the superposition of radial and non-radial pulsation modes. But it was only about 1980 that *pure* non-radially pulsating B stars were discovered. This discovery came about in two ways. One was the detection of absorption *line profile variations* in B stars, which was made possible by electronic (Reticon and CCD) detectors on astronomical spectrographs. The other was careful photometric surveys which detected light variations in both the line–profile–variable B stars, and a new class of photometric variables with periods of typically one to two days, which were called Slowly Pulsating B stars (SPB stars) or LBV in the GCVS (an unfortunate term, since LBV usually refers to a 'luminous blue variable'). The strategy for discovering Beta Cephei stars had been to observe them for a few hours during the night, and look for the three to six-hour variations. SPB stars vary only slightly in a few hours; observations over many hours must be carefully combined and analyzed, allowing for the 'alias' effects which would be found in the power spectrum. The *Hipparcos* satellite photometry, despite its unusual distribution of observing times, was useful in identifying and studying both SPB and Beta Cephei stars. SPB stars have since been found in several clusters, and in the Large Magellanic Cloud. SPB stars, with their multiple periods, are good targets for asteroseismology, though their relatively long periods mean that many years (or decades) of data are required.

At about the same time, line profile variability and short-period photometric variability was discovered in several Be stars; it had actually been discovered several decades earlier, but its significance was not recognized. Most astronomers believe that this variability is also caused by non-radial pulsation. The theory of pulsation in rapidly rotating stars is very complex, however, so this belief has not yet been fully tested by modelling.

The SPV stars, however, are not especially rapidly rotating, and theory has been very successful in reproducing the observed instability strip with high-order g modes (figure 6.29).

Epsilon Persei is a bright, remarkable example of a non-radially pulsating star. This bright star exhibits absorption line profile variations (figure 6.30) which have been interpreted (Gies and Kullavanijaya, 1988) in terms of non-radial pulsations with $l = 3$, 4, 5, and 6, with periods ranging from 2.3 to 4.5 hours.

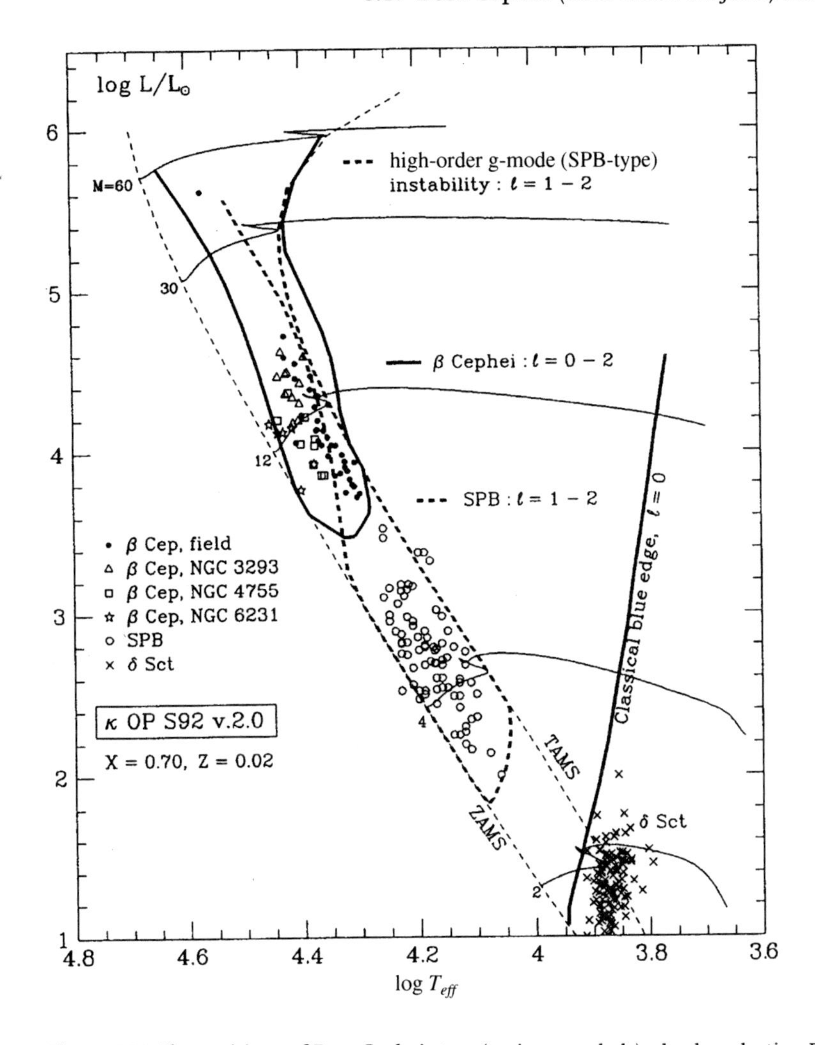

Figure 6.29 The positions of Beta Cephei stars (various symbols), slowly pulsating B stars (SPB: open circles), and Delta Scuti stars (x) in the H–R diagram. These are compared with theoretical regions of instability to radial modes (solid boundaries) and non-radial g modes (dashed boundaries). The classical blue edge of the Cepheid instability strip is also given. (From Pamyatnykh, 1999.)

As with other non-radially pulsating stars, the brightness variations are very small – only a few hundredths of a magnitude.

6.15.2 *Subdwarf B (sdB) stars*

This is a group of B stars which are not young stars on the main sequence, like the variables described above, but are 'passing through' near

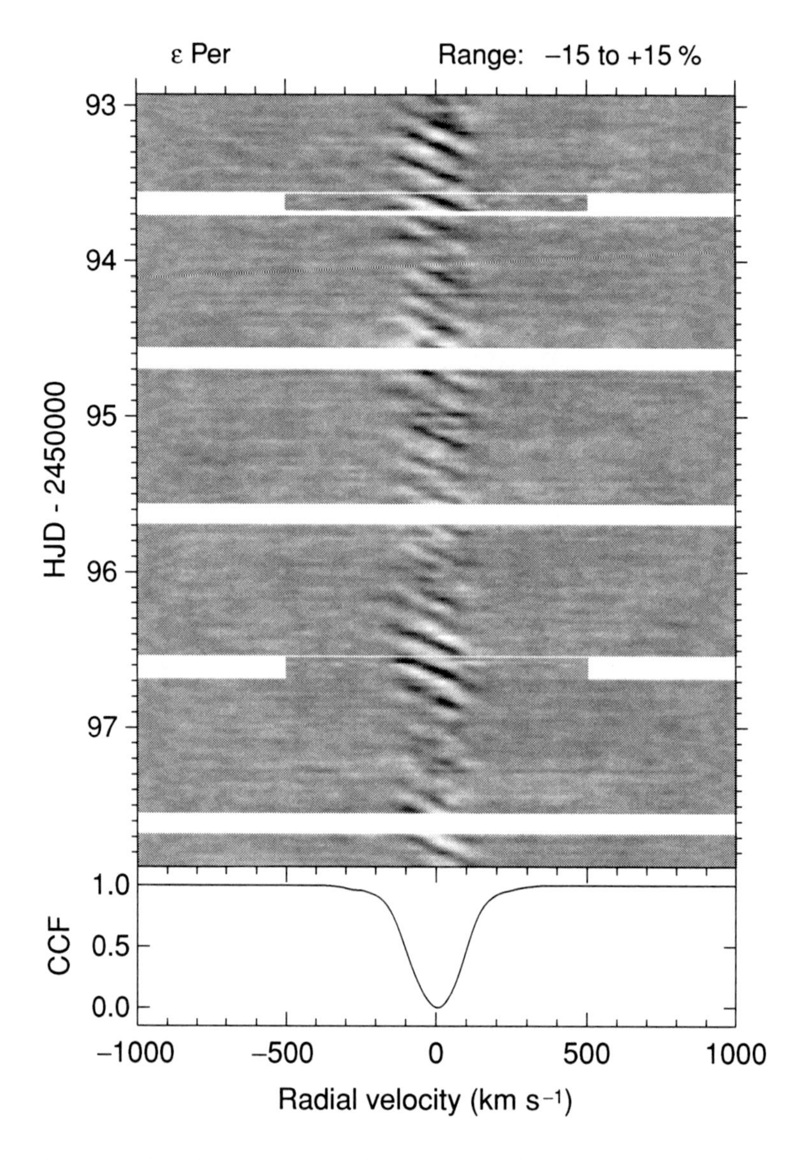

Figure 6.30 The absorption-line profile variations of ϵ Persei. At the bottom is the average profile of the absorption line at wavelength 455.2 nm; at the top is a gray-scale plot, with time increasing downward, of the deviation of the line profile from the average. The pattern of diagonal bright and dark bands represents 'bumps' running across the profile, caused by non-radial pulsation waves. (From Geis and Kullavanijaya, 1988.)

the end of their lifetime. They are extreme horizontal branch (EHB) stars: low-mass stars, with a helium-burning core surrounded by a hydrogen-rich shell, but whose outer hydrogen-rich layers were too thin (0.02 $M_{\odot}$ or less) for them to evolve to the asymptotic giant branch after helium is exhausted in their core. They lie on the extreme hot end of the horizontal branch on the H–R diagram, with B spectral type (temperatures of 20 000 to 40 000K), but 'subdwarf' in the sense that their size and luminosity are lower than for a normal B star. They will eventually end their lives as low-mass white dwarfs.

In a remarkable turn of events, a group of theorists tested the stability of detailed state-of-the-art structural and evolutionary models of such stars, and predicted that they should pulsate (see Charpinet *et al.*, 2001, for a review). Independently, a team of observers at the South African Astronomical Observatory discovered real pulsators of this type; they have been named *EC 14026* stars, after the prototype. They have temperatures of 28 000–36 000K, and periods of a few hundred seconds – non-radial *p*-modes. Some have a dozen periods or more.

In 2003, another team of astronomers found that some subdwarf B stars pulsate in non-radial gravity modes, with periods of an hour or two – ten times longer than the periods of the EC 14026 stars. The stars are referred to as PG1716+426 (PG1716) stars, after the prototype (or more colloquially as 'Betsy stars' after their discoverer). The driving mechanism, for both groups, is partially ionized iron atoms, deep in their envelopes.

The current challenge is to identify the modes in these complex pulsators, and photometric and spectroscopic observing campaigns of up to 400 hours are being used to do this. The payoff is asteroseismological data – masses, internal structure, and other properties of these stars. In the case of the sdB star PG1219+534, this has led to a mass of 0.457±0.012 $M_{\odot}$, and an estimate of the thickness of the star's hydrogen-rich outer shell, which confirms the picture mentioned above.

6.16 Pulsating red giants (PRGs)

As stars expand and cool to become red giants or asymptotic giant branch (AGB) stars, they become pulsationally unstable; virtually every star in the upper right portion of the H–R diagram is variable. Several studies have independently shown that microvariability (amplitude a few hundredths of a magnitude) sets in at mid-K spectral type, and larger-amplitude variability (amplitude 0.1 magnitude or more) sets in at spectral type M0. Stars with these small amplitudes are very numerous, but their variability was not well studied until

Table 6.3. *Bright and/or interesting Beta Cephei and related stars*

Star	HD	V Range	Period (d)	Spectrum	Comments
ϵ Per	24760	2.88–3.00	several	B0.5V	large-amplitude lpv*
53 Per	27396	4.81–4.86	several	B4.5V	prototype lpv
λ Eri	33328	4.22–4.34	0.701538	B2IVe	prototype nrp† Be star
β CMa	44743	1.93–2.00	0.25003	B1II-III	prototype
γ^2 Vel	68273	1.81	rapid?	WR	brightest WR star
β Cru	111123	1.23–1.31	0.2365072	B0.5III-IV	in Southern Cross
α Vir	116658	0.95–1.00	0.1737853	B1III-IV+B2V	stopped pulsating
β Cen	122451	0.61–0.66	0.157	B1III	brightest β Cep star
σ Sco	147165	2.86–2.94	0.246839	B2III+O9.5V	spectroscopic binary
ζ Oph	149757	2.56–2.58	–	O9.5Ve	prototype 'moving bumps' nrp star
λ Sco	158926	1.62–1.68	0.2136966	B1.5IV	–
P Cyg	193237	3.0–6.0	40-100	B2pe	S Dor star; pulsating?
BW Vul	199140	6.52–6.76	0.20104117	B1-2IIIe	largest-amplitude β Cep star
β Cep	205021	3.16–3.27	0.1904881	B2IIIe	prototype

Notes: *lpv = (absorption) line-profile variable.
†nrp = non-radial pulsator.

long-term photoelectric photometry was done. They have now been found in globular clusters as well as in field stars. Stars with mid-M spectral type tend to have amplitudes which are large enough to be studied with visual and photographic photometry, and these stars have been well known for a century. The coolest M stars tend to have amplitudes of several magnitudes, and are called *Mira stars*. Pulsating red giants are complex objects, and among the most heterogeneous and bizarre of all variables, but they are quite numerous among the bright stars, so they cannot be ignored. They are a challenge to theorists; their structure is dominated by convection, which is a poorly understood process in astrophysics.

We can now make several general statements about PRGs:

- Giants cooler than spectral type K5III tend to be variable in brightness.
- Such stars make up about 10 per cent of the stars in the *Yale Catalogue of Bright Stars*.
- The cooler and larger the giant, the larger the amplitude of variability; the rare, large-amplitude *Mira stars* are the most extreme.
- The cooler the giant, the longer the period; the periods are consistent with low-order radial pulsation.

- Pulsating red giants – especially the smaller-amplitude ones – tend to be semi-regular, though there are usually one or more strict periods present; indeed, many small-amplitude stars, and larger-amplitude SR variables are multiperiodic.
- Up to half of the small-amplitude PRGs have long secondary periods, typically ten times the primary (radial) period; the cause of these long secondary periods is not known, despite considerable effort (Wood *et al.*, 2004).

Figure 3.7 shows the interesting behaviour of the pulsating red giant R Doradus. It changes pulsation modes (and therefore periods) on a time scale of a decade or so.

6.16.1 *Classification of pulsating red giants*

Classification has historically been done on the basis of the period, amplitude, and regularity of the light curve. According to the GCVS, *Mira variables* are defined as having light amplitudes greater than 2.5 magnitudes, periods of 80–1000 days, and well-pronounced periodicity. *SRa variables* have amplitudes less than 2.5, and persistent periodicity. *SRb variables* have poorly expressed periodicity. *Lb variables* are slow and irregular. There are separate classes for red supergiants: SRc for the semi-regulars (and SRd for yellow semi-regular supergiants which we have already discussed) and Lc for the irregulars. The division between semi-regularity and irregularity is rather arbitrary: it takes considerable data, and careful analysis, to determine how much periodicity is present in a star. Likewise, the division between Miras and semi-regulars (an amplitude of 2.5 magnitudes) is rather arbitrary; there is a smooth gradation of properties between them, and there are stars whose amplitudes may vary from more than 2.5 magnitudes to less.

Olin Eggen was one of the pioneers in the photoelectric monitoring and analysis of the much more numerous small-amplitude pulsating red giants. He developed a general classification for PRGs: large, medium, and small-amplitude red variables (LARV, MARV, SARV), plus Sigma Librae variables for those with amplitudes less than 0.2 in *V*. I prefer to classify PRGs simply by their amplitude and period(s).

Small-amplitude pulsating red giants are very common. Figure 6.32 shows the light curves of several of them, on both short and long time scales. See also figure 3.4, figure 3.5, and figure 3.6 for a light curve, power spectrum, and self-correlation diagram for the 'prototype' small-amplitude pulsating red giant EU Delphini.

Box 6.6 Star sample – BW Vulpeculae

BW Vulpeculae (HD 199140, B2III, V = 6.54) would seem to be a rather routine variable. To me, it stands out as the largest-amplitude Beta Cephei star, and one of the most vigorous of pulsators! During its 0.20104117-day period, it expands and contracts at over 200 km s^{-1}. Consider that an Earth satellite orbits at about 10 km s^{-1} and speedy Mercury orbits the sun at 'only' 45 km s^{-1}.

BW Vulpeculae is about ten times larger than the sun, about 15 times more massive, and about 10 000 times more luminous. Its surface temperature is about 20 000 K. Its 'large' amplitude is only 0.2 magnitude in visual light, but, if you had ultraviolet eyes, its amplitude would be over a magnitude. That is because this hot star radiates most of its energy in the ultraviolet, and the pulsational changes in temperature would have the greatest effect there.

BW Vulpeculae is also interesting because it shows how the $(O - C)$ method of period–change analysis can demonstrate the evolution of the star. Most recently, Christiaan Sterken (1993) has collected together all observed times of minimum brightness (the time of minimum is better defined than the time of maximum in this star). The $(O - C)$ diagram is shown in figure 6.31. The period is increasing by 2.8 s/century. This is consistent with the star's evolutionary stage and with theoretical models; it is nearing exhaustion of the hydrogen fuel in its core, and preparing to use the hydrogen around the core.

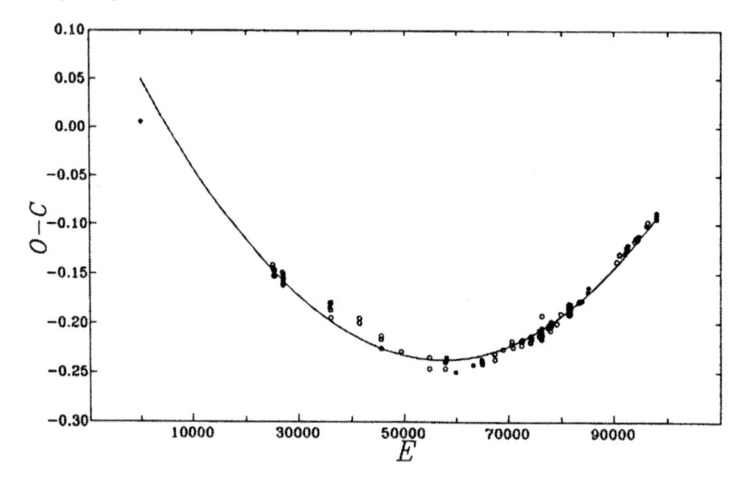

Figure 6.31 The $(O - C)$ diagram for BW Vulpeculae, covering 100000 cycles. The diagram is well fit by a parabola, indicating a period increase of 2.8 s/century. The cause of the deviations from the parabola is not clear. (From Sterken, 1993.)

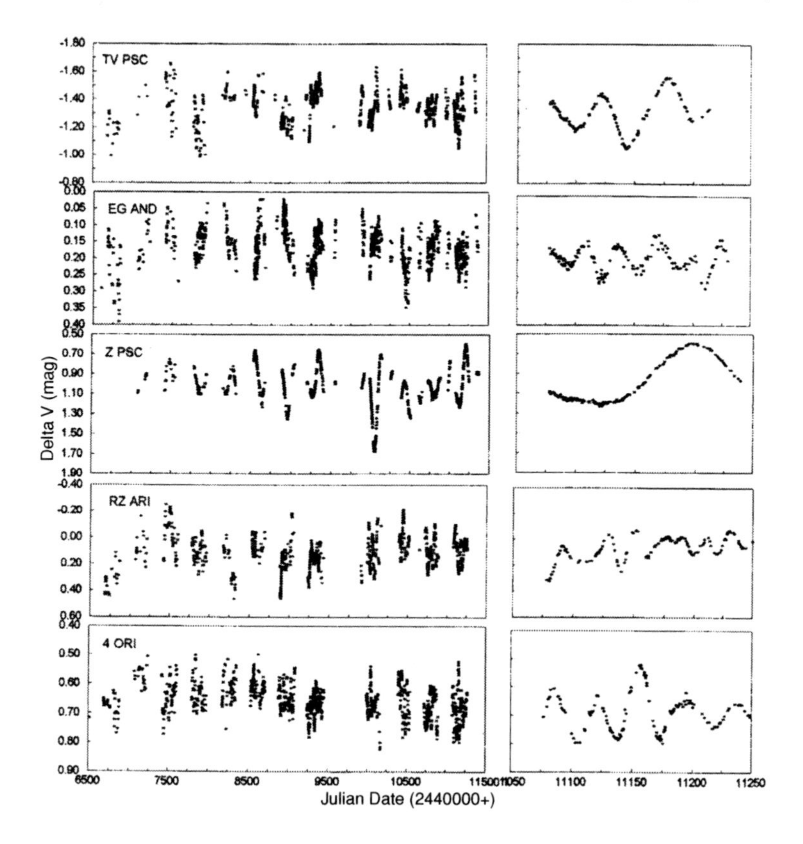

Figure 6.32 Light curves of several small-amplitude pulsating red giants. These are by far the most numerous of the pulsating red giants. The light curves on the right show the variability on a time scale of 200 days; those on the left show the variability on a time scale of 5000 days. RZ Arietis and 4 Orionis have at least two pulsation periods. This leads to the complex light curves shown at the right. (From Percy, Wilson and Henry, 2001.)

Figure 6.33 shows the light curve of the bright-red semi-regular variable Z Ursae Majoris, based on visual observations from the AAVSO International Database. Its amplitude is greater than that of stars like EU Delphini, but not great enough for it to qualify as a Mira star. Note that, as with R Doradus, there are two time scales, 100 and 200 days, that differ by a factor of two.

An important new development is the discovery of thousands of PRGs in Local Group galaxies, through large-scale surveys (section 3.2). In the period–magnitude relation (figure 6.34), there are several sequences. Since the stars are at roughly the same distance, the apparent magnitude is directly related to the absolute magnitude or luminosity. So figure 6.34 shows that a PRG with a specific luminosity can pulsate with different periods: these correspond to the

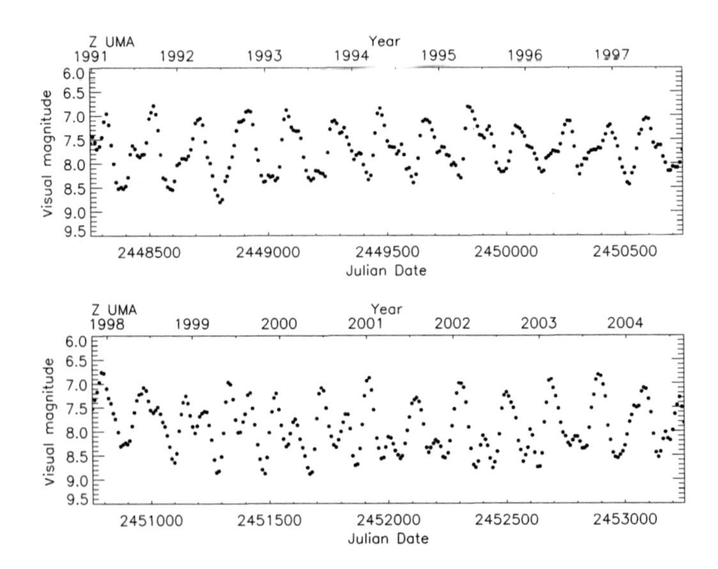

Figure 6.33 The light curve of the semi-regular pulsating red giant Z Ursae Majoris, based on measurements from the AAVSO International Database. Note the many interesting complications: the presence of variations with periods of approximately 100 and 200 days. There are even occasional intervals when the pulsations seem to disappear. (From AAVSO.)

fundamental, first overtone, and second overtone of radial pulsation. Similar results are found for bright nearby PRGs whose distances and luminosities are known from their parallaxes.

The red variables are further divided into three subclasses based on chemical composition: M – normal, with spectra dominated by strong bands of the oxides of Ti, Sc, and V, with bands of water vapour in the near-IR and continuous absorption by H-; C – carbon-rich, called R if CN is strong and N if C_2 is strong; S – oxygen-rich, showing lines from the oxides of Zr, Y, Ba, La, Sr, Nb, and Tc. In the C and S stars, elements synthesized by nuclear processes, deep within the stars, have been brought to the surface by the deep convection zones in the star, in a process known as *dredge-up*. The relative amount of C and O dredge-up determines the spectroscopic type of the star – M, C, or S. The most dramatic example of dredge-up is an isotope of technetium which is radioactive, with a half-life of four million years. The technetium which appears in the spectrum of these stars must have been produced only a few million years ago!

Population II Miras have generally low abundances of all the heavy elements. Since most of these variables show emission lines in their spectra, the spectra are actually labelled Me, Ce, and Se. There may be the usual spectral subtypes, indicating the temperature.

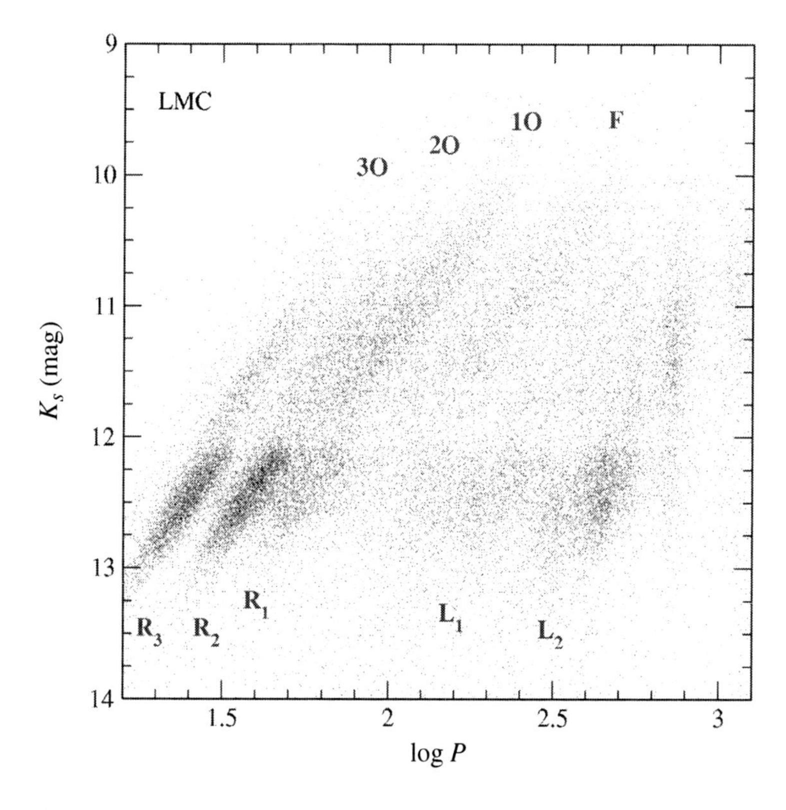

Figure 6.34 Variable red giants in the Large Magellanic Cloud, plotted in an apparent K magnitude *versus* log(period) graph. Because the stars are all at approximately the same distance, the apparent K magnitude reflects the absolute K magnitude, so this is a *period–luminosity relation*. The sequences 3O, 2O, 1O, and F (asymptotic giant branch stars) and R3, R2, and R1 (mostly red giant branch stars) correspond to stars pulsating in the third overtone, second overtone, first overtone, and fundamental mode. Sequences L1 and L2 are the 'long secondary periods' whose cause is not known. The larger numbers of points below magnitude 12 are because the red giant branch stars are more numerous than the asymptotic giant branch stars. (From L. L. Kiss, private communication.)

6.16.2 *Mira variables*

Miras have stable periods of 100–1000 days, with most between 150 and 450 days. The visual amplitudes are greater than 2.5 by definition, and can be up to ten magnitudes (figure 6.35). Being so conspicuous, they are the most numerous variables in the GCVS, but, with modern surveys being able to detect smaller-amplitude variables more easily, other types are catching up. Miras are highly evolved stars, with masses between about 0.6 and a few solar masses. They have radii of up to several hundred solar radii; if placed in the solar system where

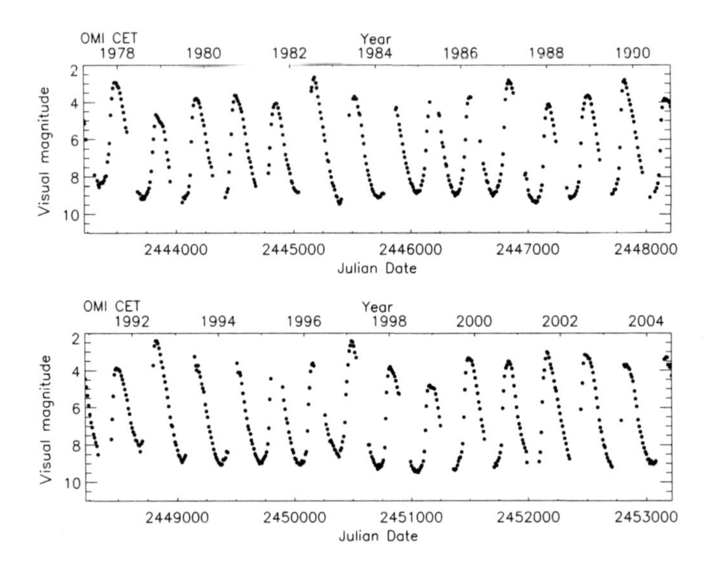

Figure 6.35 The 1978–2004 light curve of Mira, based on visual measurements from the AAVSO International Database. Note the variable magnitude at maximum. The magnitude at minimum is approximately constant because of the presence of the close hot binary companion. (From AAVSO.)

the sun is, they would extend beyond the earth's orbit! For these stars, however, the concept of 'radius' is complicated. The atmosphere is very extended. And the star will look much larger in regions of the spectrum (such as V) at which the atmosphere is opaque. Some Miras are Population I stars, others are Population II; some of the latter are found in globular clusters,

The large amplitudes of these variables make them especially suitable for visual observation by amateur astronomers. It is from such observations, over many years or decades, that the periods are found. Evolutionary period changes have been detected both for several individual rapidly evolving stars (the period of BH Crucis has increased from 420 to 540 days) and statistically for more slowly evolving stars. The complication is that these stars show random cycle-to-cycle period fluctuations of up to 5 per cent. Their slight irregularity, in both phase and maximum and minimum magnitude, adds interest, and makes systematic visual observation even more important. Figure 6.35 shows a 25-year light curve of Mira itself, showing the irregularity in the magnitude at maximum. Visual observers play a special role in providing professional astronomers with information on the current brightness of Miras; this helps in planning ground-based or space observations. For instance, visual observers provided the *Hipparcos* satellite scientists with such information, which was absolutely essential for this satellite's measurements of Mira stars. A special issue of the *Journal of the AAVSO* (volume 15, #2 (1986)) marks the 400th anniversary of the discovery of Mira's variability.

Table 6.4 includes some of the brighter Miras, as well as some red semi-regular and irregular variables. The AAVSO is the best source of information on light curves and times of maxima of these stars, both past and present; its International Database includes several hundred Miras, observed for up to a century or more.

Variability

In Mira stars, the variations in V are much greater than those in the near-IR magnitude, or in the bolometric magnitude, so their variability would be much less conspicuous to visual observers if they had infrared eyes! Since the bulk of the star's energy is emitted in the near-IR – typically 1-3 μm – the I magnitude, for instance, is far more representative of the star's true variability than the V magnitude. This is one of several advantages of observing Miras at near-IR wavelengths. At longer IR wavelengths, emission from dust around the star complicates the picture.

There are two reasons for the large V variations in Miras. First, as the star becomes fainter, it also becomes cooler, and a smaller fraction of the total energy is emitted in the visual portion of the spectrum. Second, and more important, as the star becomes cooler, TiO molecules form (figure 6.36), and they absorb light preferentially in the visual part of the spectrum – especially in the V band.

The visual amplitude tends to be largest in cooler stars, which also have longer periods. In a given star, however, V can vary from cycle to cycle (figure 6.35). Mira is a good example.

The colour variations in Miras are large, but difficult to interpret. The $(B–V)$ and $(U–B)$ variations may be in opposite senses: when one says that the star is getting hotter, the other may be saying that the star is getting cooler! But these colour indices are not very appropriate for stars which emit most of their energy in the IR. And they are affected by the strange and variable spectra of these stars, as noted above. In many Miras, for instance, there is a pronounced 'line weakening' which may be caused by the dilution of the line absorption by continuous absorption – by the H^-ion for instance.

The most conspicuous feature of the spectrum is hydrogen line and continuum emission (figure 6.36). This emission is most pronounced just after maximum brightness, and is associated with the pulsation of the star. The emission is caused by *shock waves* which travel outward through the atmosphere at supersonic velocities (though, even at 10 km s^{-1}, it takes the waves 100 days to traverse the deep atmosphere!). The radial velocity variations in Miras are rather difficult to interpret, because, at different phases, the temperature and the opacity of the

Table 6.4. *Bright and/or interesting pulsating red giants and supergiants*

Star	HD	V Range	Type	Spectrum	Period (d)	Comments
EG And	4174	7.08–7.80	Z And	M2IIIep	symbiotic	
RZ Ari	18191	5.62–6.01	SRb	M6III	30	multiperiodic SARV
R Hor	18242	4.7–14.3	Mira	M5-8eII-III	407.6	–
ρ Per	19058	3.30–4.00	SRb	M4IIb-IIIa	50	bright SARV
o Cet	33196	2.0–10.1	Mira	M5-9e	331.96	prototype
α Ori	39801	0.00–1.30	SRc	M1-2Ia-Ib3	2335	Betelgeuse
η Gem	42995	3.15–3.90	SRd + EA	M3IIIab	232.9	bright SARV
μ Gem	44478	2.91±	–	M3III	–	bright SARV
L^2 Pup	56096	5.10± SR?	M5IIIe	–	–	
R Car	82901	3.9–10.5	Mira	M6.5IIIpevar	308.71	–
R Leo	84748	4.4–11.3	Mira	M6-9.5e	309.95	–
S Car	88366	4.5–9.9	Mira	K5-M6e	149.49	–
Z UMa	103681	6.2–9.4	SRb	M5IIIe	195.5	well-known SR
R Vir	109914	6.1–12.1	Mira	M3.5-8.5e	145.63	–
FS Com	113866	5.30–6.10	SRb	M5III	58	multiperiodic SARV
R Hya	117287	3.5–10.9	Mira	M6-9eS(Tc)	388.87	–
σ Lib	133216	3.20–3.46	SRb	M3.5III	20	prototype VSARV
α Sco	148478	0.88–1.16	Lc	M1.5Iab-Ib	–	Antares
g Her	148783	4.3–6.3	SRb	M6III	89.2	–
α^1 Her	156014	2.74–4.00	SRc	M5Ib-II	–	–
η Sgr	167618	3.05–3.12	Lb	M3.5III	–	–
R Lyr	175865	3.88–5.00	SRb	M5III	46	–
R Aql	177940	5.5–12.0	Mira	M5-9e	284.2	–
CH Cyg	182917	5.60–8.49	Z And + SR	M7IIIab + Be	–	symbiotic
R Cyg	185456	6.1–14.4	Mira	S2.5,9e-S6,9e(Tc)	426.45	–
δ Sge	187076	3.75–3.83	Lb:	M2.5II-III + B9V	–	–
χ Cyg	187796	5.2–14.2?	Mira	S7,1e:-S10,1e:	406.84?	–
EU Del	196610	5.79–6.90	SRb	M6.4III	59.7	prototype SARV
μ Cep	206936	4.04±	SRc:	M2Iae	–	very red!
β Gru	214952	2.0–2.3	Lc	M3-5II-III	–	–
β Peg	217906	2.31–2.74	Lb	M2.5II-IIIe	–	–
χ Aqr	219576	5.01± –	M3III	–	–	
Z And	221650	8.0–12.4	Z And	M2III + B1eq	–	symbiotic
R Aqr	222800	5.8–12.4	Mira	M5-8.5e + pec	386.96	symbiotic; radio jet
R Cas	224490	4.7-13.5	Mira	M6-10e	430.46	–

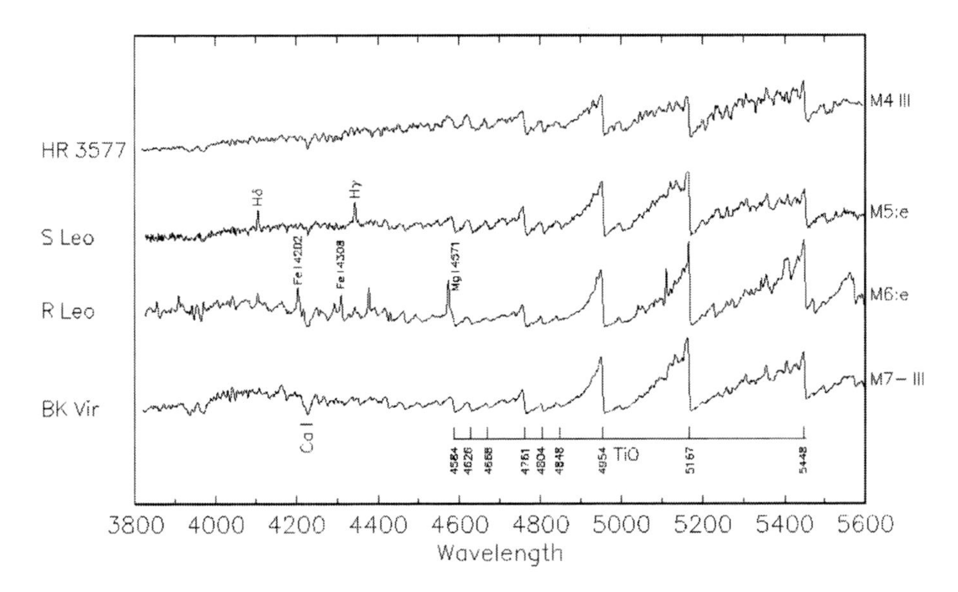

Figure 6.36 The spectra of four pulsating red giants. HR 3577 (FZ Cancri) has a very small amplitude. BK Virginis has an amplitude of 1.5, and R Leonis and S Leonis are Mira variables with large amplitudes. Note the emission lines in the two Mira variables, and the TiO molecular bands in all four of these cool stars. (From Richard O. Gray.)

atmosphere change, and we see to different depths or layers in the atmosphere. Since the velocity of the atmosphere varies with *depth* as well as with *time*, it is not always clear whether the pulsational velocity variations are real or apparent.

Related to the problem of the thickness of the atmosphere, and of the radial velocity variations, is the problem of the radius and its apparent variations. If the star is truly pulsating, then the radius should increase and decrease. But how does one define the radius? At visual wavelengths, where the atmosphere is opaque because of TiO absorption, the star looks large. At near-IR wavelengths, the star looks smaller. Thus, when the radius is measured at visual wavelengths, using interferometry, lunar occultations, or the Barnes–Evans method, for instance, large values are obtained. There is an alternate way of measuring the radius: if the radial velocity of a single layer of the atmosphere could be followed through a pulsation cycle (a difficult thing to do!), then the gravity of the star (GM/R^2) could be measured from the way in which that layer 'bounces'. Using the estimated mass of the star (0.6 to a few $M_\odot$), the radius can be determined. It turns out to be relatively small, in agreement with the value determined from studies at near-IR wavelengths. The average radius of a Mira, in solar units, is approximately equal to its period in days.

Period–luminosity–colour relation

The period of a Mira star depends primarily on its radius, secondarily on its mass, and – in the most extreme stars – on how the mass is distributed within the star. The masses of Miras are in the range of 0.6 to a few solar masses, so the period can be considered, to a first approximation, to depend on radius only. But the radius is a function of the luminosity and the effective temperature, according to Stefan's law. There is also a relation between luminosity and temperature, because Mira stars lie on a specific locus on the H–R diagram – the AGB. So a period–luminosity or period–luminosity–colour relation should be expected.

As with other types of pulsating variable stars, there are several ways of determining and calibrating the period–luminosity relation. Some Miras belong to globular clusters (none belong to galactic clusters); these are the Population II Miras. Statistical parallaxes can be determined from the proper motions and radial velocities; until recently, this was the most effective approach. The resulting P–L–C relation was used extensively to map out the structure of our galaxy. More recently, Miras have been found and studied in the Magellanic Clouds, especially using near-IR photometry; the distances of the Magellanic Clouds can be measured, independently, in other ways. And precise parallaxes and proper motions were determined for many Miras by the *Hipparcos* satellite. As mentioned above, amateurs' monitoring of the Miras was essential for this process.

The most effective period–luminosity relations, for Miras, are those that use IR magnitudes. Feast (2004) has developed the following

$$M_K = -3.47(\pm 0.19) \log P + 1.00(\pm 0.08) \tag{6.14}$$

For many years, there was uncertainty about whether Mira stars were pulsating in the fundamental mode, or the first overtone. Several lines of evidence, including direct observations with interferometers, confirm the fundamental mode.

Masses

The mass of a star is its most fundamental property. Masses can be determined directly for some stars in binary systems. Miras and their relatives are frequently found in binary systems, but they are usually complex, interacting systems such as symbiotic systems (section 7.3.4). See Mikolojewska (2002) for a list of 16 symbiotic stars with mass determinations for the red-giant component; the masses are 1–3 $M_\odot$. There are at least two Miras in long-period visual binary systems, but the periods are so long (Mira: about 400 years; X Oph: about 550 years) that only estimates of the masses can be obtained; they are about one solar mass. Miras in globular clusters must have masses less than the sun's, since the globular clusters are too old for more massive stars to survive. And we can estimate, for other Miras, that their masses must be no more than a few $M_{\odot}$, from evolutionary models of the stars. My students and I have determined 'pulsation masses' for a few *small-amplitude* pulsating red giants by determining the period and pulsation mode, inferring the radius from the observed luminosity and temperature, and using the pulsation constant Q determined from models. We find masses similar to that of the sun.

The nature of Miras

The nature of Miras can be deduced from their position in the H–R diagram, as well as from a general consideration of their properties and behaviour. They are stars which have completed their hydrogen-burning, moved up the red giant branch (and perhaps been small-amplitude pulsating red giants), gone through the helium flash, and are now at the end of the asymptotic giant branch phase of evolution. They are the coolest, largest, most luminous red giants. They have a dense core of carbon, surrounded by a helium-burning shell, surrounded by a region of inert helium, surrounded by a thin hydrogen-burning shell. To put this in scale: if the core of a Mira was represented by a grape, the atmosphere of the Mira would fill a large sports arena!

By comparing Miras with models of stellar evolution, we can deduce their ages: from three to ten billion years. The Mira stage is brief, however: the stars are not only using their limited nuclear fuel at a prodigious rate, but they are also losing their mass into space; their outer layers will dissipate within a million years. This was demonstrated theoretically by George Bowen and Lee Anne Willson, in the 1980s. They showed that the pulsations in the outer layers of the Mira (which are driven by the hydrogen ionization zone) produce shock waves which increase the density of the atmosphere (figure 6.37). Dust particles condense. They absorb visible and IR radiation, and are pushed outward, carrying gas with them. The shock waves are visible in the spectra of the stars, and the dust and gas, driven off the star to form circumstellar shells, can be directly observed at IR and radio wavelengths. Molecules such as OH and water emit radio radiation at specific wavelengths; these can be observed, and used to detect the presence of the molecules and, by the Doppler effect, how fast they are moving outward. Eventually, the emitted gas is excited by the UV radiation from the exposed core of the Mira star, and we see a beautiful *planetary nebula*.

Study of the radio emission from these molecules yields an interesting conclusion: they do not emit as if they were simply part of a warm gas. Rather, IR

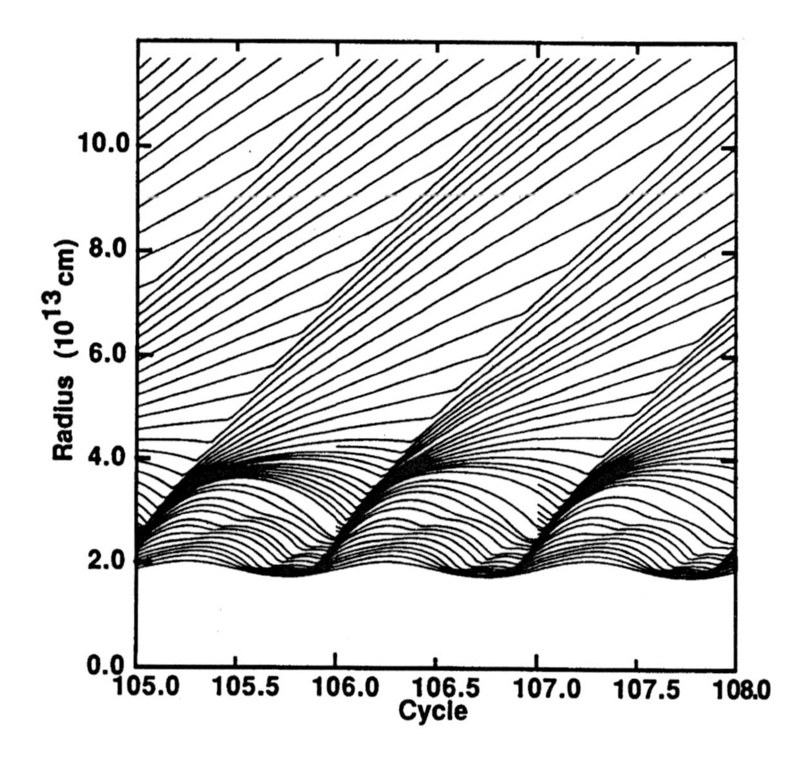

Figure 6.37 Radius *versus* time for selected layers (solid lines) in the atmosphere of a Mira star, based on theoretical models by Bowen (1988). The radius is the distance from the centre of the star. The lowest layer is the photosphere. It expands and contracts because of the pulsation mechanism in the star. Notice that the outer layers are ejected from the star as *mass loss*. (From Bowen, 1988.)

radiation from the star amplifies the radio emission at certain wavelengths. They act as natural celestial *masers* – a term which stands for 'microwave amplification by stimulated emission of radiation'.

The largest, most luminous Mira stars are so heavily obscured by gas and dust that they are almost invisible in the optical region of the spectrum, but they are very conspicuous in the IR. The OH emission is also strong, so they are called *OH/IR stars*. They have pulsation periods up to 2000 days.

Mira stars and OH/IR stars are a major source of interstellar dust – silicates in the case of the oxygen-rich stars, and graphite in the case of the carbon-rich stars. Along with supernovae, they also produce elements heavier than helium, and eject these into the interstellar material, where they become part of new generations of stars and planets.

This 'picture' of Mira stars, and their dynamic behaviour, can now be studied by high-resolution IR spectroscopy. At IR wavelengths, the star is brightest, and the photometric variations smallest. These studies are challenging, however, because of the long time scales involved, the irregularity of long-period variables, and the diversity of their properties. The work of Ken Hinkle, Tom Lebzelter, and their collaborators (e.g. Lebzelter *et al.*, 2005) is a good example. It confirms that the deep layers of Mira stars pulsate regularly, with typical velocity amplitude of 25 km s^{-1}. Above these layers, shock waves move upward through the atmosphere. The absorption lines in the spectra are split into two components, because more than one shock wave may be visible at any time. There may be other non-symmetric, non-periodic processes, such as convective motions and non-radial pulsation. The outer atmosphere is expanding at typically 10–15 km s^{-1}.

Semi-regular (SR) variables have smaller amplitudes, no shock waves, and little or no mass loss. But the difference between Miras and SRs is complicated, and may depend on pulsation mode, amplitude, and luminosity. IR spectroscopy can also address the mystery of the long secondary periods in pulsating red giants. These periods are *not* due to binarity, in most cases.

One complication in our understanding of Miras is the fact that, in most of the star, energy is transported outward by convection – a very poorly understood process in astrophysics. In particular, the interaction between convection and pulsation is unclear. Convection may explain some of the cycle-to-cycle irregularities in Miras. Another complication is that, in AGB stars, the nuclear-burning shells occasionally become unstable and produce *shell flashes* – a sort of 'burp' in the energy-generation processes. The star evolves to hotter temperatures within a few thousand years, then returns to the AGB state. This may explain the small fraction of Miras which show very rapid changes in their pulsation period.

Miras in globular clusters represent a fruitful area for comparison between observation and theoretical models. Until recently, globular clusters were seldom monitored regularly enough to follow the longer-period variable stars within them, but studies of the photometric and velocity variations of such stars are now being carried out, with interesting results.

Figure 6.38 summarizes the process by which red giants, of various masses, become Mira stars, and lose mass. The initial masses are on the left-hand scale. The stars evolve to higher luminosities; see scale at the bottom. As they become Mira stars, they lose mass at increasingly high rates (labelled diagonal lines), reaching the final masses shown along the curved line at the bottom.

6.17 Red supergiant (SRc) variables

We now turn to a group of variables which are significantly different from other red variables. They are massive, young, extreme Population I stars, whereas most red giant variables are less massive, older, disc Population I or

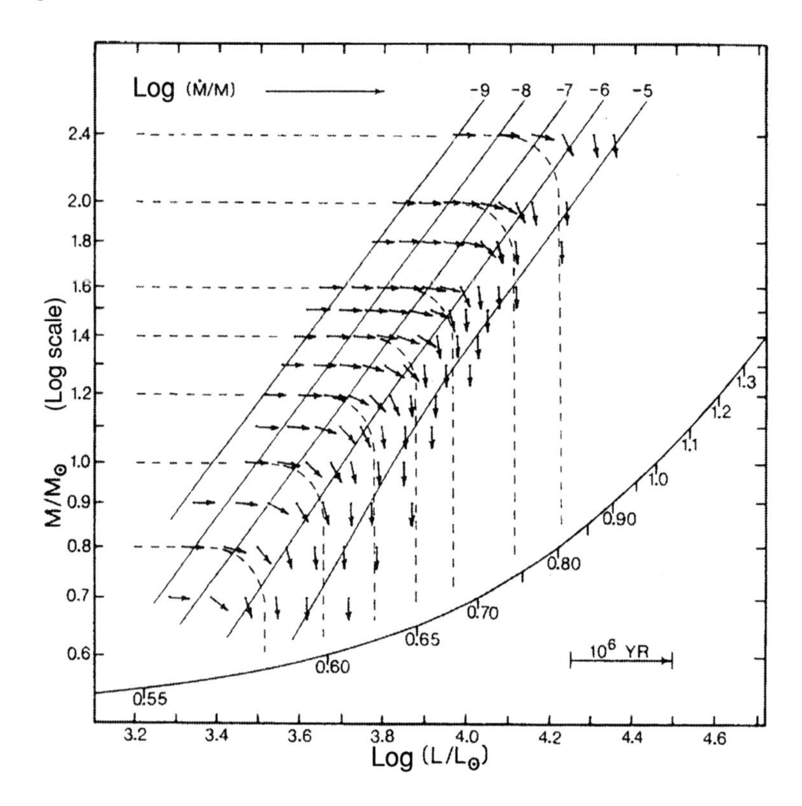

Figure 6.38 'Lemming' diagram for Mira stars. Stars of different masses (vertical coordinate) evolve to the right, to higher luminosity. As they do, their rate of mass loss increases as a result of the effects of pulsation. The rates of mass loss are indicated by the diagonal lines. As the star loses mass, it moves downward to the curve, which indicates the remaining mass of the star and its core. (From Bowen and Willson, 1991.)

Population II stars. Over 50 SRc variables are listed in the GCVS, but only about two dozen SRc variables are well studied. The best-known examples are Betelgeuse (α Orionis) and μ Cephei. Virtually every red supergiant is variable. They are so luminous that they can be seen in many other galaxies.

The peak-to-peak amplitude of variability ranges up to 5 magnitudes in V, and is greater for more luminous stars. The variability is semi-regular at best, but there are characteristic time scales of 250–1000 days. These are correlated with the luminosity of the star, as would be expected if the variations were caused by pulsation. Following earlier work by Richard Stothers, which established that SRc variables were pulsating, Stothers and Leung (1971) derived a semi-theoretical relation

$$M_{bol} = -7.20 \log P + 12.8 \quad (6.15)$$

Box 6.7 Star sample – Mira

Mira (o Ceti, HD 14386, SpT M5e-M9e, V = 2.0 to 10.1) is the prototype of the most extreme pulsating red giants – the coolest, largest, most luminous, longest period, and largest amplitude – greater than $\Delta V = 2.5$ by definition. Its period is 331.96 days, but fluctuates. Its maximum magnitude also varies, from 2 to 5. Its minimum magnitude is more or less constant at about 9.5, because of the presence of an usually unresolved close companion (figure 6.39).

The variability of o Cet was discovered in 1596 by David Fabricius. Its 11-month period was determined in 1638 by Johann Holwarda; it was the first known periodic variable. The name *Mira* – the wonderful – was bestowed on it in 1642 by Johannes Hevelius. Since then, it has been observed by thousands of variable star observers, and it has been the subject of hundreds of research papers. Many of these discuss its spectrum, which is complex and varies noticeably from one cycle to the next.

It has been known since the 1920s that Mira has a 9.5-magnitude hot dwarf companion, subsequently named VZ Ceti, about 0.85 arc sec away; it varies between magnitude 9.5 and 12, on time scales of minutes to hours. The orbital period is about 400 years. Because the companion is hot, it is best seen in the UV, and when Mira itself is near minimum brightness, of course. A recent HST image by Margarita Karovska (figure 6.39) shows evidence for a mass stream in the vicinity of Mira itself. Mira is losing mass at a rate of about one Earth mass per year, as a result of its pulsation-driven wind: the shock waves enhance the density of the atmosphere; dust condenses in these density enhancements; the dust grains absorb IR radiation, which pushes them outward; they carry the gas in the atmosphere with them. Some of that gas forms an accretion disc around the white dwarf companion, making the system a mild example of a symbiotic binary star. Like many symbiotic stars, it has also been detected as a radio source.

Mira and its companion have also been observed with the *Chandra* X-ray satellite. X-ray emission from the hot accretion disc around the companion was not surprising. But an X-ray flare was detected from Mira itself – possibly a result of the shock waves travelling through its atmosphere.

The $(O - C)$ diagram for Mira can be explained as the result of random cycle-to-cycle fluctuations. These unfortunately mask any evolutionary period changes, even though the star is evolving, as a result of its mass loss, on a time scale of a few hundred thousand years. At that time, the white dwarf core of Mira will be revealed, and it will light up the surrounding gas as a planetary nebula.

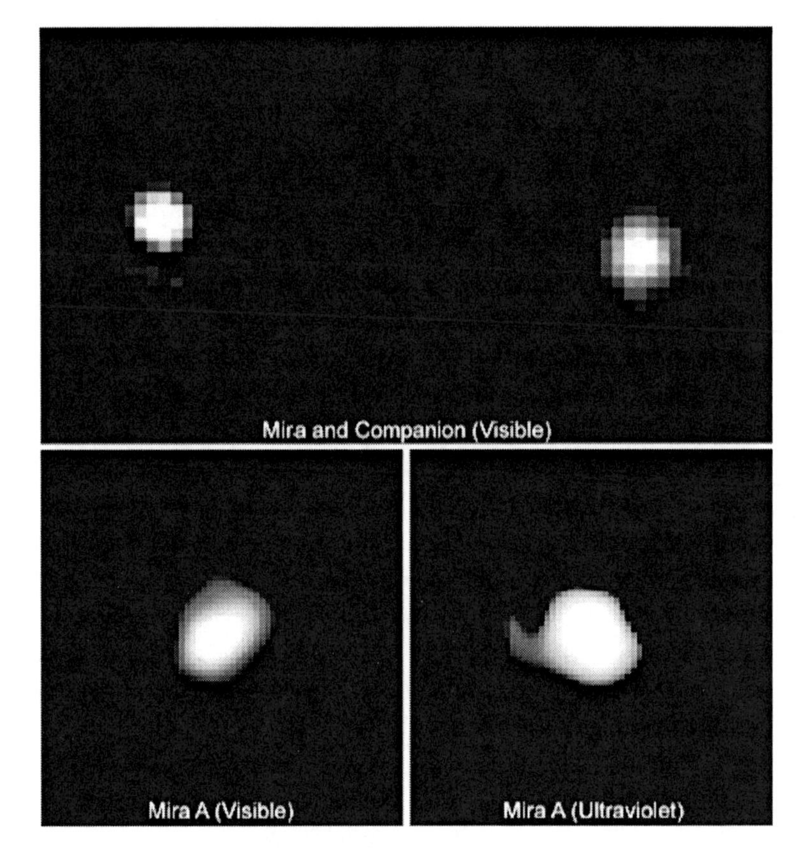

Figure 6.39 HST image of Mira and its hot companion. Top and left: visible light; bottom right: ultraviolet light. There is evidence for a mass stream from the cool to the hot component. (From M. Karovska/NASA.)

Feast *et al.* (1981) studied SRc variables in the Magellanic Clouds, at infrared wavelengths, where their flux is maximum, and the effects of reddening, absorption, metallicity, and duplicity are all minimized; they derive

$$M_{bol} = -8.6 \log P + 16.4 \quad (6.16)$$

which gives similar results to the relation derived by Stothers and Leung. There are also longer-term variations. It was initially thought that these might be caused by the rotation of the star, and the fact that the bright and dark convection cells in its outer layers might be large and few; it would thus be a rotating variable. More recent observational and theoretical studies suggest that the convection cells are small and many, and would not produce significant rotational variability. It is possible that the 'long' and 'short' periods correspond to

Box 6.8 Star sample – Betelgeuse

Betelgeuse [α Orionis, HD 39801, M1-M2Ia-Ibe, V = 0.2 to 1.3] is an especially popular star with students and the general public because it is bright, it is red, it is part of a familiar constellation, it is a potential future supernova, and its name can be mispronounced as 'beetlejuice'. Even in an urban environment, it can be seen, and its brightness can be measured relative to that of Aldebaran and Procyon. These measurements can be compared with archival measurements, for instance those in the AAVSO International Database.

Betelgeuse is one of the few stars, other than the sun, that has been imaged (figure 6.40). Efforts to 'resolve' the disc of the star go back to 1920, when Albert Michelson used an optical interferometer on the 2.5m (100') Mt. Wilson telescope to measure a disc diameter of 0.044 arc sec. More recently, the HST obtained a 'picture' of the disc, showing a bright spot which may be a large convection cell (Gilliland and Dupree, 1996). It is clear that the surface of this star is far from simple and symmetric, so it is not surprising that the variability is complex.

Figure 6.40 HST image of Betelgeuse, showing the size of the star as compared with the size of the earth's orbit. (HST/NASA.)

Betelgeuse can be imaged because of its large size, and the relatively close distance of 425 LY. On the other hand, the definition of 'size' is rather fuzzy, because the star is! At wavelengths at which the star's atmosphere is more opaque, the star looks larger. Still, if Betelgeuse were placed at the centre of our solar system, its outer atmosphere would stretch far into the realm of the gas giants.

The variability of Betelgeuse was discovered by John Herschel in 1836. It is a bright example of an SRc variable, a red supergiant with appreciable periodicity. It actually varies on two time scales, one of about a year, and one of about five years. While the shorter time scale is probably caused by pulsation, the cause of the longer time scale is still uncertain – as is the case with the long secondary periods of other pulsating red giant and supergiant stars (Goldberg, 1984).

fundamental and first overtone pulsation, even though they and their ratio seem too large. Theoretical pulsation models are uncertain for such extreme stars. Keep in mind that other types of pulsating variables have long secondary periods, whose cause is uncertain.

The challenge in understanding these variables is the long time scales involved, and the irregularity. Decades of systematic observations are necessary. For some stars, there are long datasets of visual observations, and/or measurements from photographic sky surveys. These can be used to determine the dominant period in the stars, and may possibly resolve the nature of the long secondary periods, and the irregularity. Another issue is the relationship between the SRc variables and the Lc variables, which are irregular red supergiant variables. The SRc and Lc variables may, in fact, be part of a 'spectrum', ranging from more periodic to more irregular.

SRc variables are M1–M4 supergiants, with temperatures of 3000–3500K, absolute magnitudes (determined from membership in galactic clusters and associations) of −5 to −7, corresponding to bolometric magnitudes of −7 to −9. They are among the most luminous stars known. Their masses were initially 15 to 30 solar masses, but they will have lost significant mass, so the present values will be less – typically 5 to 20 $M_{\odot}$.

Since the absolute magnitudes of the brightest M supergiants are fairly uniform: -8.0 ± 0.2, with little or no dependence on the type or luminosity of their parent galaxy, they can be used as distance indicators. The power of this technique is only slightly lessened by their small variability.

In addition to Betelgeuse, which is highlighted in a sub-section of its own, μ Cephei is a famous pulsating red supergiant. This extremely red star was referred to as a 'garnet star' by William Herschel. Its variability was discovered in 1848 by John Russel Hind. Analysis of measurements from 1848 to the present give periods of 850 and 4400 days. One AAVSO observer monitored it for 60 years! Like Betelgeuse, it is ejecting a wind of gas and dust at a rate of up to one earth mass a year. It is surrounded by gas and dust shells which suggest that the mass loss is enhanced every thousand years or so. Like Betelgeuse, an attractive star to observe in the night sky!

Recently, Kiss *et al.* (2006) have carried out the most comprehensive study of SRc variables: an analysis of AAVSO visual observations of 48 stars with a mean time-span of over 60 years. In addition to 'short' pulsation periods, they find long secondary periods in a third of the stars. They also find period irregularity which they ascribe to the effect of convection cells in the outer regions of the stars.

7

Eruptive variable stars

Eruptive variables are considered to include any variables which flare up relatively quickly, and fade more slowly. They include flare stars which brighten in seconds, and some types of novae and symbiotic stars which may take months or even years to brighten. This class sometimes even includes R Coronae Borealis stars, which are the inverse of eruptive variables! In the GCVS4, this class includes many types of pre-main sequence stars, and also S Doradus, Gamma Cassiopeiae, and erupting Wolf–Rayet stars. In eruptive variables, there is generally a sudden input of energy into a star, or part of a star, and we see the star's response – a violent outburst.

7.1 Flare stars

Flare stars, also known as UV Ceti stars, are dwarf K and M stars (mostly the latter) which randomly and unpredictably increase in brightness within seconds to minutes, by up to several magnitudes, then slowly return to normal (figures 7.1, 7.2). In the GCVS4 they are classified as UV, or UVN if they are associated with pre-main sequence stars, or RS if they occur in an RS Canum Venaticorum binary system. These flares are qualitatively similar to those on the sun. The flares are one aspect of *activity* on these stars; emission lines in both the visible and ultraviolet spectra, and X-ray emission from a hot (up to 10 000 000 K) corona are others. So flare stars are generally classified as dMe, the 'e' referring to the presence of emission lines in the spectrum. Some flare stars are also BY Draconis variables (section 4.5); this is yet another manifestation of their activity.

Flare stars are the most numerous variables in our galaxy. That is because dwarf stars, cooler than the sun, constitute over 90 per cent of all stars. In a

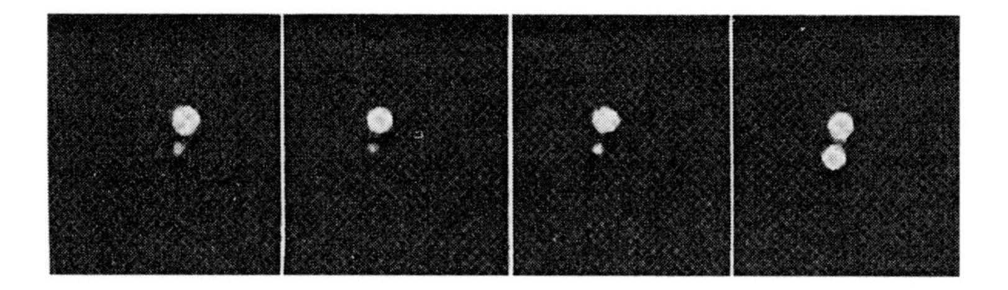

Figure 7.1 Right: a flare on the fainter component of the visual binary star system Krüger 60. The images are taken over an interval of a few minutes. (Sproul Observatory photograph.)

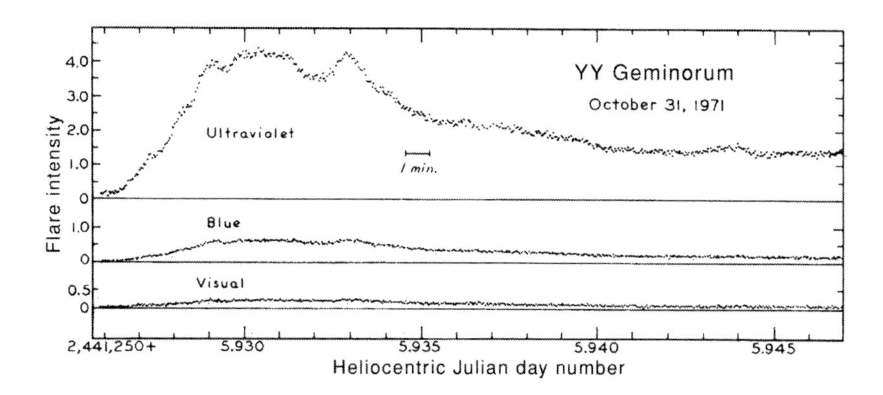

Figure 7.2 A flare on YY Geminorum, the companion of Castor. The light curve covers 30 minutes. The flare intensity is greatest in the ultraviolet, because the flaring region is very hot. (From Moffett, 1974.)

sense, the nearest flare star is the sun! Its flares are localized in sunspot regions, and can easily be seen with a small telescope. They would be much more difficult to see if the sun was a distant star because, in the sun's case, the total brightness increases only slightly. The nearest other flare star is the nearest star in the night sky – α Centauri C, or Proxima Centauri. The prototype UV Ceti is our tenth-nearest neighbour. The literature on flare stars is very extensive, consisting mainly of papers on the discovery and monitoring of these stars, especially in clusters.

7.1.1 *Variability*

Because flare stars are found within a specific type of star (dwarf K and M stars with emission lines), it is easy in principle to discover them. One merely needs to survey them, patiently, using visual, photoelectric, or imaging techniques. Visual techniques are adequate, though tedious, for some aspects of the study of flare stars, and many amateurs search for and monitor flare stars visually. Photoelectric techniques – specifically high-speed photometry – are essential for detailed physical study of these stars. CCD techniques now combine the advantages of efficiency (wide-field imaging of many flare stars within the same cluster) with photometric accuracy. The flares are most pronounced in the short-wavelength part of the spectrum, partly because the emitting material is very hot, and partly because the star itself is less bright at shorter wavelengths. In any one star, large flares are rare; small ones are more common. And flares occur more or less randomly.

Figure 7.2 shows the brightness changes which occur in a typical flare. The rise time is a fraction of a minute; the decay time is a few minutes, though it may take hours for the star to return completely to its normal brightness. The total energy released is 3×10^{28} to 3×10^{34} ergs; the luminosity of the star itself is 10^{28} to 10^{33} ergs/sec. The flares occur at random intervals, and the logarithm of the probability of a flare with an energy greater than some given value E_0 is proportional to $-\log E_0$. This is another way of saying that large energy flares are improbable.

Bursts of radio energy often accompany the optical flares. The simultaneous arrival of the radio and optical flares is a dramatic illustration of the fact that radio signals travel at the same speed as light.

7.1.2 *Physical properties*

Flare stars are dKe and dMe stars – dwarf K and M stars with emission lines. It is possible that flares occur on G or even F type stars on the main sequence, as they do on the sun, but the flares would be too faint, relative to the brightness of the underlying star, to be seen from a distance. Actually, there are sporadic reports of flares on stars as hot as type B, but the reality of these flares is open to question; they may be instrumental artifacts. As Carl Sagan said, 'extraordinary claims require extraordinary evidence'. The dKe and dMe stars have absolute magnitudes fainter than +7, masses and radii less than 0.4 times those of the sun.

7.1.3 *Flare mechanism*

The optical flare is best explained by the sudden release of energy at the surface of the star. The energy is equivalent to that of billions of atomic bombs, released over a few minutes. This creates an outward-moving shock wave, which heats and ionizes the gas. After the shock wave passes, the ions recombine with the electrons, giving off light and creating the bright emission lines seen in the spectrum. The white-light flare occurs first, then the emission lines. Since the initial heating produces a very hot gas, the flare is most pronounced at short wavelengths – violet, ultraviolet, and X-ray. In 2004, the GALEX UV satellite observed a 9-magnitude UV brightening, within six minutes, on the red dwarf Gliese 3685A. The emission lines can be quite broad, testifying to the large velocities (up to hundreds of km s^{-1}) of the shocked gas.

Understanding of solar flares has recently been greatly advanced by the Reuven Ramaty High Energy Solar Spectroscopic Imager (RHESSI) satellite, which has observed thousands of solar flares in the X-ray and gamma-ray regions of the spectrum. There are some practical reasons for wanting to understand (and even possibly predict) solar flares: the energetic particles that are produced arrive at earth a few days later, where they may threaten satellites, radio communications, and the electrical power grid. For more information about solar flares, see:
http://hesperia.gsfc.nasa.gov/sftheory/

The radio flares on flare stars bear a strong resemblance to so-called Type III solar radio bursts, except that they are 10^3 to 10^5 times more energetic. They are strongly polarized (90 per cent circular, 20 per cent linear), indicating that magnetic fields and synchrotron radiation are involved. (Synchrotron radiation is produced by charged particles, spiralling in a magnetic field.) Lower-frequency radio waves emerge at later times, as the radio-emitting plasma rises to higher, more rarified layers in the star's atmosphere; lower layers are opaque to lower-frequency waves.

But what causes the sudden release of energy in the first place? Presumably the same process as on the sun – the release of magnetic energy when magnetic field lines twist, break, and reconnect. The magnetic field lines are formed and twisted by a stellar dynamo effect, which results from rotation and convection in the star – two processes which play an even more crucial role in these stars than they do in the sun. These processes are extremely complex and still not fully understood; they involve *magnetohydrodynamics*, a theory which is as complex as the name is long.

Why? As in the BY Draconis stars, the physically important property of a flare star is its relatively rapid rotation. Rapid rotation tends to occur in stars which are *relatively* young (i.e. they have not lived more than a small fraction of their lifetime), or which are intrinsically rapidly rotating, or which have somehow failed to lose their rotational energy with age, or which are members of close binary systems. In the first case, the rapid rotation is a result of the star's recent formation; it has not had time to 'spin down'. In the last case, the rapid rotation is due to tidal effects of the companion star: it tends to 'spin up' the star to synchronize the rotation with the revolution. This is the case in RS Canum Venaticorum stars (section 4.4). The other crucial property of K and M dwarfs is their deep outer convection zone – significantly deeper than in the sun. The interiors of the coolest red dwarfs are completely convective.

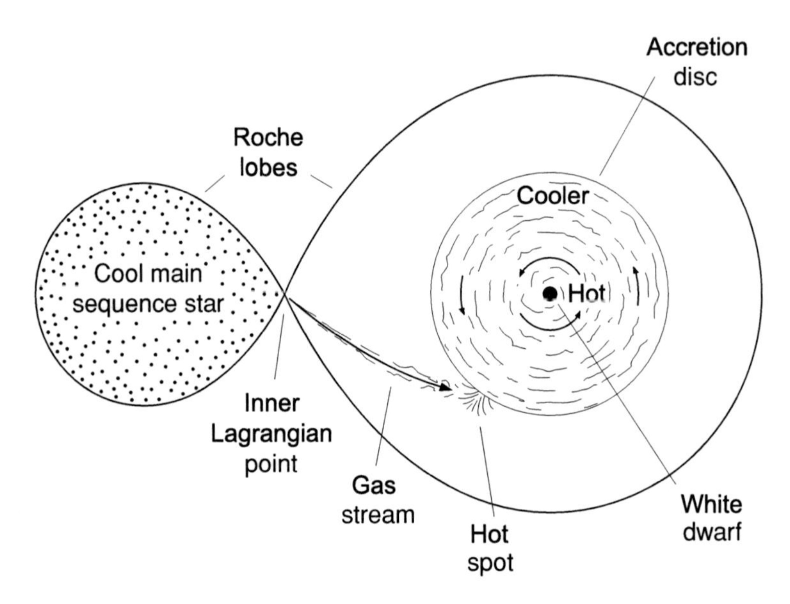

Figure 7.3 The generic model of a cataclysmic variable: the primary star is a white dwarf; the secondary star is a cool main sequence star (or in the case of some recurrent novae, a red giant); mass transfer through the inner Lagrangian point feeds an accretion disc, with a hot spot at the point of impact. The accretion disc is hottest near the star. (Jeff Dixon Graphics.)

7.2 Cataclysmic variables

The term *cataclysmic variable* (CV) includes a broad range of eruptive variables which share a common model: they are close binary systems in which one star – usually a cool (G to M type) main sequence star – fills its Roche lobe and is transferring matter through the inner Lagrangian point to an accretion disc, circling around a compact object – a white dwarf, according to the standard definition (figure 7.3). To quote Brian Warner (2000): 'The exchange of matter, from donor to receiver, in a close binary star leaves the receiver pregnant with possibilities.' These include rapid, small-amplitude flickering as matter impacts the disc, flare-up of the accretion disc due to thermal instability, thermonuclear explosion as hydrogen-rich material accumulates on a stellar corpse, and even the explosion of the corpse if it is a white dwarf which accumulates enough mass to exceed the Chandrasekhar limit. If the compact star is a neutron star or black hole, or if it has a very strong magnetic field, then the possibilities can be even more extreme and interesting. If the magnetic field is weak, then mass transfer through the accretion disc dominates; if the magnetic field is strong, then mass transfer is dominated by the effects of the magnetic field; in intermediate cases, the situation is equally complex and interesting. CVs have provided astronomers with a better understanding of mass transfer and accretion discs in the many environments in which they exist in the universe, from protostars to the cores of galaxies. It has also provided them with a better understanding of the collapsed objects which occur in such binary systems – white dwarfs, neutron stars, and black holes.

CVs vary on a wide variety of time scales, with a variety of amplitudes, as a result of a variety of geometrical and physical processes in the different parts of the CV system. Amateur astronomers contribute to our understanding of CVs, using visual, photoelectric, and CCD techniques, on time scales from seconds to decades.

There is an on-line catalogue of cataclysmic variables at:
http://icarus.stsci.edu/~downes/cvcat/

Novae are variables which unpredictably brighten by 6–19 magnitudes to $M_v \sim -9$ within a few days, then slowly fade. In the GCVS4 they are sub-classified as fast (NA), slow (NB), and very slow (NC). The brightness changes are accompanied by unique and spectacular spectroscopic changes as the nova throws off a shell of matter, and by changes in the ultraviolet, infrared, and radio emission. The nova is the result of a thermonuclear runaway explosion of hydrogen-rich material, accreted on to the surface of the white dwarf. There are also *recurrent novae* (NR in the GCVS4), which have erupted more than once; *dwarf novae* of several kinds, which brighten by a few magnitudes every few weeks; and *nova-like objects* (NL in the GCVS4), which look like novae or DNe before or after outburst, but are not known to have erupted themselves.

Hundreds of CVs are bright enough to be monitored by visual observers with small telescopes, and visual observers have played an important role in our understanding of these stars. One of the remarkable aspects of variable star astronomy is that the demand for visual observations through backyard telescopes has actually *increased* by a factor of 25 since the birth of space astronomy around 1970. One reason is that visual observers have been able to support spacecraft observations of CVs, by detecting the onset of their outbursts, and monitoring the visual brightness during the spacecraft observations.

7.2.1 *Novae*

Novae unpredicably brighten by up to 15 magnitudes in only a few days, often attaining naked-eye visibility if they are close enough within our galaxy (figures 7.4 and 7.5). Nova Aquila 1918 (V603 Aquila), at brightest, was outshone only by Sirius. Thus they look like a 'new star', hence the name *nova*, meaning new. Only recently has it been possible to look and see if any star existed at the position of the nova prior to its outburst. On survey images, such as the photographic *Palomar Observatory Sky Survey* (POSS), one often finds the pre-nova:

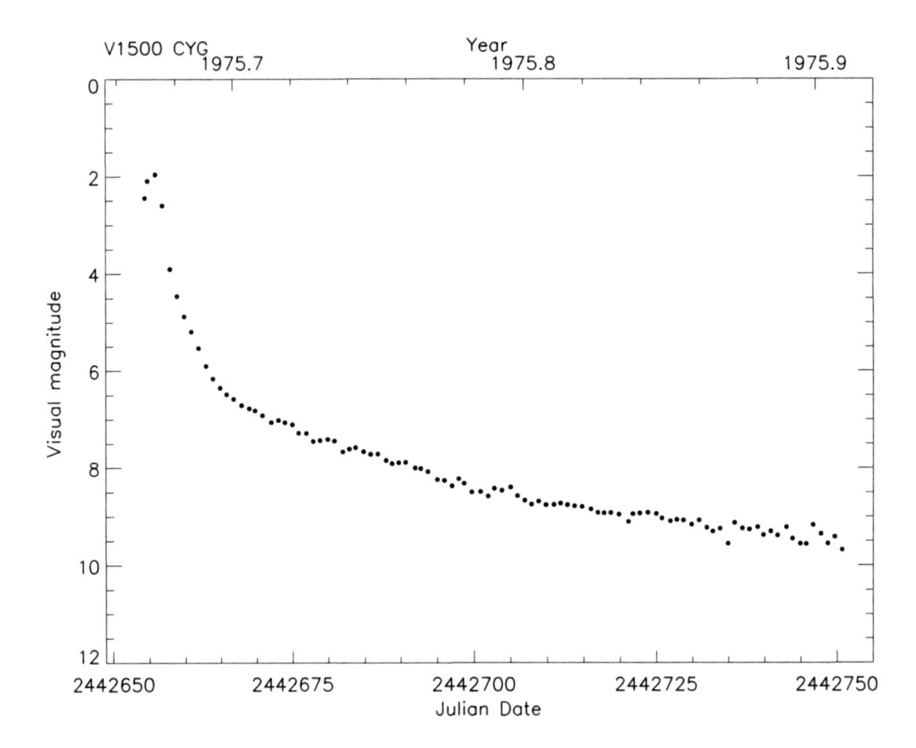

Figure 7.4 The light curve of the Nova Cygni 1975 (V1500 Cygni), based on visual measurements from the AAVSO International Database. The nova was independently discovered by dozens of amateur astronomers, including the author's 11-year-old daughter. Note the amplitude, duration, and very smooth shape of the light curve. (From AAVSO.)

a faint star which is not new at all, but is approaching the end of a long and eventful life.

As many as 30 novae may erupt in our galaxy each year, but most of them are never seen, being obscured by distance or by absorption by interstellar dust. Many novae – especially the brighter ones – are discovered by amateur astronomers, systematically and patiently monitoring the sky by eye or camera, though automated surveys are increasingly usurping this role. Novae have been discovered in this way for over 2000 years, as evidenced by the records kept by Oriental astronomers. Until the 1930s, novae and supernovae were not distinguished, but there is no longer any confusion. A few novae are discovered spectroscopically, when their unique emission-line spectrum is accidentally recorded on a wide-field *objective prism* photograph.

The situation is somewhat different in nearby galaxies such as the Magellanic Clouds, M31 and M33. Regular wide-field images will pick up most of the novae which erupt in these galaxies each year. Since we do not have to look edgewise

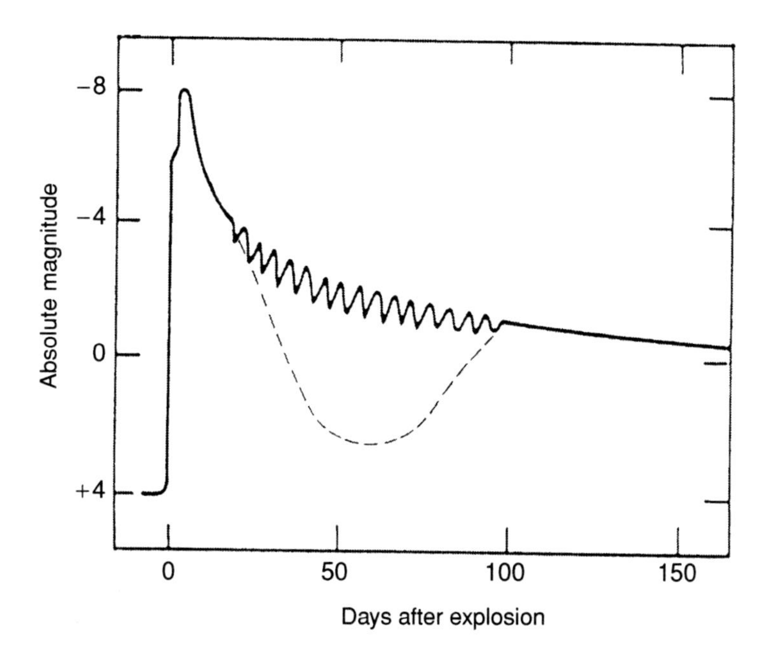

Figure 7.5 A schematic light curve of a nova, showing various features which might be found: the rapid rise to maximum, the increasingly slow decline, and the fading or oscillation about 50 days after maximum. (From Kaler, 1997.)

through absorbing dust clouds, absorption is less of a problem. On the other hand, we can only observe them near maximum light. It is difficult to study them spectroscopically, and nearly impossible to see them at minimum light. Nothing beats a bright, nearby nova in our own galaxy!

Most galactic novae belong to intermediate or old Population I, but that is a biased sample. One of the most complete extragalactic samples is that of Neill and Shara (2004) for M81; it clearly shows that the M81 novae are primarily Population II. A few galactic novae have been identified in globular clusters, and are therefore also Population II. When plotted on the sky, they are concentrated to the disc and to the nucleus of our galaxy. They have ages of up to billions of years and (as discussed later in this section) they are the end product of the evolution of a close binary system of stars.

Novae are designated by their constellation and year of discovery, and later receive an official variable star designation; Nova Cygni 1975 is also V1500 Cygni.

Light curves

The schematic light curve of a typical nova is shown in figure 7.5. This shows several forms of behaviour which can be exhibited by a nova on its fade from maximum.

The pre-nova stage is poorly understood. At best, we can determine the magnitude of the pre-nova from survey photographs. It is not clear whether there are 'advance symptoms', months, years, or even decades ahead of time, that would signal that an outburst was about to occur. Nor is it clear how long it takes for a nova to return to a final, constant, post-nova stage. Theory suggests that it takes 10 000–100 000 years for the star to be ready for another outburst. Nova Cygni 1975 appears to have taken several months to prepare for its outburst. Only in the case of recurrent novae do we have any real idea about the spectrum of a nova before eruption.

The rise to maximum brightness, and the pre-maximum halt are only slightly better understood. Novae are usually discovered at or after maximum. Occasionally, there are sky patrol photographs which record the nova's rise to maximum. In the case of Nova Cygni 1975, an amateur astronomer Ben Mayer was conducting a systematic photographic sky patrol for meteors when he recorded several images of the nova's rise.

After maximum, the initial rate of decline may be fast, slow, or very slow (sometimes classified as A, B, or C). Nova Cygni 1975 (V1500 Cygni) was among the fastest; Nova Delphini 1967 (HR Delphini) was among the slowest. The standard parameter is t_3, the time in days required to fade from maximum by three magnitudes. This may be days, weeks, or months, depending on the speed of the nova. The parameter t_3 (in days) is correlated with the absolute magnitude at maximum: the brighter the nova, the faster the decline. For novae with smooth, reasonably fast declines: $M_v = -12.25 + 2.66 \log(t_3)$; otherwise, $M_v \sim -6.4$. Once this relation has been established, for instance, by observing novae in nearby galaxies of known distance, it provides an effective method for determining the distance of a nova. From the initial rate of decline, the maximum M_v can be determined. This, when combined with the maximum apparent magnitude, and corrected for interstellar absorption, gives the distance. This relation also implies that the M_v is about −5.5, 15 days after maximum, independent of speed class.

Well after maximum, during the *transition region* of the light curve, the brightness changes may vary from one nova to another. The brightness may decline smoothly. There may be periodic 'oscillations' superimposed on the slow decline; these may be due to thermal instabilities in the star, or to processes in the expanding shell that alternately create dust, and destroy it. The brightness may decline precipitously, then recover somewhat, as dust forms in the expanding gas around the star, then disperses.

Eventually, the nova reaches the post-nova stage which – if all novae are recurrent – eventually becomes the pre-nova stage again! The post-nova stage also demonstrates, of course, that the star survived the cataclysm.

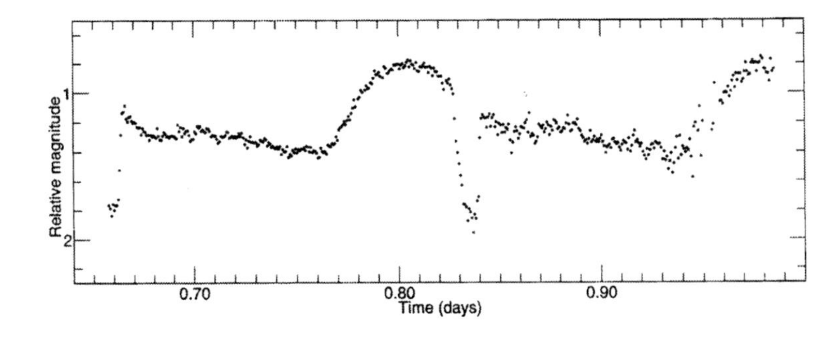

Figure 7.6 The light curve of the CV U Geminorum, showing the eclipse of the hot spot, and the brightening when the hot spot is most easily seen. It is from light (and velocity) curves like these that the nature of CVs can be deduced. (From Hellier, 2001.)

Novae and their relatives also show characteristic brightness fluctuations at minimum light. This variability is rapid, and often of small amplitude. It was first discovered, with difficulty, using conventional photometry, but has been studied in much more detail using *high-speed photometry* – an adaptation of photoelectric photometry which makes it possible to make observations every second or so. The variability is characterized by:

- Intense flickering on a time scale of minutes.
- Brief periodic minima, such as would be expected from an eclipse.
- Humps, or relatively slow increases and decreases in brightness.

Figure 7.6 shows the brightness variability of Z Chamaeleontis at minimum. The variability is periodic; the period is 0.074 day. The significance of this variability will be discussed in connection with the close binary model of CVs, in the next section.

Figure 7.5 shows a schematic light curve of a nova, which can be compared with the spectroscopic variations discussed later in this section.

Ultraviolet and infrared variability

Visual observations by themselves can often be misleading. Such is the case with novae because, if the UV and IR luminosities are added to the visual luminosity, then the brightness of the nova remains relatively constant for several weeks after visual maximum. It then declines slowly. The star pours out energy at the rate of about 3×10^4 $L_\odot$. This is equivalent to the *Eddington luminosity* – the maximum luminosity which a star of a given mass can have before the outward radiation pressure exceeds the inward gravitational force. But an increasingly small fraction of this energy appears in the visible portion of the spectrum. Immediately after the outburst begins, the temperature of the star is about 10 000K, and most of its energy is released in the visible. Later, the temperature of the star increases considerably as we see deeper into its expanding photosphere, and most of its energy is released in the UV, where it can be detected by UV telescopes in space. The IR energy comes, not from the star, but from dust which forms in the expanding gas around the star, absorbs the visible and UV radiation from the star, and re-emits it as IR radiation.

Radio emission also arises from the ejected material, specifically by free–free emission from the gas. The IR and radio emission provide useful ways of studying the temperature and density of the ejected material.

Spectroscopic variability

Novae undergo a unique and spectacular pattern of spectroscopic variability, first discovered by visual spectroscopy in 1866. This variability arises from the ejection of a shell of matter with a mass of about 0.0001 $M_\odot$ and a velocity of several hundred or more km s^{-1}. The following spectroscopic stages may be present:

- *Pre-nova.* Except in the case of recurrent novae; this is deduced from the post-nova spectrum – a deduction which may or may not be justified. The spectrum shows absorption lines of a G-M type star, plus broad emission lines of H, He I and He II, and Ca II, and a strong blue continuum coming from the accretion disc.
- *Pre-maximum.* Absorption lines similar to those of a B-A type supergiant star, blue-shifted by several hundred km s^{-1}. Novae with faster declines from maximum are found to have faster expansion velocities – up to 4 000 km s^{-1}.
- *Maximum.* Absorption lines similar to those of an A-F type supergiant, blue-shifted by several hundred km s^{-1}.
- *Principal* (0.6 to 4 mag below maximum light). Wide emission lines with blue-shifted absorption lines ('P Cygni profiles') of H, [O I], [N II], and other metals. The [ ] signify *forbidden lines* which are formed in very low-density environments.
- *Diffuse enhanced* (1.2 to 3.0 mag below maximum light). Absorption and emission lines similar to those of the principal spectrum, but blue-shifted by up to 2000 km s^{-1}.
- *Orion* (2.1 to 3.3 mag below maximum light). Absorption lines of He I, N I, O II (like an Orion B star, hence the name), with hazy emission. The former spectrum appears to be that of the star, the latter that of the ejected material.

- *λ4640* (3.0 to 4.5 mag below maximum light). Emission lines of He I and N III.
- *Nebular* (4 to 11 mag below maximum light). Emission lines of [O III], [Ne III] etc., as in a gaseous nebula, energized by the radiation from the hot star in the system.
- *Post-nova*. Similar to pre-nova.

It should be emphasized that several of these stages may be present simultaneously, and that each stage may have multiple components. The spectrum can be complex indeed! In fact, these divisions are quite arbitrary; there is a smooth transition from one stage to another.

The ejected material

It is apparent from the blue-shifted features in the spectrum of a nova that it is ejecting material at high velocities. The ejected material can eventually be seen through its optical emission lines, and later through its IR and radio emission. After many years, it can be seen on direct images, as wispy material surrounding the central star. The clumpy nature of the ejecta remind us that stellar winds and outbursts are rarely spherically symmetric (figure 7.7).

As the expanding shell is gradually resolved in images, its rate of angular expansion can be compared with the linear rate, determined from the spectrum. This leads to an estimate of the distance by the *expanding photosphere method* – a variant of the Baade–Wesselink and Barnes–Evans methods.

The study of the ejected material provides important information about the cause of the nova outburst – a runaway thermonuclear reaction in the outer layers of the star, caused by a buildup of material transferred from a binary companion. The ejected material has been processed by this thermonuclear reaction; its composition therefore reflects the products of the reaction. There is also some suggestion that the existence and nature of the reaction depends on the *initial* composition of the material in the star. The composition of the ejecta might shed some light on this suggestion.

In the photographic era, it was difficult to study the spectra of faint objects. Photographic plates were inefficient and hard to calibrate, so it was almost impossible to determine the absolute strength of the emission lines from the ejecta, and hence determine their quantitative composition. Electronic detectors on large telescopes have alleviated this problem.

The emission lines in the optical spectrum of a nova are caused by gas, excited by the ultraviolet radiation from the central star. Its temperature is about 50 000K, but the radiation is increasingly diluted, further from the star.

Figure 7.7 The recurrent nova T Pyxidis, showing discrete clouds of ejected material. (HST/NASA.)

There is also evidence for regions of even higher temperature – 10^6 K – perhaps caused by shock waves produced by the turbulent motions of the gas.

The chemical composition of the gas is still uncertain, but the optical results, so far, indicate the following: (i) helium is normal to high; (ii) carbon is probably high; (iii) nitrogen is up to 30 times normal; (iv) oxygen is 10–100 times normal; (v) metals heavier than oxygen are probably normal. UV satellite studies of the ejecta from several novae provide solid support for the thermonuclear runaway hypothesis by confirming the predictions based upon it.

Although the observational results – and the theoretical predictions – are uncertain, the situation is improving, and a meaningful comparison between the two is now possible. The most highly resolved image of nova ejecta is for the recurrent nova T Pyxidis (figure 7.7).

The total amount of ejecta can be estimated in a number of ways: from the spectrum of the expanding photosphere of the star; from the emission lines in the spectrum of the nova; and from the IR and radio emission from the ejecta. The results are in the range 10^{-4} to 10^{-3} $M_{\odot}$. The mass of the ejecta seems to be about the same for fast, medium, and slow novae, but the total energies are greater for fast novae. Although the ejected shell is usually assumed to be spherical, this is for simplicity and convenience only. There is every reason to believe that the shell may be asymmetric – due to the rotation, magnetic field, and/or duplicity of the star – and clumpy – due to instabilities in the flow. The simplest models of the IR and radio emission suggest that the density in the shell decreases as $1/r^2$, where r is the distance from the centre of the star.

Box 7.1 Star sample – Nova Cygni 1992

Nova Cygni 1992 (V1974 Cygni) was an especially important and well-studied nova. It was discovered on 19 February 1992 by amateur astronomer Peter Collins. Within hours, the *International Ultraviolet Explorer* (IUE) satellite observed it, as did a wide range of other ground-based and space telescopes. At the same time, the nova provided a testing ground for many areas of theoretical astrophysics, from thermonuclear energy generation, to the hydrodynamics of a thermonuclear runaway, to the complex atomic processes going on in a hot expanding plasma.

The nova brightened, as the fireball swelled, but it was important to observe it at *all* the wavelengths, from the visible to the X-ray, at which the nova was emitting. As the ejected material thinned, UV and X-ray telescopes detected the very hot surface of the white dwarf, and they followed it as it gradually cooled over the next few months. The runaway nuclear burning had ceased. The ejecta, too, were cooling. Clumpy debris was detected and studied at both UV and radio wavelengths, confirming that nova shells were not smooth and spherically symmetric. Eventually, the debris could be 'finger-printed' for composition, through visual and UV spectroscopy, to reveal elements from both the envelope and core of the white dwarf – especially oxygen, neon, and magnesium.

The underlying system is now visible; it has an orbital period of 0.0183 ...days; there is also a slightly longer superhump period (see section 7.2.3).

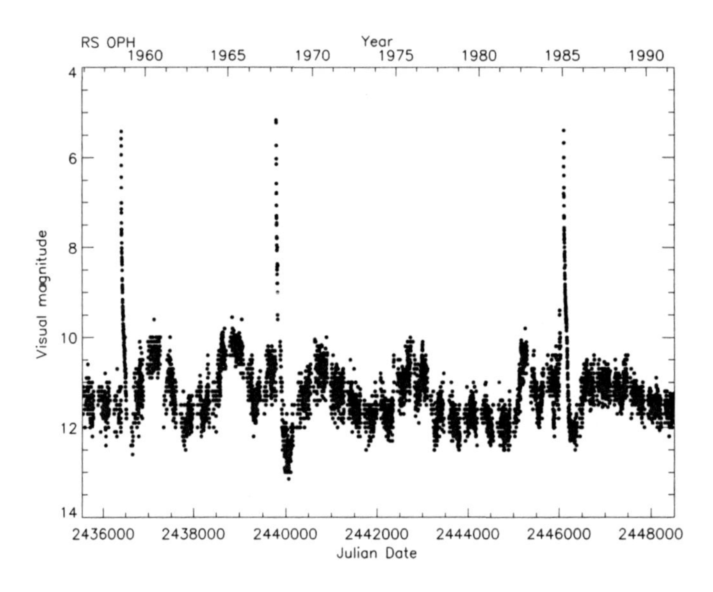

Figure 7.8 The light curve of RS Ophiuchi, a recurrent nova, from 1957 to 1991, based on visual measurements from the AAVSO International Database. Three outbursts are visible, of up to eight magnitudes. There are also interesting variations, of one or two magnitudes, at minimum. (From AAVSO.)

7.2.2 *Recurrent novae*

Recurrent novae are novae which are known to have erupted more than once. As of 2001, there were eight recurrent novae known in our galaxy (Warner, 2001): T Coronae Borealis (1866, 1946), V394 Coronae Australis (1949, 1987), RS Ophiuchi (1898, 1933, 1958, 1967, 1985; this star subsequently erupted in early 2006), T Pyxidis (1890, 1902, 1920, 1944, 1966), V3890 Sagittarii (1962, 1990), U Scorpii (1863, 1906, 1936, 1979, 1987), and V745 Scorpii (1937, 1989); note that the recurrence times are 10–100 years. The decline times t_2 (in this case, the time to fade two magnitudes from maximum) are days, except T Pyxidis which was much slower; the amplitudes are 4–9 magnitudes (less than for ordinary novae); and the absolute visual magnitudes at minimum are 0–1 (much brighter than for ordinary novae). It is possible, of course, that outbursts of these, or other possibly recurrent novae, have been missed.

Recurrent novae are important, of course, because they allow us to see a type of nova both before and after the fact. Also, since amateur astronomers regularly monitor known recurrent novae, they – and therefore professional astronomers – have a greater chance of observing the early stages of an eruption.

Four recurrent novae – T Coronae Australis, RS Ophiuchi, V3890 Sagittarii and V745 Scorpii – have red giant secondaries, with orbital periods of months. T Pyxidis is unique, among the recurrent novae, in its slowness.

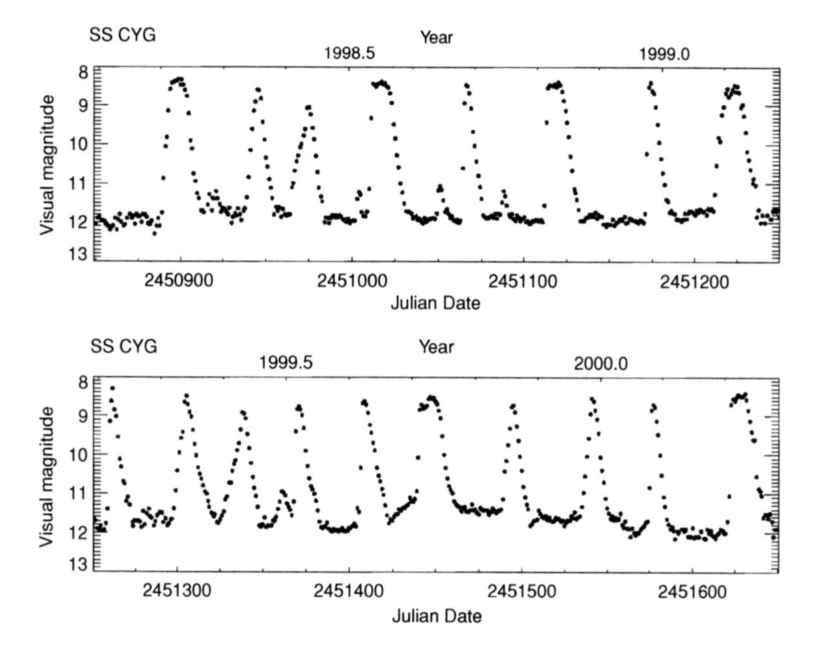

Figure 7.9 The light curve of SS Cygni, the prototype U Geminorum type of dwarf nova, based on visual observations from the AAVSO International Database. Note the amplitude, duration, spacing, and shape of the outbursts. (From AAVSO.)

Figure 7.8 shows the 50-year light curve of RS Ophiuchi. Three outbursts are visible. Another outburst occurred in early 2006. Note that, after an outburst, the star fades to a minimum, then slowly recovers.

7.2.3 *Dwarf novae*

Dwarf novae (DNe) are defined as hot dwarf variables which suddenly brighten by up to six magnitudes. They do so repeatedly, at irregular intervals of a few weeks. Although there is a reasonably clear-cut distinction between recurrent novae and DNe, there is at least one object – AI Com – which has a range of nine magnitudes, and a repetition time of about three years. The first DN – U Geminorum – was discovered by John Russell Hind on 15 December 1855. SS Cygni (figure 7.9) is another well-known CV. It was discovered at the Harvard College Observatory in 1896 by Louisa Wells, one of the many women employed by HCO Director, E. C. Pickering. It is probably the most extensively observed CV – about a quarter of a million measurements have been made by AAVSO observers alone. Several hundred DNe are now known, but only a few dozen have been studied in any detail. Table 7.1 lists several whose properties at minimum are reasonably well known.

Table 7.1. *Bright and/or interesting cataclysmic and related variable stars*

Name	Type	V Range	$P_{orbital}$ (d)	Comments
RX And	UGZ	10.3–14.0	0.209893	frequent outbursts
GK Per	NA + XP	0.2–14.0	1.9968	nova 1901
VW Hyi	UGSU	8.4–14.4	0.074271	close, bright, well-studied
U Gem	UGSS + E	8.2–14.9	0.176906	prototype 'normal' DN
SU UMa	UGSU	10.8–14.96	0.07635	prototype DN with superoutbursts/humps
Z Cam	UGZ	10.0–14.5	0.289841	prototype DN with 'still-stands'
T Pyx	NR	6.3–14.0	0.076223	outbursts 1890, 1902, 1920, 1944, 1966
ER UMa	UG:	12.4–15.2	0.06366	co-prototype: DN with short outburst cycles
SW Sex	E/WD+NL	14.8–16.7B	0.134938	prototype: edge-on CVs?
RZ Leo	NR	11.15–17.5P	0.0760383	co-prototype: DN with short outburst cycles
AM CVn	NL/AMCVn	14.10–14.18	0.011907	prototype helium-rich CV
UX UMa	EA/WD+NL	12.57–14.15	0.196671	prototype DN in permanent outburst
T CrB	NR	2.0–10.8	227.5687	outbursts 1866, 1946
U Sco	NR	8.7–19.3	1.230552	-
RS Oph	NR	4.3–12.5	455.72	outbursts 1898, 1933, 1958, 1967, 1985
DQ Her	NB+EA	–	0.193621	prototype intermediate polar
AM Her	AM/XRM + E	12.3–15.7	0.128927	prototype polar
WZ Sge	UGSU + E + ZZ	7.0–15.53B	0.056688	large amplitude, short period
V1974 Cyg	N	4.2–17.5V	0.08126	nova 1992
AE Aqr	XP	10.4–12.56B	0.411656	unique 'propellor star'
V1500 Cyg	N	1.69–21.0V	0.139613	nova 1975
SS Cyg	UGSS	7.7–12.4	0.27513	best-known DN
RU Peg	UGSS + ZZ:	9.0–13.2	0.3746	-
IP Peg	UG + E	12.0–18.6B	0.158206	-
VY Scl	NL	12.5–18.5P	0.1894	prototype NL with low states

Many of the features of the brightness variability of DNe can be gleaned from light curves such as shown in figure 7.9. In fact, most of what we know about the long-term photometric behaviour of CVs – such as frequency, amplitude, spacing between outbursts, and correlations between these quantities – comes from the work of amateur astronomers. CVs have characteristic ranges (3.5 magnitudes for SS Cygni) and characteristic intervals between outbursts (about 50 days for SS Cygni). There is a very approximate relation between these, in the sense that the longer the characteristic time between outbursts, the larger the outbursts. It is interesting, in fact, that DNe and recurrent novae fit a single relation between amplitude and time between outbursts: $A = 0.80+1.667\log(t)$ where t is in days. This relation, when extrapolated to the amplitude of novae, gives $T = 10^4$, which

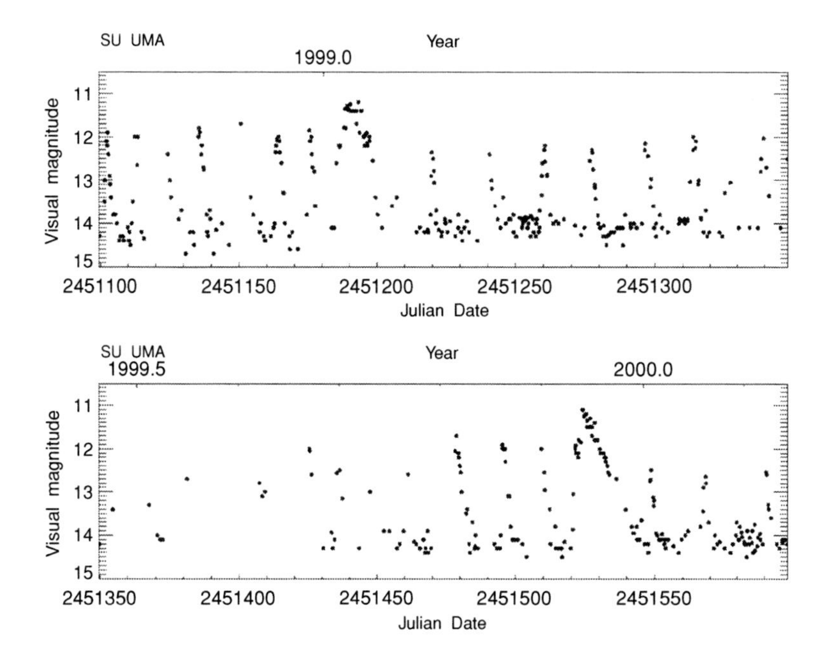

Figure 7.10 The light curve of SU Ursae Majoris, based on visual observations from the AAVSO International Database. SU Ursae Majoris is the prototype of dwarf novae which have occasional superoutbursts. Two are visible here, at approximate JD 2451200 and 2451530. Note the amplitude, duration, spacing, and shape of the normal outbursts and superoutbursts. (From AAVSO.)

is not inconsistent with current estimates. This may be coincidental, however, since the causes of the outbursts are totally different.

It should be emphasized that the outbursts are in no way periodic. In SS Cygni, for instance, the time between outbursts typically ranges from 30 to 70 days. The outbursts of SS Cygni (figure 7.9) are numbered consecutively (to our knowledge, none has been missed for many decades) and are classified A, B, or C according to the rapidity of rise to maximum and 1, 2 ... 10 according to the length of time the star spends above minimum during that outburst. There is some tendency for long outbursts to be followed by long intervals before the next outburst.

Stars which exhibit this behaviour are usually referred to as the U Geminorum subclass of DNe (figure 7.9) (though SS Cygni is sometimes considered the prototype, the classification of CVs is as complex as for other types of variables); in the GCVS4, DNe in general are classified as UG, and 'normal' DNe as UGSS.

A second subclass is the Z Camelopardalis stars (UGZ). These are much less common (by a factor of ten) than the U Geminorum stars. Their defining characteristic is a 'still stand' or halt in their light curve on the way down from maximum. The star may remain at this intermediate magnitude for many weeks. Z Camelopardalis stars also tend to have smaller amplitudes, and shorter times between outbursts, but this is only a tendency, not a rule.

There is a third subclass of DNe: the SU Ursae Majoris stars (UGSU: figure 7.10). Normally, their outbursts are similar to those of the U Geminorum stars, though the interval between outbursts is shorter than average (16 days for SU Ursae Majoris itself) and the orbital periods tend to be shorter (about 100 minutes) – except in SU Ursae Majoris itself. The defining characteristic of the SU Ursae Majoris stars is *supermaxima*: brighter, longer, rarer, and more regularly spaced maxima interspersed among the normal ones. In SU Ursae Majoris itself, the supermaxima are about a magnitude brighter than the regular maxima, last about 20 days instead of a week, and recur about every six months. Superimposed on the supermaxima are *superhumps*, which have periods which are 10 per cent greater than the orbital period. The fraction by which the superhump period exceeds the orbital period is a smooth function of the orbital period, which suggests that the superhumps are connected with orbital or dynamical phenomena. The superhumps appear to be associated with a precessing bright spot on the accretion disc around the star, though a universally accepted explanation of the supermaxima and superhumps is not yet available. The superhump period may change, and can be studied by the $(O - C)$ technique. If it is connected with a bright spot in the accretion disc, then the period might be expected to change if the accretion disc expands or contracts.

SU Ursae Majoris stars are sometimes divided into subtypes, including the WZ Sagittae stars which are extreme SU Ursae Majoris stars, and the RZ Leo Minoris or ER Ursae Majoris stars, which display extremely short superoutburst cycles, and spend up to half their time in outburst. These subtypes seem to differ in their average rate of mass loss. WZ Sagittae stars are sometimes called TOADS: Tremendous Outburst Amplitude Dwarf Novae. WZ Sagittae's giant outbursts are 20–30 years apart, typical of a recurrent nova; it is also the nearest CV at 43.5 parsecs.

Curiously, normal DNe such as SS Cygni (figure 7.9) have two distinct kinds of outbursts: more energetic ones and less energetic ones, but there is no evidence for superhumps on the former. Some of the brightest normal DNe (U Geminorum, RU Pegasi, SS Cygni, KT Persei) also show quasi-periodic oscillations during outburst, but these are rapid ($\sim$ 30 seconds) and transient.

In general, the importance of DNe is that they provide direct information about the nature of an accretion disc, and its behaviour during an outburst.

In addition to these rapid variations, and the variations associated with the outbursts, there are long-term variations associated with the magnetic cycles on the cool main sequence component. These cycles will slightly affect the rate of mass transfer which, in turm, will affect the brightness of the system at minimum, the interval between outbursts, and even the orbital period. Small changes in the latter can be monitored by precise timing of the eclipses, if eclipses occur.

One of the main reasons why the demand for AAVSO visual observations has *increased* by a factor of 25 since the birth of the space age is the contribution that visual observers can make in the discovery and study of dwarf nova outbursts. This is illustrated by figure 7.11. Visual observers noted an outburst in SS Cygni, and notified AAVSO headquarters which, in turn, notified astronomers using the EUVE and RXTE satellites, which observe in the far UV and the X-ray regions, respectively. The visual observers continued to monitor the star. The results (figure 7.11) lead to an understanding of how the outburst progresses, and why it occurs in the first place.

Spectroscopic variability

The spectroscopic variability of DNe resembles that of novae except that, at maximum light, their spectra tend to lack emission lines. The spectra at minimum, however, are similar to those of novae. Some of the most important spectroscopic studies of DNe have been made in the ultraviolet. During outbursts, the DNe show P Cygni line profiles, indicating that there is a high-velocity wind flowing at right angles to their accretion disc, and that the DN is losing mass. Spectroscopically, a dense accretion disc can mimic an A-type giant or supergiant star.

7.2.4 *Nova-like objects*

There are many objects whose spectra resemble those of post-novae, but which are not known to have undergone an eruption. Furthermore, these stars show the same types of brightness variability which novae and DNe show *at minimum*. Unfortunately, the class of nova-like objects is poorly defined, simply because it lacks any specific defining property such as an outburst. It occasionally includes the symbiotic stars, for instance. Conversely, many of the nova-like objects could be considered symbiotic stars.

Nova-like objects can be divided into a number of subclasses:

- *UX Ursae Majoris stars.* These are non-eruptive stars which photometrically and spectroscopically appear like DNe in a permanent state of eruption, or like Z Camelopardalis stars at standstill; the emission is probably powered by continuous high mass transfer from the companion star into a steady-state accretion disc, causing the disc

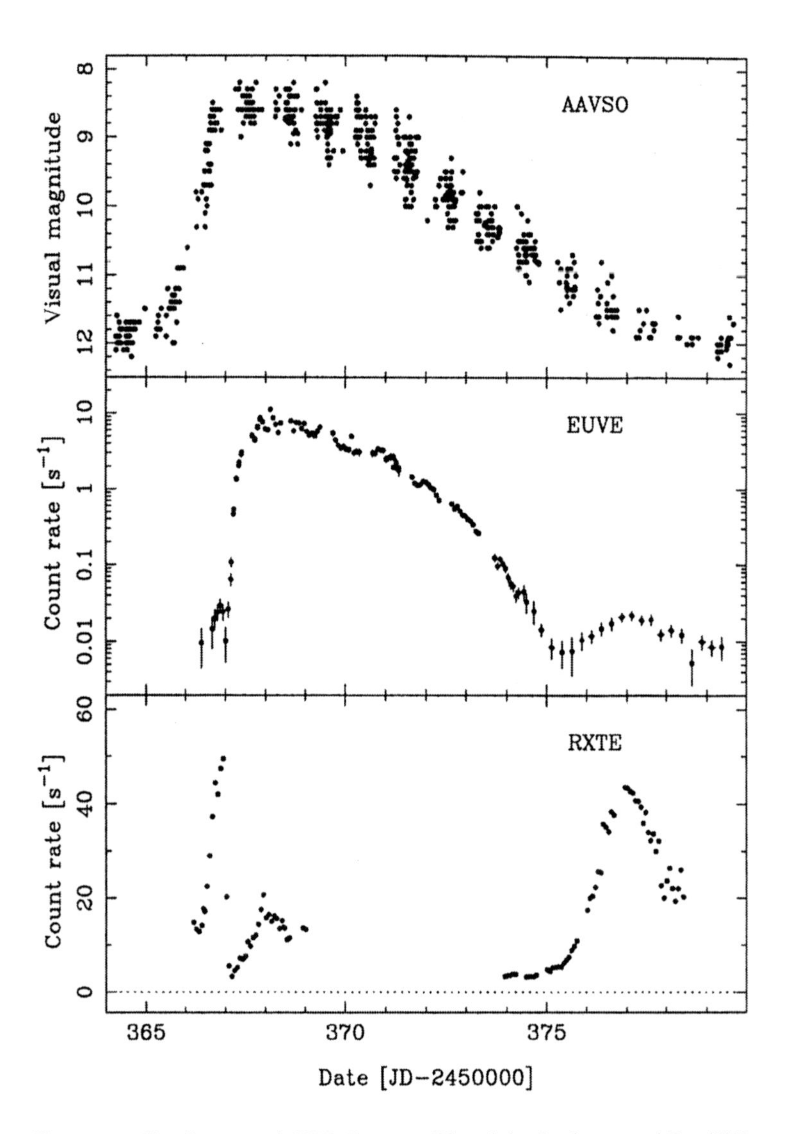

Figure 7.11 Simultaneous AAVSO, Extreme Ultraviolet Explorer, and Rossi X-Ray Timing Explorer observations of SS Cygni through outburst. The AAVSO observations made it possible for EUVE and RXTE to study the progress of the outburst from the beginning. (From J. A. Mattei, private communication.)

to appear very bright. They are generally classified as NL or ISA in the GCVS4. At present, however, the UX Ursae Majoris classification is not sufficiently precise to distinguish between other nova-like variables, and should perhaps be used to denote nova-like variables in general.

- *VY Sculptoris stars.* These show occasional deep minima or 'low states', lasting from weeks to months, apparently caused by a temporary halt in the mass transfer. Are they UX Ursae Majoris stars in which the mass transfer turns off? It is interesting to note that VY Sculptoris was once classified as an R Coronae Borealis star!

- *AM Herculis stars or polars.* These are the most complex, bizarre, and energetic of the nova-like objects. They are characterized by magnetic fields of millions of Gauss (over 200 MGauss in AR Ursae Majoris), which completely control the flow of matter in these stars. The matter flows from the companion star but, rather than forming a disc around the white dwarf, it flows down *accretion columns* on to the white dwarf's magnetic poles. The X-ray emission varies with the orbital period, as the accreting magnetic pole moves into and out of the field of view. The light emitted from these accretion columns is highly polarized, which provides a distinguishing characteristic for these objects. These stars are also sources of hard X-rays, produced by matter screeching to a halt at the bottom of the accretion column. In fact, they were discovered (in 1976) as a result of their strong X-ray emission. The long-term light curves of AM Herculis stars show 'high states' and 'low states' as in the VY Sculptoris stars.
- *Intermediate polars*, formerly DQ Herculis Stars. DQ Herculis was the first example discovered. In these systems, the magnetic field of the white dwarf is strong enough to affect the flow of matter, but not strong enough to control it entirely, except between the disc and the star, where the magnetic field tends to dominate. Presumably there is a continuous spectrum of objects from those in which the magnetic field has no discernable effect, to those in which it is dominant.

  DQ Herculis stars show coherent optical variability with periods of about 30 seconds. The coherence requires that the variations be caused by either pulsation or rotation; the latter turns out to be the case. The variations are sufficiently coherent that the spin-up or spin-down of the white dwarf can be measured by the $(O - C)$ technique. Note that the prototype is an old nova; it is the coherent variations in the star (discovered many years ago) which lend its name to this subtype of nova-like objects. Rapid coherent oscillations are also present in two UX Ursae Majoris stars, including the prototype, and in about a dozen dwarf novae during outburst.
- *Asynchronous polars* are systems in which the spin period of the white dwarf is slightly different from the orbital period. V1500 Cygni is a famous example.
- *AE Aquarii stars* or *propellor stars* rotate so fast that the spinning magnetic field actually expels the accretion stream from the system.

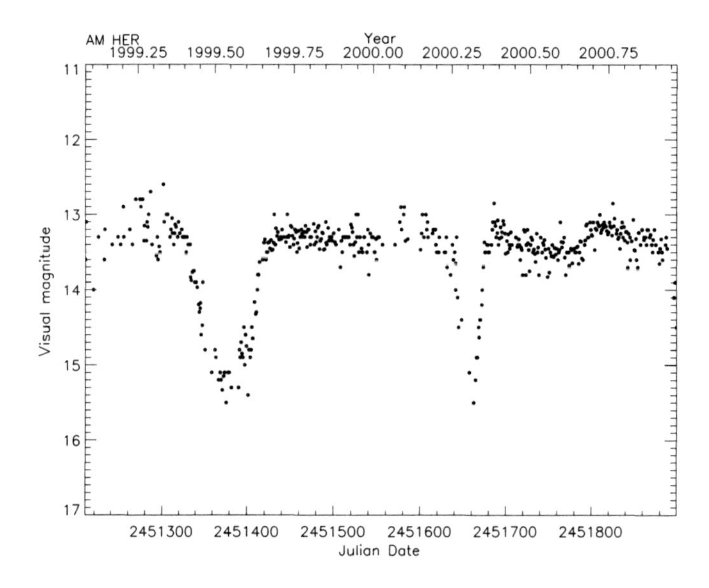

Figure 7.12 The light curve of the polar AM Herculis, based on visual observations from the AAVSO International Database. Note the amplitude, duration, spacing, and shape of the declines in brightness. (From AAVSO.)

- *AM Canum Venaticorum stars* are systems consisting of two white dwarfs, or a white dwarf and a helium star.

Polars are of special interest, because the gas flow is completely different from that in normal CVs. Figure 7.12 shows the light curve of a polar, including two minima. Figure 7.13 shows the currently accepted model for a polar.

7.2.5 *Rapid brightness variability in cataclysmic variables*

As we have seen, CVs display a remarkable variety of types of rapid variability. Virtually all of them flicker randomly, not only in light (figure 7.6) but in X-rays. This is mostly due to interaction between the mass transfer stream and the hot spot on the accretion disc. During outburst, many DNe oscillate coherently with periods of 10–30 seconds. The actual period may vary by several per cent, but the oscillations remain coherent for hundreds of cycles. Two UX Ursae Majoris stars show similar 30-second oscillations; this is not surprising, since UX Ursae Majoris stars resemble erupting DNe in other respects. Dwarf novae may show another type of variability during outburst: *quasi-periodic oscillations, or QPOs*: slower, more long-lasting, but less coherent than the above. Finally: the well-known ex-nova DQ Herculis displays a remarkably stable (one part in a trillion) 71-second oscillation, which, due to its stability, must be caused by either rotation or non-radial pulsation. No universally accepted explanation for

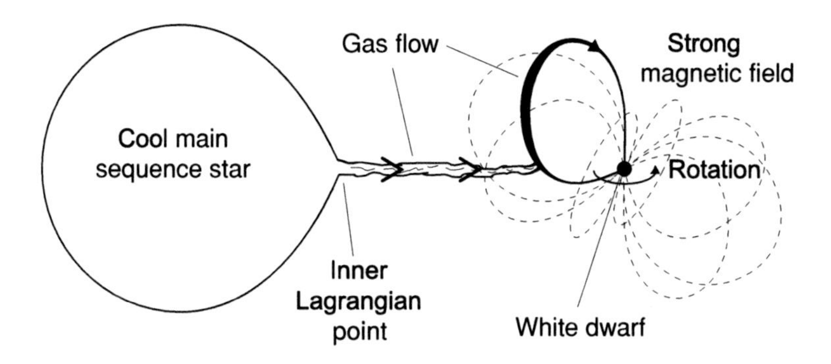

Figure 7.13 A schematic model for a polar. Material is transferred from the cool companion. It encounters the white dwarf's strong magnetic field, and is guided to the magnetic polar regions of the white dwarf. (Jeff Dixon Graphics.)

the 10–30 second oscillations is available, though the short periods require that they be produced on or near the white dwarf. The less coherent oscillations may be due to rotational effects (such as orbiting blobs) in the inner accretion disc. And many of the white dwarf components may be pulsating ZZ Ceti stars, though these variations may not be visible if the white dwarf's contribution to the brightness of the system is small.

Much important information about the many complex forms of CV variability have come from the *Center for Backyard Astrophysics*, a multi-longitude network of small telescopes, mostly operated by amateur astronomers. The co-ordinator is Joseph Patterson, Columbia University, USA. One of the most recent publications from the group (Patterson *et al.*, 2003) presents the results on 20 stars – a veritable factory!

7.2.6 *The Close Binary Model for Cataclysmic Variables*

The fundamental discovery regarding the nature of CVs was made by Merle F. Walker in 1954; he discovered that DQ Herculis was a binary system with a period of 4.6 hours. Further work by Robert P. Kraft and others in the 1960s suggested that *all* CVs were binary systems with the following set of properties:

- Periods of mostly 1–12 hours but with a marked deficiency in the 'period gap' of 2.2–2.8 hours.
- The primary component is a white dwarf with a radius about 0.01 $R_\odot$, and a mass of about 0.5 $M_\odot$.
- The secondary component is a G-M type star, usually near the main sequence, with a mass of 0.1–1 $M_\odot$, which fills its Roche lobe, is therefore slightly distorted from a spherical shape, and transfers matter through the inner Lagrangian point towards the primary, at a

rate of about 10^{-9} to 10^{-10} $M_\odot$ a year. The G-M type component *must* fill its Roche lobe; this requires that the period is short, and also results in a relationship between the period, and the mass and radius of the secondary.

- An *accretion disc* of material around the white dwarf, produced by the material (and its angular momentum) flowing through the inner Lagrangian point.
- A *hot spot* on the accretion disc, produced by material impacting the accretion disc. The accretion disc and the hot spot produce most of the light from the system, including H, He, and Ca II emission lines which are seen in CVs when they are not erupting.
- X-ray and extreme ultraviolet emission, due to the high temperature of the material falling into a strong gravitational field. Most of this emission comes from the inner edge or *boundary layer* of the accretion disc, near the white dwarf.

The close binary model for CVs is shown schematically in figure 7.3.

It is interesting to follow the reasoning which led up to this picture. The late-type component is usually visible in the post-nova spectrum. The presence of the white dwarf is inferred from the radial velocity variations of the late-type component and of the 'hot component' – the source of the emission lines and the blue continuum, which turns out to be the accretion disc. The white dwarf is not easily visible in the *visual* spectrum of the system, because it is faint, but it often dominates the *UV* spectrum. Its nature is also revealed by the rapid oscillations which are seen in the light from some CVs (the DQ Herculis stars); such oscillations could only come from a white dwarf.

Some CVs show periodic brightness variations at minimum. These include a periodic hump, and a periodic minimum which is interpreted as an eclipse – not of the white dwarf, but of the hot spot on the accretion disc. As with 'standard' eclipsing binaries, these eclipsing CVs provide much information about the geometry of the accretion disc as the limb of the cool star scans across the system: the disc is very thin, especially near the white dwarf; it is hottest and brightest in the centre and at the hot spot; it is coolest and faintest at the edge. Astronomers use increasingly sophisticated methods for modelling these eclipsing systems. The accretion disc is the source of most of the emission lines of H, He, and Ca II, and of most of the blue continuum. The rest comes from the hot spot. The hot spot is formed by a shock front on the accretion disc, formed where matter impacts the disc on its way from the inner Lagrangian point. The hot spot is thermally unstable, and flickers on a time scale of minutes. When the hot spot is presented so that it is most visible to the observer, its brightness is greatest, and the flickering is most intense; this is the origin of the hump in the light curve. The eclipse provides further evidence for the hot spot, since it is the hot spot which is eclipsed. During the eclipse, the flickering ceases, since the source of the flickering is in eclipse. *SW Sex variables* are nova-like systems which are seen edge-on, so that eclipse phenomena are especially noticeable.

The close binary model seems to apply to all classes of CVs. There appear to be no systematic differences between the properties of the binary systems in novae, U Geminorum and Z Camelopardalis stars, except that SU Ursae Majoris and AM Herculis stars seem to have significantly shorter periods than other subtypes. What, then, determines the type of variability which will occur in a given close binary system? Various properties of both the primary and secondary components are undoubtedly involved, but the exact combination has not yet been identified.

To understand the origin of CVs, it is well to review the discussion of equipotential surfaces, Roche lobes, and the evolution of binary systems in general. The white dwarf component of the CV has already gone through its evolution. The late-type component fills its Roche lobe, perhaps due to evolutionary expansion, but also due to the strangling effects of *magnetic braking* and *gravitational radiation* and other losses of angular momentum in the system. This release of energy and angular momentum acts to shrink the binary system, and eventually to destroy it. Thus, as John Faulkner has stated, a CV consists of 'two stars in mortal embrace'.

At this point, it is worth asking how two stars ended up in this predicament. First of all, the majority of stars are members of binary systems. As long as the separation is less than a few hundred solar radii then, as soon as the more massive star evolves and swells up into a red giant, it will start to transfer mass to its companion, the Roche lobes will overfill, the stars will effectively orbit within a common envelope, and friction will cause them to spiral inward. By this point, the originally more massive star is a white dwarf, and its companion is still a more-or-less normal star.

If the separation is now such that the period is greater than half a day, then the cool star is detached, and nothing interesting will happen – at least for the present. If the separation is smaller, then there will be Roche lobe overflow, mass transfer, an accretion disc, and CV phenomena will occur. Meanwhile, the cool companion will be affected by *magnetic braking*, as has occurred in stars such as the sun. The orbit will gradually shrink, and the period will decrease, producing CVs with periods of 12 down to three hours. At that point, something seems to happen (or not happen), because we observe a large deficit of CVs in the *period gap* from three to two hours (figure 7.14). Apparently the cool companion

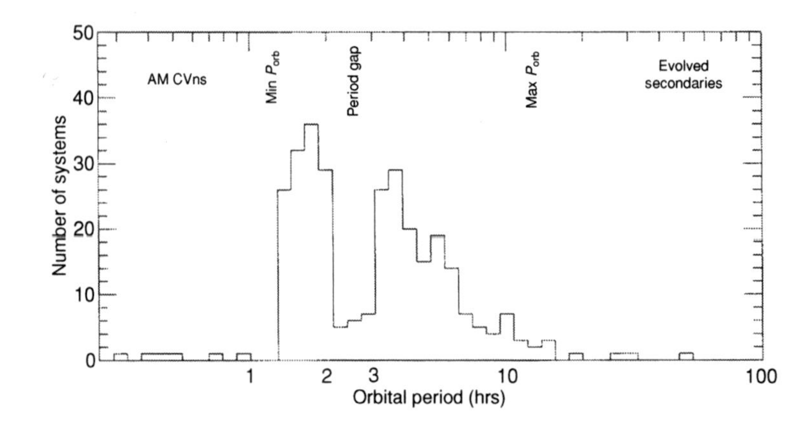

Figure 7.14 The distribution of the orbital periods of CVs, showing the long-period cutoff at about ten hours, the period gap between two and three hours, and the low-period cutoff at about 80 minutes. (From Hellier, 2001.)

adjusts its interior structure, and Roche overflow does not occur again until the orbit has shrunk to a two hour period. Now, the shrinkage of the orbit is driven primarily by *gravitational radiation*. The orbital motion of these close masses produces waves in the fabric of space-time; these waves carry off energy at the speed of light. (Gravitational radiation was first detected, indirectly, from the 'double pulsar' (section 4.7).)

By the time that the orbit has shrunk to a two hour period, the internal structure of the cool star has changed: it has become degenerate like a white dwarf. White dwarfs become *smaller* when they become more massive; at the *Chandrasekhar limit* of 1.44 solar masses, their radius approaches zero. So as the mass of the cool star decreases through mass loss, its radius *increases*, leading to a slightly longer period. But by this time, the mass of the cool star approaches 0.01 solar mass, gravitational radiation and mass transfer wane, and the system becomes faint and almost undetectable.

An exception are the *AM Canum Venaticorum stars*, which can have even shorter periods. In these systems, the companion is not a normal star but a helium star, which can become even smaller and more compact than a normal star.

7.2.7 *Novae and Dwarf Novae: the cause of the cataclysm*

We now have a good idea of the nature of the stars which are cataclysmic variables. But what produces the cataclysm?

In the case of the novae, the cause is reasonably well understood: it is a runaway thermonuclear reaction in the outer layers of the white dwarf. Hydrogen-rich material is transferred from the late-type component through the inner Lagrangian point. The material strikes the accretion disc (which probably has a mass of about 0.0001 $M_{\odot}$) and eventually spirals down on to the equatorial regions of the white dwarf.

The white dwarf consists of a thin layer of hydrogen rich material, overlying a core of heavier elements which are degenerate (their electrons are no longer attached to atoms, and they fill all available states) and isothermal (they have a constant temperature throughout). As more hydrogen-rich material is added to the surface of the white dwarf, the temperature at the base of the hydrogen-rich layer climbs to 2×10^7 K, as the accreted material compresses and heats the gas. This value is reached after about 0.0001 $M_{\odot}$ of material has been transferred; this requires 10^5 to 10^6 years. At this temperature, a runaway thermonuclear reaction begins. The nature of the reactions depends on the temperature, and on the abundances of C, N, and O in the surface layers of the star. Hydrogen reacts with carbon, and a series of reactions occur in which various isotopes of C, N, O, and F are produced. At the peak of the runaway, energy is produced at a prodigious rate, resulting in a total luminosity of close to 100 000 $L_{\odot}$ – precisely the value which is observed. The speed of the runaway is limited by the speed at which hydrogen can be cycled through the series of reactions, which eventually produces He and great quantities of energy. As a result, only a small fraction of the thermonuclear energy in the envelope material is used before it is blasted into space.

The runaway proceeds quickly enough, however, so that the outer layers of the star initially have no time to expand – they heat instead to 10^5 K. They then expand explosively, as more and more energy is fed into them from below. The intense radiation from the hot surface of the star aids in the expansion. Matter streams off the surface at several hundred km s^{-1} and the photosphere of the star – the layer at which the gas becomes thick enough that we cannot see below it – expands from a radius of 0.01 to about 20 $R_{\odot}$, now mimicking a supergiant star but with a rapidly expanding surface. Most of the energy of the nova is contained in this ejected material; only a small fraction appears in the form of radiation.

The nova continues to generate and release energy at close to 100 000 $L_{\odot}$ as the surface layers of the star stream away. As more and more of the surface layers peel off, the photosphere of the star appears to recede (though in fact its gases are streaming rapidly outward), and the star appears to get smaller and fainter. After most of the hydrogen-rich material (about 0.0001 $M_{\odot}$) on the surface of the white dwarf has blown away, the thermonuclear reactions eventually shut down. This may take up to ten years.

Attempts are being made to explain the different speed classes of novae. According to theoretical calculations, a fast nova requires that the white dwarf be massive (about one solar mass) and that the accreted material be enriched in

C, N, and O to provide a catalyst for the rapid burning of H via the CNO cycle. The observational results are not inconsistent with this prediction, but there are so many factors which may affect the outburst – mass and composition of the white dwarf, mass transfer rate – that it is difficult to draw any general conclusion.

The explanation for the outbursts of DNe is rather different. Dwarf novae appear to have smaller mass transfer rates (10^{-10} $M_\odot$/year), so, in a fraction of a year, barely 10^{-11} $M_\odot$ is transferred – much too much to ignite a thermonuclear reaction. We also observe that little or no mass is ejected in a DN outburst.

The outbursts of DNe are due to thermal instabilities in the accretion disc. The late-type component, and probably the white dwarf, can be ruled out as the site of the outburst. John Faulkner has pursued the idea that the outbursts are due to the effects of mass transfer on to the accretion disc. He finds two forms of behaviour, depending on the mass transfer rate: a 'cycling' behaviour typical of U Geminorum stars, and a steady-state behaviour typical of the stillstands in the light curves of Z Camelopardalis stars (and of the normal state of UX Ursae Majoris stars). In the U Geminorum stars, the thermodynamic properties of the gas in the accretion disc can change, such that the stored gas (and its gravitational potential energy) in the accretion disc can be dumped rapidly on to the white dwarf (Osaki, 1974). This is yet another example of how the microscopic properties of atoms can result in macroscopic effects such as the eruption of a dwarf nova.

In the Z Camelopardalis stars, the periods are 3–10 hours, and the rate of mass transfer is close to the boundary between steady accretion, and cyclic outbursts, so the star can alternate between these behaviours.

The cause of the outbursts of recurrent novae is less clear. Ronald Webbink has suggested that the outbursts may be due to discrete mass transfer events caused by thermal instabilities in the late-type component, which in recurrent novae is usually a giant (hence the brighter absolute magnitudes of these objects). Michael Shara has proposed an alternative mechanism based on the ignition of He on a high-mass, low-luminosity white dwarf. In view of the heterogeneous nature of recurrent novae, it is possible that all of these hypotheses – and others – may be correct! The most recent and comprehensive series of models of nova outbursts (Yaron *et al.*, 2005) shows that it *is* possible to produce a recurrent nova in only 10–20 years, via accretion and a thermonuclear runaway. In this way, the recurrent novae would be close cousins of ordinary novae.

The SU Ursae Majoris phenomenon has been ascribed to resonance effects in the disc. This takes place in systems with very short orbital periods

(1–3 hours). If mass and angular momentum build up in the disc, it expands until the outer part of the disc is in a resonance with the orbiting secondary. The disc then collapses, dumping an unusually large amount of mass and energy on to the white dwarf. In the process, the accretion disc becomes eccentric, and precesses or rotates in space. This leads to a variation in the amount of energy produced as the gas stream impacts the disc at the hot spot. Because the disc is precessing, the period of the superhumps is slightly different than the orbital period.

Finally, we should note that some stars can be both a dwarf nova *and* a nova, and may exhibit a wide range of different forms of variability!

All in all, cataclysmic variables provide a fascinating and important picture of the accretion process – a process which also occurs in forming stars and, on a grander scale, in active galactic nuclei. Spectroscopic and photometric techniques such as Doppler and eclipse mapping have revealed the detailed structure of the stars and especially the accretion disc.

7.2.8 *Symbiotic stars*

A symbiotic star is one whose spectrum shows the simultaneous presence of features from a cool star (such as absorption lines from molecules) and from a hot object (such as high-excitation emission lines). Observed at short wavelengths, they appear as hot stars; observed at longer (near-infrared) wavelengths, they appear as cool stars! The first examples (CI Cygni, RW Hydrae and AX Persei) were noted in 1932. Symbiotic stars turn out to be binary systems consisting of a cool giant (usually M type) and either a hot main sequence star, or more usually a white dwarf with an accretion disc. There is a tendency to include, in this class, any interacting binary with a hot component and a cool one; this certainly complicates the discussion. Variable symbiotic stars (and almost all of them are variable) are called *Z Andromedae* stars. The variability of this star was discovered around 1900. The variability of the Z Andromedae stars is so diverse and complex that it could be discussed in several different chapters of this book: there may be eclipses of one component by the other (so they are related to the VV Cephei stars); the M giant may be a pulsating variable (so EG Andromedae, which I consider to be a garden-variety pulsating red giant, which happens to be a binary, is included in this class); there may be flickering or eruption from the accretion disc as in a CV, if mass transfer is taking place from the cool star to the hot object. In fact, the recurrent nova T Coronae Borealis is often listed as a Z Andromedae variable. Table 7.2 lists a few of the brighter Z Andromedae variables. Figure 7.15 shows

Table 7.2. *Bright and/or interesting symbiotic stars*

Name	V Range	Spectral Type	Period (days)
Z And	10.53 (7.0–12.0)	M2III + B1eq	756.85
EG And	7.23 (7.08–7.80)	M2IIIe	482.57
R Aqr	7.69 (5.8–12.4)	M7IIIpe + pec	44yr?
T CrB	–	–	–
CH Cyg	8.84 (5.6–8.49)	M7III + Be	5750
CI Cyg	11.1 (9.9–13.1)	M5III + Bep	855.25
AR Pav	10.57 (7.4–13.62)	M3III + shell + cont.	605
AG Peg	8.65 (6.0–9.4)	M3III + WN6	816.5
BL Tel	7.20 (7.09–9.41)	M + F5Iab/b	778.6
RR Tel	6.5 (6.5–16.5)	Mira? + pec	

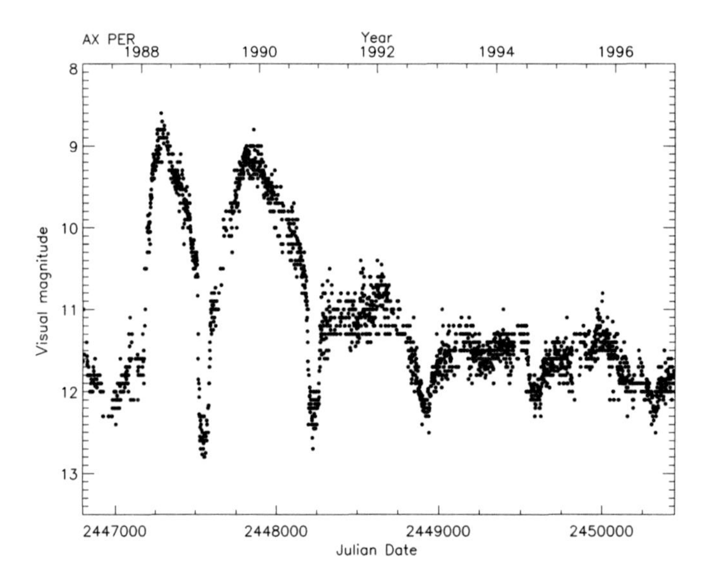

Figure 7.15 The light curve of AX Persei, based on visual observations from the AAVSO International Database. Two aspects of the variability can be seen: a pronounced brightening at the start, and eclipses separated by the binary period of 682.1 days. (From AAVSO.)

the light curve of a symbiotic star – a complex interplay between binarity and eruption.

There is also an interesting class of *symbiotic novae* – eruptive variables consisting of a red giant (often a pulsating variable) and a hot, compact object, usually a white dwarf. They erupt every few centuries, then decline over a period of years to decades. Examples include V1016 Cygni, HM Sagittae, FG Serpentis, RT Serpentis, RR Telescopii, and most recently PU Vulpeculae.

7.3 Supernovae

A supernova is a star which, within a few days, brightens by 10 to 20 magnitudes, reaching a peak absolute magnitude of -15 to -20 or brighter, then slowly fades. In the process, the star is irreversibly transformed into a rapidly expanding shell (a *supernova remnant*) and/or a collapsed core – a neutron star or possibly a black hole. Despite a superficial resemblance between novae and supernovae, the eruption mechanism – and the consequences – are totally different. A supernova explosion 'is forever'.

The supernova is perhaps the most important and complex phenomenon in stellar astrophysics, and it would be difficult to do justice to all its facets in this short section. The supernova is the terminal phase in the evolution of the rare, massive stars. The core of the star, exhausted of nuclear fuel, collapses to nuclear density. The collapse takes a second or two but, during this time, gravitational energy is being released at a rate which exceeds the luminosity of all the other stars in the visible universe. In some binary systems containing a white dwarf, a supernova may occur if the white dwarf accretes enough mass to approach the Chandrasekhar limit; a runaway thermonuclear explosion then leads to the demise of a less massive star as well. Supernovas are the source of *pulsars* (section 4.7), which are rapidly spinning neutron stars which emit pulses of electromagnetic radiation; the pulsars also accelerate some of the *cosmic rays*, high-energy particles which fill our galaxy and impinge on our upper atmosphere. Supernovae disperse heavy elements, synthesized in the precursor star or in the supernova explosion, into interstellar space, where they can become part of the next generation of stars and planets, and – in the case of our planetary system – of life. The supernova remnant, expanding at thousands of km s^{-1}, churns up the interstellar medium and helps to trigger the formation of stars and planets, as part of 'galactic ecology'. Supernovae, at maximum brightness, are the brightest 'normal' stellar objects. They can be seen, even in distant galaxies and, since their maximum absolute magnitude can be calibrated, they can be used as distance indicators, far out into the universe. As standard candles, they were instrumental in one of the most exciting and profound discoveries of the last decade – the discovery of the accelerating universe, and of *dark energy*.

We shall concentrate on the brightness, colour, and spectroscopic variability of supernovae, and on the processes which cause them. We should not, however, lose sight of the many other aspects of supernovae, and the many ways in which they are important in astronomy

There is a catalogue of supernovae at:
www.sai.msu.su/sn/sncat/

7.3.1 *Discovery*

The brightness changes in a supernova are so striking that they would be easy to detect and identify, provided that the supernova was bright enough, and the observer knew where to look. There is some difficulty in distinguishing novae from supernovae from light curves alone, but the spectra are quite different, as are the absolute magnitudes – fainter than −10 for novae, and brighter than −15 for supernovae. Nevertheless, there are a few peculiar objects such as η Carinae which tend to blur the division – at least observationally.

Supernovae can be conveniently divided into *Galactic* supernovae (in our galaxy) and *extragalactic* supernovae. There is probably a supernova in our galaxy a few times each century, but most of them are obscured by distance and by interstellar absorption. Those that are visible at all have a good chance of reaching naked-eye brightness, but no supernova has been visually observed in our galaxy since 1604. In that year, Johannes Kepler discovered and studied a 'new star' which rose to rival Venus in brightness. A few years previous, in 1572, Tycho Brahe had discovered and studied a similar object, with similar brightness. Tycho's and Kepler's supernovae were instrumental in overthrowing the Aristotelian view of the universe, which included the belief that the stars were unchanging; they came along at just the right time! Prior to 1572, several 'guest stars' had been recorded by astronomers in East Asia. Some of these discoveries can now be identified with known supernova remnants, the most famous being the Crab Nebula. The supernova giving rise to this nebula was observed by astronomers in East Asia in 1054 AD; its magnitude was about −4. An even brighter supernova (magnitude −9) was observed in 1006 AD. It too was not recorded in Western records. The interpretation of the records of these 'historical supernovae' is a fascinating and challenging task, both for historians and for astronomers. SN 1987A, in the Large Magellanic Cloud, reached naked-eye brightness in 1987.

The mode of detection of the next Galactic supernova depends on when it appears. It might be discovered by an amateur astronomer, in a systematic visual search for novae or supernovae. However, sophisticated survey telescopes, looking perhaps for near-earth comets or asteroids, are quickly taking over. If the supernova was distant and heavily obscured, it might be discovered (probably accidentally) through its radio emission. With the advent of neutrino observatories such as the Kamiokande and Sudbury Neutrino Observatories, it might be discovered through the burst of neutrinos that it emitted.

The first extragalactic supernova to be recorded was S Andromedae, in 1885. Located in M31, the Andromeda galaxy, it was first thought to be a nova since, at that time, there was no obvious distinction between objects such as Tycho's and Kepler's stars and ordinary novae. If it were a nova, its absolute magnitude would be -10 at most, and its distance would be a few thousand parsecs. This episode helped to mislead astronomers into believing that objects like M31 were nearby objects in our own galaxy, rather than galaxies in their own right.

By the 1930s, the difference between novae and supernovae was becoming clear, and a group led by Fritz Zwicky (who coined the term 'supernova' and pioneered the understanding of them) began a systematic search for supernovae: a photographic survey of 4000 galaxies using a 0.45m Schmidt telescope. Any supernova discovered could then be studied spectroscopically using the 2.5m Mt. Wilson reflector. After World War II, the survey continued with the 1.2m Schmidt telescope on Mt. Palomar. By 1967, a total of over 200 supernovae had been discovered (at an average cost of about $1000 in 1967 funds), most of them by Zwicky's group. Some galaxies are especially rich in supernovae; NGC6946 has had four.

Interest in supernovae heightened in the 1980s – partly because of the widespread availability of CCDs, and partly because of the excitement surrounding the appearance of SN 1987A. It was also increasingly apparent that one type (Ia) of supernova had a very uniform peak luminosity, so they could be used as standard candles. The availability of CCDs made it possible to discover and observe these supernovae, at greater distances (or with smaller telescopes). Supernova hunting with CCDs became popular with amateurs, and with educators looking for exciting student projects. It became possible to compare a CCD image of a galaxy directly with an image stored in the memory of a computer. Supernova searches could be completely automated. Newly discovered supernovae could then be observed spectroscopically with large telescopes, to confirm their nature. These amateur discoveries continue; for instance, Tim Puckett (USA) and his team have discovered 109 supernovae, as of 2005.

Nevertheless, the most remarkable supernova hunter was not an automated telescope with a CCD camera, but a visual observer – Rev. Robert Evans, in Australia. With small telescopes, he monitored hundreds of galaxies visually, and discovered about two dozen supernovae. Because his observing procedure was systematic and well-documented, his *negative* observations (i.e. the absence of supernovae in the galaxies, as well as the presence) could be used to determine the frequency of occurrence of supernovae in galaxies of different types.

More recently, supernova discovery and observation has occurred in three different distance ranges. One of the key projects of the HST was to determine the distance scale of the universe. For relatively nearby galaxies, that could be done using Cepheids and other kinds of stellar standard candles. For more distant galaxies, supernovae had to be used, since they are the brightest stellar objects known. But supernovae had to be discovered and measured in the nearby galaxies whose distances could be independently determined (using Cepheids, for instance), in order to determine their peak absolute magnitudes. This could be done with relatively small telescopes. The distances of more distant galaxies, combined with their radial velocities, also enabled the Hubble constant to be determined. By the 1990s, HST had determined the Hubble constant to be about 73 km s^{-1} Mpc^{-1}, with an accuracy of better than 10 per cent.

The third step was to discover and measure even more distant supernovae. The expansion of the universe should be slowly decelerating (so it was thought), due to the gravitational force of galaxies on each other. This should be measurable by looking back in time by looking far out into space – to the most distant supernovae accessible to the largest ground-based telescopes, and to HST. Astronomers had discovered that the small differences in peak luminosity of Type Ia supernovae were correlated with the shapes of their light curves, so, by observing the light curves over several weeks, the peak apparent magnitudes could be determined or inferred. Then the absolute magnitude could be inferred from the calibration, and the distance could be calculated from the inverse-square law of brightness. Type Ia supernovae became the ultimate 'standard candle' – bright and, with the 'stretch correction' applied, uniform in peak brightness.

This was not an easy task. The supernovae were distant, faint, rare, and transient. Much telescope time had to be expended in searching for the supernovae, using the largest possible telescopes, and the widest-field detectors possible. The discoveries had to be followed up with frequent observations, to determine the light curves (photometrically), and to classify the supernovae (spectroscopically). They were beyond the reach of small telescopes and amateurs – even Rev. Evans.

But the results of this work were remarkable: as a result of two independent search projects – the High-Redshift Supernova Project (e.g. Schmidt *et al.*, 1998), and the Supernova Cosmology Project (e.g. Perlmutter *et al.*, 1998) – it turned out that the expansion of the universe was *accelerating*, not decelerating. There must be some process which exerts a repulsive force, proportional to the separation of the galaxies. According to theoretical physics, there are several possibilities. These range from Einstein's 'classical' cosmological constant, to possible effects of parallel universes in the new brane-world theory. New observations of very distant supernovae are rapidly converging to a solution – and undoubtedly to new questions. The understanding of the dark energy that drives the acceleration is one of the hottest topics in cosmology today.

For this kind of research, it is not actually necessary to calibrate the peak absolute magnitude of the supernovae, and determine their distance. It is

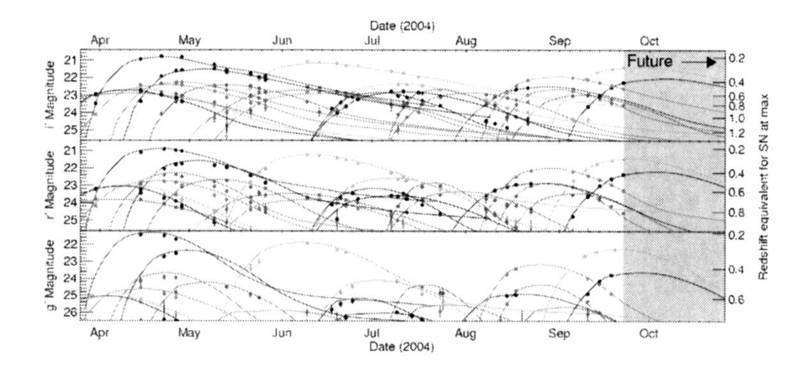

Figure 7.16 Light curves of a large number of supernovae in the *gri* photometric system, discovered by the CFHT Supernova Legacy Survey. Regions of the sky are sampled regularly, enabling reasonably complete light curves to be obtained. These can be used to classify the supernovae, and to determine the apparent magnitude at maximum. As supernovae are discovered photometrically, spectra can also be obtained to confirm the classifications. (From Mark Sullivan and the Supernova Legacy Survey.)

sufficient to compare the peak *apparent* magnitudes of a specific type of supernova (usually Type Ia) at small and large redshifts.

Now surveys of supernovae are being extended to the faintest, most distant supernovae, in order to better understand the cause of the deceleration. The largest is part of the Canada–France–Hawaii Legacy Survey; it is on the way to discovering about 700 distant supernovae (figure 7.16). A similar but somewhat smaller survey is *ESSENCE* – so-called after the name of one possible form of 'dark energy'.

7.3.2 *Brightness variability*

Supernova light curves are generally characterized by a rapid rise, a rounded maximum, and a more-or-less linear decline in magnitude. Based on the detailed shape of the light curve, supernovae can be divided into Type I and Type II. For supernovae of a given type, the light curves are very similar, especially Type I. In fact, if the type is known, fragmentary observations can be 'fitted' to a standard light curve. This is helpful, for instance, if a supernova is discovered after maximum, and the observer wishes to determine, in retrospect, the magnitude at maximum.

A few supernovae have light curves which do not appear to fit these two types, and Zwicky has classified them as Types III, IV, and V, but it is not clear whether these types have any fundamental physical significance. Type V, typified by SN 1961V, was probably a luminous blue variable like η Carinae, and Types III and

IV were probably peculiar Type II supernovae – ones now called Type IIn with narrow hydrogen emission lines. Given the wide range of possible properties of stars which could explode as supernovae, it is not surprising that there is a range of light curve shapes.

Initially, there was evidence that Type I supernovae could be divided into 'fast' and 'slow', perhaps representing different sub-populations of stars, and that Type II supernovae could be divided into those with a plateau on the decline portion of the light curve, and those without. However, it was difficult to draw any firm conclusions, especially when the differences between the subtypes were subtle. Spectroscopic observations are necessary for a meaningful classification.

The linear decline of the magnitude is of some interest because, since magnitude is a logarithmic scale, a linear decline in magnitude corresponds to an exponential decline in luminosity ($L \sim L(0)e^{-t/\tau}$, where τ is the *e-folding time* which is related to the *half life* $\tau_{1/2}$ by $\tau = 1.443\tau_{1/2}$). For Type II supernovae it is 100–200 days. Astronomers have long suspected that this exponential decline occurs because, long after maximum, the luminosity of the supernova is powered by the decay of radioactive isotopes. Spectroscopic observations of SN1987A confirmed that this is the case.

The colour variability of a supernova is even more complicated. At maximum brightness, the supernova radiates more or less like a normal B or A type supergiant star, so (B–V) is about −0.2 to 0.0, plus any interstellar reddening. After maximum, the spectrum of the supernova is a complex superposition of emission lines, which ultimately determine the colour. And different types of supernovae have different kinds of emission lines. It is also important to know both the observed and the intrinsic colour of the supernova, because the amount of reddening by dust can be used to determine the amount of absorption or dimming. This is essential, if the supernova is to be used for distance determination. And if the supernova is very distant, the spectrum will be reddened by the cosmological red shift, and this will also affect the colour.

7.3.3 *Absolute magnitude at maximum*

Perhaps the most important property of the light curve is the magnitude at maximum, because the *absolute* magnitude at maximum is roughly similar for supernovae of the same type, so measurement of the apparent magnitude at maximum establishes the distance.

The absolute magnitude at maximum is best calibrated by using supernovae observed in relatively nearby galaxies whose distance can be determined by other means – Cepheid variables, for instance. For a few more supernovae, the observed properties of the supernova (including the light curve) can be fit to theoretical models, or a variant of the Baade–Wesselink method can be used. This method, called the *expanding photosphere method*, has been applied to about two dozen supernovae. The spectrum provides an estimate of the expansion velocity in linear units. The observed flux and the temperature (from the colour), measured over at least two epochs, provides an estimate of the expansion velocity in angular units. The ratio of these gives the distance.

For supernovae in more distant galaxies, the distance can be obtained from the Hubble relation, which states that, for extra-galactic objects, the distance D is related to the velocity of recession V by $D = V/H$ where the distance is in megaparsecs, the velocity is in km s^{-1}, and H is in km s^{-1} Mpc^{-1}. H is now known to be about 73 ± 4 in these units, thanks in part to a key project of HST. (Technically, this is H_0, the present value of H, since H is variable with time.)

Current values for the absolute magnitude at maximum, in B, are about -19.3 ± 0.1 for Type Ia supernovae (section 7.3.5).

There is some scatter in the values of the absolute magnitude at maximum, but, as mentioned above, this is correlated with the shape of the light curve – specifically with the rate of decline after maximum (figure 7.17) – so the peak absolute magnitude can actually be determined to an accuracy of about 0.17 mag. The uncertain effect of dust adds another 0.11 mag to the uncertainty. If supernovae are to be used for 'precision cosmology', then it is important to know the absolute magnitude as precisely as possible, including how it depends on the properties of the progenitor, and the nature of its surroundings. Great progress has been made in this area.

7.3.4 *Frequency and distribution*

Up to 1980, about 500 supernovae had been discovered, but only about a quarter of them had been classified. These formed the basis for the first statistical analyses designed to determine the number of supernovae per year of each type, in galaxies of different types and luminosities. Such analyses are fraught with problems. They assume that a complete sample of galaxies has been searched, and that no supernovae have been missed due to obscuration – a doubtful assumption, especially if the galaxy is seen edge-on. Supernovae may also be missed if they are seen against the bright nucleus of the parent galaxy; this was especially true for photographic searches. The outcome of these searches is given, for instance, by van den Bergh and Tammann (1991), in units of supernovae per century per 10^{10} solar luminosities. It is noted that Type Ia supernovae occur in all types of galaxies, though more frequently in spiral and irregular galaxies; type II supernovae occur only in spiral and irregular galaxies, especially those which are most gas-rich. This is consistent with the general belief that Type Ia supernovae occur among older populations of stars, whereas Type II supernovae occur among massive young stars. Very recent results suggest that

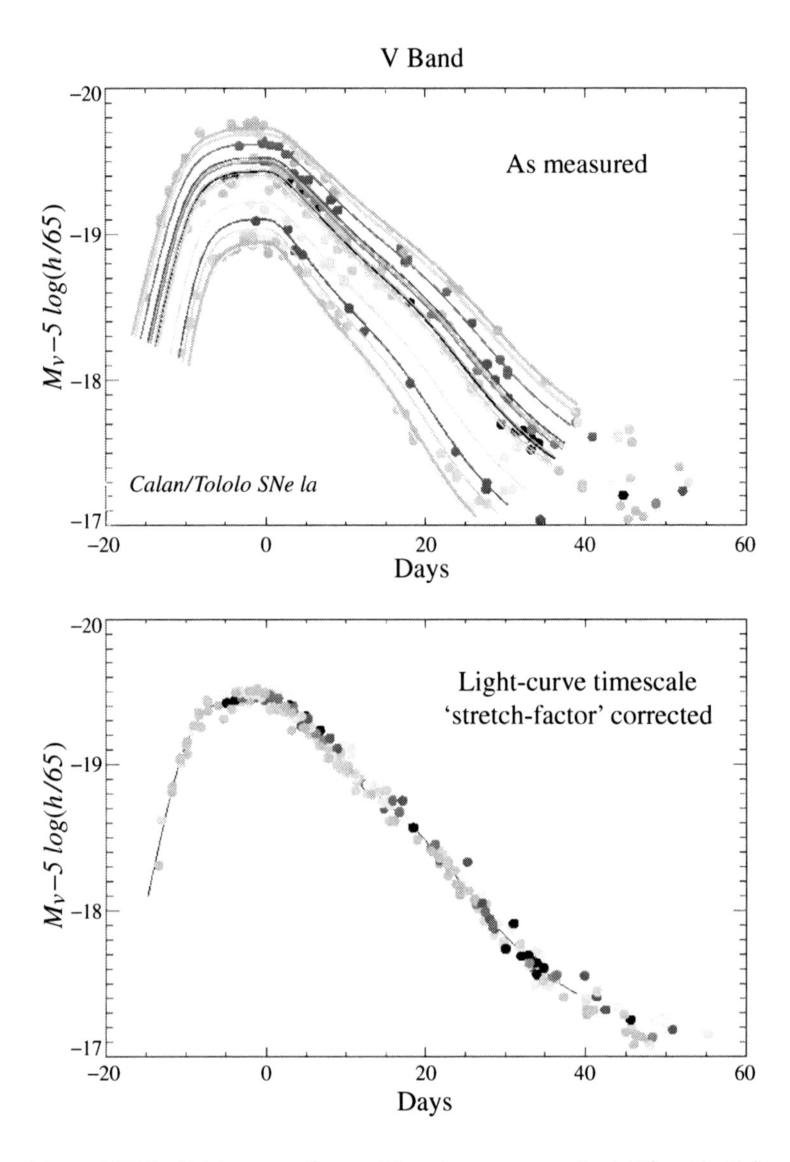

Figure 7.17 The light curves of several Type Ia supernovae (top). When the light curves are corrected for the 'stretch factor' (below), they are almost identical, and can be used to determine the distance of the supernova. (From Saul Perlmutter and the Supernova Cosmology Project.)

there may be two populations of Type Ia supernovae, one associated with the total mass of the galaxy, and one associated with the rate of star formation. As of late 2005, 3348 supernovae had been discovered and catalogued.

Within our own galaxy, there are several ways in which we could estimate the frequency of supernovae: (i) from surveys of other galaxies like our own; (ii) from the observed frequency of supernovae in our galaxy since record-keeping began (correcting for supernovae which were distant and/or obscured, or in the southern and daytime sky); (iii) from the number and age of supernova remnants in our galaxy; (iv) from the number and age of pulsars in our galaxy. These methods give reasonably consistent results: 2 - 14, 2.2 $\pm$ 1.3, 1.3 - 5, and 10, per century, respectively (i.e. a handful per century). In any case, it appears that supernovae are more frequent in our galaxy than visual observations would suggest! Recently, X-ray telescopes have proven useful in discovering the hot remnants of supernovae in our galaxy. There are also major projects to survey the plane of our Milky Way galaxy at many wavelengths; one project is the *Canadian Galactic Plane Survey*, and its southern and international counterparts. These are also helping to complete the census of the supernovae which have occurred in our galaxy in the recent past.

7.3.5 *Spectra*

The spectrum of a supernova is that of a stellar photosphere, expanding at several thousand km s^{-1}. The study of supernova spectra has benefitted greatly from the development of powerful spectrographs, with sensitive detectors, on very large telescopes. Observations of supernova spectra provide two kinds of information: the continuous spectrum gives a measure of the temperature of the radiating layers, and the line spectrum gives information about their temperature, composition, velocity, and excitation. The temperature can also be measured (crudely) from the photometric colours - at least for Type II supernovae; values of 12 000–20 000K at maximum, declining to 6000–7000K, are found.

The spectra are now classified, at the most basic level, as hydrogen-rich (Type II), and hydrogen-deficient (Type I). Type I spectra are subclassified as having prominent Si II λ6355 absorption (Type Ia), or having prominent He lines but not Si II absorption (Type Ib), or having neither (Type Ic). (The Type II supernovae are further classified on the basis of their light curves.) The spectra provide information about the composition of the star that exploded: those with Type Ib and Ic had lost most of their hydrogen-rich envelopes before they exploded; those with Type Ic had lost most of their helium-rich regions also.

The line spectra of Type II supernovae are reasonably well understood. There are emission lines of H, He I, C III, and Fe II, along with absorption lines of Ca II, Na I, Mg I, and the elements mentioned above. These show *P Cygni profiles*: broad emission lines with absorption lines Doppler-shifted to shorter (more violet) wavelengths, indicating expansion velocities of up to 10 000 km s^{-1}. The chemical composition is reasonably normal.

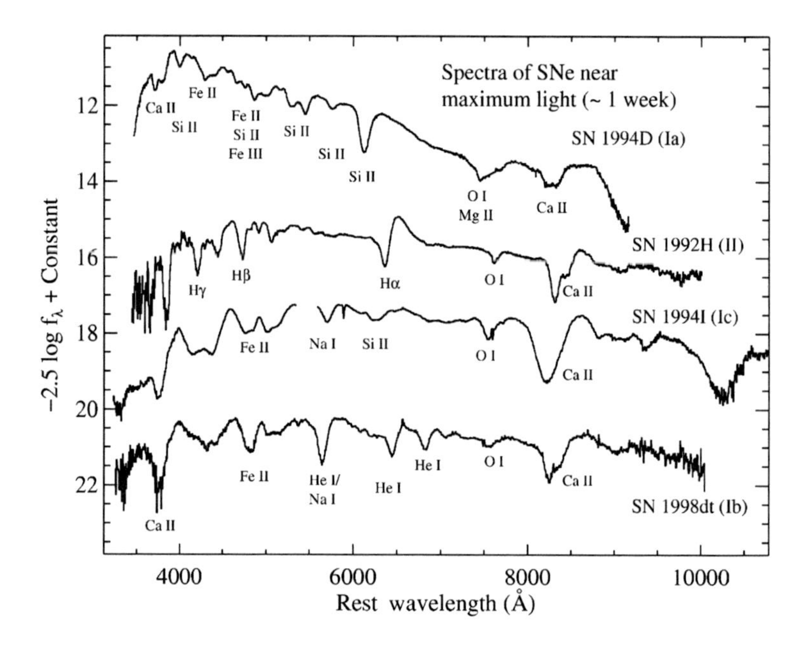

Figure 7.18 Spectra of Types Ia, Ib, Ic, and II supernovae, near maximum light. The elements responsible for the absorption lines are indicated. Note that the Type I supernovae do not contain hydrogen. (From A. Filippenko and T. Matheson, private communication.)

The line spectra of Type I supernovae are much more complex. Elements such as Si, Ca, and Fe (in Type Ia) and Ca, Na, He, and Fe (in Type Ib) are seen in the 'early' spectrum, and Fe, Ni, and Co in the 'late' spectrum – but no hydrogen! The total mass of material above the photosphere is of the order of 0.30–0.6 solar masses. Figure 7.18 shows the spectra of supernovae of Types I and II.

On the basis of the photometric and spectroscopic observations, we can begin to 'picture' the supernova event. Material is streaming off the star at up to 10 000 km s^{-1}. The photosphere, which is the layer at which the gas becomes transparent enough so that the photons can escape, has a radius of 50 000 times that of the sun, as deduced from the luminosity and temperature at maximum.

This picture is undoubtedly oversimplified. The supernova process is unlikely to be spherically symmetric: the original star was rotating, had a magnetic field, and may have been a binary. The explosion itself was probably not symmetric either.

Supernova 1987A was discovered by Ian Shelton in February 1987. It was the brightest supernova seen from earth in 400 years! Fortunately, it occurred after the development of UV and X-ray telescopes in space, and large telescopes with electronic detectors on the ground, so it was more comprehensively studied than any other supernova.

7.3.6 *Outburst mechanism*

This brings us to the question of the cause of the supernova outburst. We now know that Type II supernovae result from the collapse of the core of a massive star which has exhausted its nuclear energy. Type Ia supernovae are often thought to be white dwarfs that, due to mass transfer from a binary companion, have exceeded the Chandrasekhar limit and collapsed. In fact: as accretion brings them close to the Chandrasekhar limit, the temperature rises to the point at which the carbon undergoes runaway thermonuclear fusion. The resulting explosion destroys the star. But this knowledge was many decades in the making.

The outburst releases about 10^{51} ergs of energy, mostly in the form of the kinetic energy of the expanding material and, for the core-collapse supernovae, in the form of neutrinos. Only a small fraction of the energy appears as visible radiation. This total energy is to be compared with the total nuclear energy in a star like the sun (10^{52} ergs), and to the gravitational binding energy of a one-solar-mass main sequence star (10^{49} ergs), a white dwarf (10^{51} ergs), and a neutron star (10^{54} ergs). It is clear that the gravitational collapse of a star to the neutron star state is more than energetic enough to power a supernova explosion, and could do so in the requisite time of a few seconds. A supernova outburst is the equivalent of putting a 10^{54} erg gravitational 'bomb' at the centre of the pre-supernova star.

Stars do not normally collapse, because their thermonuclear reactions are self-regulating. Contraction of the star increases the temperature and density; the thermonuclear reactions go faster, produce more thermal pressure, and act to expand the star. Stars will collapse only if they have no further thermonuclear reactions to provide self-regulation. This occurs toward the end of a massive star's lifetime, when it has exhausted its supply of thermonuclear fuel.

The majority of stars (like the sun) contract gradually. During their lifetime, they gradually build up a dense core of inert material whose state is said to be *degenerate*. It holds itself up, not by thermal pressure but by an intrinsic quantum repulsion between closely packed electrons. If this dense core has a mass greater than 1.44 solar masses (called the *Chandrasekhar limit*), however, its electrons are unable to provide enough repulsion to balance gravity. The electrons are forced to recombine with nuclei, a process called *inverse beta decay*. Energy and pressure are removed. The collapse becomes catastrophic, and does not cease until the star reaches nuclear densities (10^{14} gm/cm^3) and becomes a *neutron star*.

It is possible that, in such a process, the collapse of a core will drive off the outer layers of the star, but the modelling of this process is notoriously difficult. Current modelling does suggest that this might happen if the collapse of the core produced a shock wave, or a burst of neutrinos which could blow off the outer layers, or if a layer of thermonuclear fuel outside the core were to detonate. This is probably the origin of Type II supernovae. Stars of 10–25 solar masses end their lives in this way, producing neutron star cores, surrounded by an expanding envelope containing a large fraction of the star's original mass. Stars with masses greater than 25 solar masses probably yield black holes, as remnants, though it is possible that some such stars avoid this fate, and produce neutron stars instead; the precise fate of very massive stars may depend on details such as their rate of rotation.

Imagine, then, the event. Within the core of an apparently normal red supergiant (like Betelgeuse!), more than 10^{51} ergs of energy are suddenly released. Shock waves cause the surface layers of the star to stream outward at 10 000 km s^{-1}. This process continues for months, forming the beginnings of a supernova remnant, and gradually revealing the neutron star at the centre.

Images of Supernova 1987A remind us that the pre-supernova star, especially if it was massive, may have emitted layers of material through stellar winds, in the years before its final collapse.

The origin of Type I supernovae was, until recently, much less clear. The solution turns out to be ... Type I supernovae do not come from single stars; they result from white dwarfs in close binary systems. In some cases, two white dwarfs may actually merge, forming a single entity, above the Chandrasekhar limit. More commonly, mass transfer on to the white dwarf from a normal star may raise its mass close to the Chandrasekhar limit. It *deflagrates*, converting the inner half of the star into nickel-56. The deflagration wave travels outward, creating lighter elements, and finally turning into a detonation which explodes the star. By monitoring the spectrum of the star over the next few weeks, astronomers can see evidence of the different layers of nuclear burning. The nickel-56 produces visible evidence of a different kind: it decays to cobalt-56 and then to iron-56, powering the star through radioactive decay. As mentioned earlier, the light curve provides a strong clue that some supernovae are powered in this way. Spectroscopic observations of SN1987A confirm this picture directly. It is interesting that the supernova – the most energetic process in stellar astrophysics, and the most spectacular phenomenon in variable star astronomy – is connected so visibly with a process which occurs at the nuclear level!

Type II supernovae also provide yet another example of the importance of close binaries in astrophysics. It is also a link between supernovae and the more mundane types of cataclysmic variables.

This is an over-simplified picture. Stars explode in a variety of environments, with a variety of possible companions, with a variety of evolutionary histories. For instance, Types Ib and Ic supernovae are believed to occur in stars which have previously shed their hydrogen-rich and helium-rich layers, respectively.

7.3.7 *Supernova remnants*

Supernova remnants (SNRs) are neither stars nor variable stars, but they result from the most extreme of variable stars – the supernovae. SNRs are the expanding shells of material which are ejected by most supernovae. The most famous is the Crab Nebula, the remnant of the supernova of 1054 AD. It is visible through a small telescope, and is #1 in Messier's famous list. Other SNRs are barely visible optically, or not at all. But most SNRs are strong radio sources. The Crab Nebula is the radio source Taurus A, which means that it is the brightest radio source in that constellation. Cassiopeia A is the brightest radio source in the sky, other than the sun. Based on its expansion rate, it must have been produced by a supernova which occurred around 1667 AD, but not recorded – perhaps because it was obscured by interstellar dust. Cassiopeia A is slowly decreasing in radio brightness, so it is a variable radio source – though not technically a variable star.

SNRs are energized by the complex interactions between the shock wave produced in the supernova explosion, the ejected material, and the gas and dust around the star – some of which may have been ejected by stellar winds, earlier in the star's life. SNRs can be very hot, and many have been discovered or observed by X-ray telescopes such as *Chandra*, launched in 1999. Such telescopes have also observed the hot neutron stars at the centres of some SNRs. Their measurements of the surface temperatures of neutron stars of known age have provided important information about the interiors of neutron stars, and the nature of matter in these extreme environments.

In 1993, astronomers had a rare opportunity to observe the formation and development of a SNR – that of SN1993J in M81, discovered by an amateur astronomer Francisco Garcia Diaz in Spain on 28 March 1993. Radio astronomers were able to monitor the expansion of the SNR, using a network of telescopes in Europe and North America. These observations indicated that the remnant was not slowing down significantly, due to interaction with circumstellar material, for instance. They also showed that, while the remnant was expanding symmetrically, the emission was stronger on one side of the remnant. Perhaps most important, from the rate of *angular* expansion measured from the radio observations, and the rate of *linear* expansion (about 6000 km s^{-1}) measured from the optical spectrum, astronomers were able to measure an accurate distance to M81 – about 11 million light years.

Box 7.2 Star sample – Supernova 1987A

On the night of 23–24 February 1987, Ian Shelton – resident astronomer at the University of Toronto Southern Observatory on Las Campanas in Chile – was photographing the Large Magellanic Cloud with a 25 cm photographic telescope called an astrograph. He wisely decided to develop his photographic plate before going to sleep. A bright 'new' star was visible in the LMC on the plate, and in the real sky when he went outside to look (figure 7.19). It was a supernova! This discovery had a broad range of implications, scientific and otherwise:

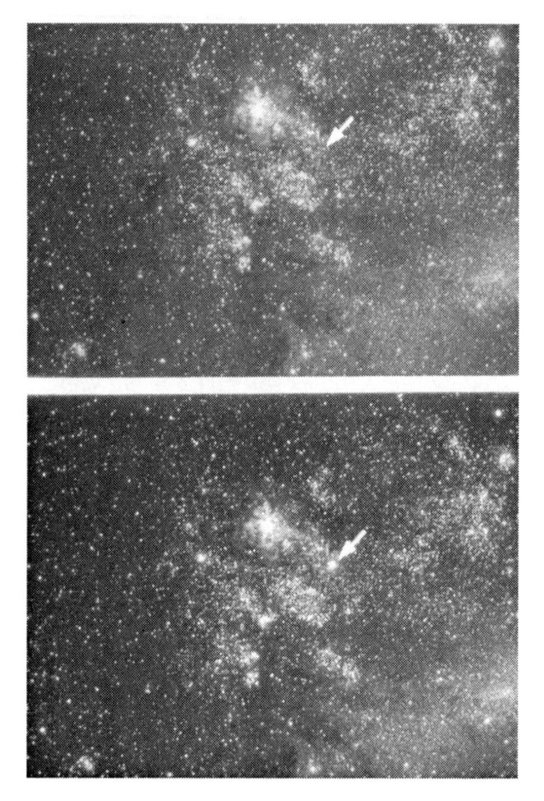

Figure 7.19 Discovery images of SN1987A. These photographs show the region around the supernova. The precursor is visible on the top panel – the first time that a previously catalogued star has been observed to explode as a supernova. The supernova was discovered by Ian Shelton, University of Toronto Southern Observatory. (University of Toronto photographs by I. Shelton.)

- The brightest supernova in 400 years, therefore historically significant.

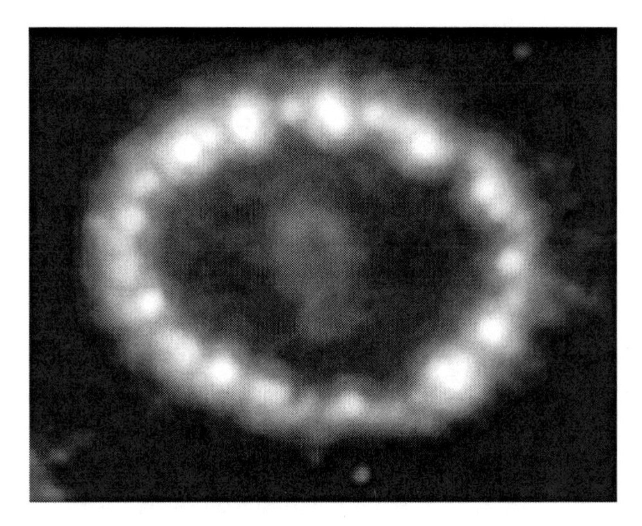

Figure 7.20 HST image of SN 1987A, showing the development of 'hot spots' in a ring of previously ejected material around the supernova, as a result of interaction with debris ejected in the supernova explosion. (HST/NASA.)

- And therefore the first supernova which could be observed with the full range of space and ground-based telescopes.
- A bright, pathological, celestial object, of great interest to amateur astronomers and the public, as well as to scientists.
- The first supernova explosion of a previously observed and catalogued star – the young, massive B3 supergiant Sanduleak -69° 202.
- The first supernova observed through the neutrinos which it emitted; these were detected by three underground neutrino observatories, confirming the mechanism by which supernovae explode (and demonstrating that neutrinos travel at the speed of light).
- The spectra of SN1987A eventually demonstrated that new elements had been created in the explosion, including radioactive elements which powered the light curve of the supernova.

The study of this 'nearby' (150 000 light years away!) supernova energized the study of supernovae in general. It also reminded us that astronomy is an observational science, and that astronomers – professional and amateur – must constantly be monitoring the sky. Astronomers continue to study the supernova, among other things waiting for a neutron-star core to appear. HST images have revealed the complex surroundings of the supernova, including mass lost from its previous evolution (figure 7.20).

7.4 Gamma-ray bursters

In October 1963, the US launched the first of a series of secret military satellites, designed to monitor a nuclear weapon test ban treaty by detecting the X-rays and gamma-rays emitted by clandestine nuclear explosions. I do not know whether any were detected, but the satellites did detect sporadic bursts of gamma-rays, lasting a few seconds or minutes. By 1972, these were being analyzed by scientists, who were able to conclude that they did not come from the earth or sun, but were of cosmic origin. Unfortunately, gamma-ray telescopes had very low resolving power, so it was not possible to know whether these *gamma-ray bursts* (GRBs) came from optically visible objects. There was no obvious correlation with visible supernovae or novae, for instance. Gamma-ray detectors were later installed on astronomical satellites (for example, IMP-6 and OSO-7); these confirmed that the 'cosmic gamma-ray bursts' were indeed cosmic, and that their energy peaked in the gamma-ray portion of the spectrum.

The next major progress came with the launch of the *Compton Gamma-Ray Observatory* (CGRO) – one of NASA's 'great observatories' series. The Burst and Transient Source Experiment (BATSE) aboard the satellite measured the approximate positions and distribution of 2704 GRBs (about one a day), and found that they were randomly distributed in direction. If GRBs were connected with stars in our galaxy, then they should have the same disc-shaped distribution as the stars (unless they are very close by), and should be concentrated to the plane of the Milky Way. They were not. If they were connected with distant galaxies, however, they should therefore have the same approximately random distribution as distant galaxies. They do.

The next breakthrough came in 1997 as a result of the combined effort of CGRO and the European *BeppoSAX* satellite, which had been launched in 1996. *BeppoSAX* was an X-ray satellite, and had much higher resolving power than CGRO. When the GRB designated GRB970228 was discovered on 28 February 1997 by CGRO, *BeppoSAX* was able to observe it and determine a precise position. Within hours, astronomers with ground-based telescopes were able to image it, and confirm that it was part of a faint galaxy. On 8 May 1997, a second GRB was passed from CGRO to *BeppoSAX* to ground-based observers, who were able to connect it with a faint galaxy at redshift 0.8.

Astronomers now believe that most 'long' GRBs are the result of superenergetic supernovae – *hypernovae* – formed in the collapse of a very massive star to become a black hole (figure 7.21). The total energies are 10^{51} ergs or more. Jets of matter and radiation are ejected along the star's rotation axes at close to the speed of light. Shock waves in these jets produce the gamma-ray bursts. As the shocks encounter material around the collapsing star, they produce the

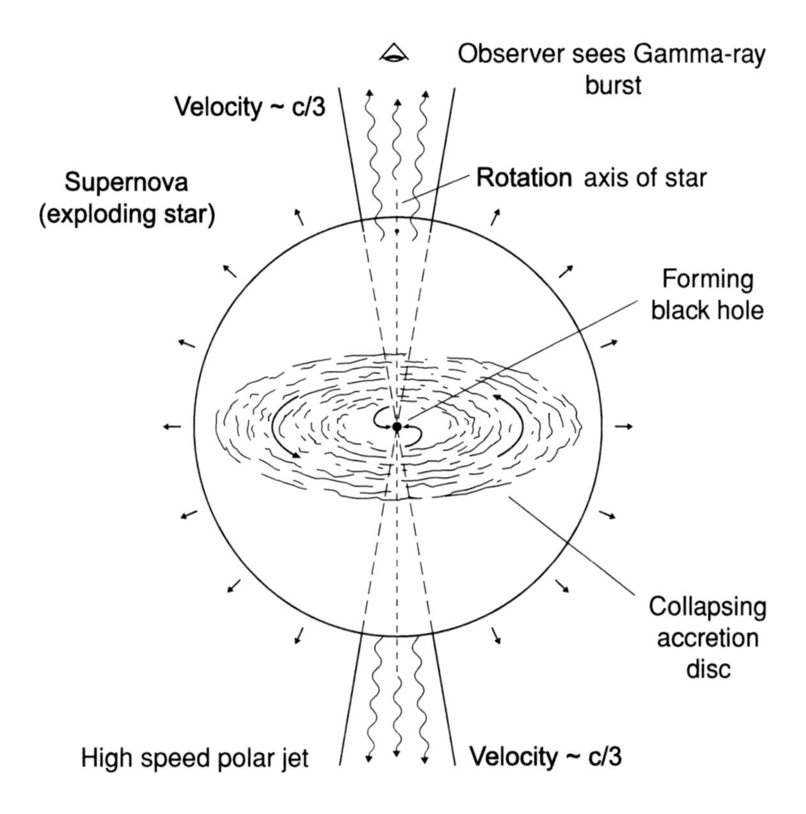

Figure 7.21 Model of a GRB. GRBs are not completely understood, but the most likely model involves a massive rotating star which collapses at the end of its life. During the few seconds of the collapse, the core of the star forms an accretion disc which collapses on to the black hole, ejecting high-velocity jets in the polar directions. In these directions, a GRB will be observed. (Jeff Dixon Graphics.)

'afterglows' which are observed at both visible and radio wavelengths, and at other wavelengths also.

A key observation came with the study of a GRB discovered by NASA's High-Energy Transient Explorer 2 (HETE-2) satellite, which was shown to be coincident with a supernova – SN 2003dh. An earlier coincidence between a GRB and SN 1998bw was suggestive, but no afterglow was observed in this case. The appearance of this phenomenon will vary, depending on the observer's orientiation, relative to the jets. Only if the jet is directed toward us – a less than one-in-a-hundred chance – will a GRB be observed. The exact orientation of the jet may produce much of the variety observed in GRBs: durations lasting from 30 ms to over 1000 s; structures ranging from simple to complex.

There are also 'short' gamma ray bursts, lasting a fraction of a second. Observations of two of these, with the *SWIFT* and *HETE-2* satellites, match well with

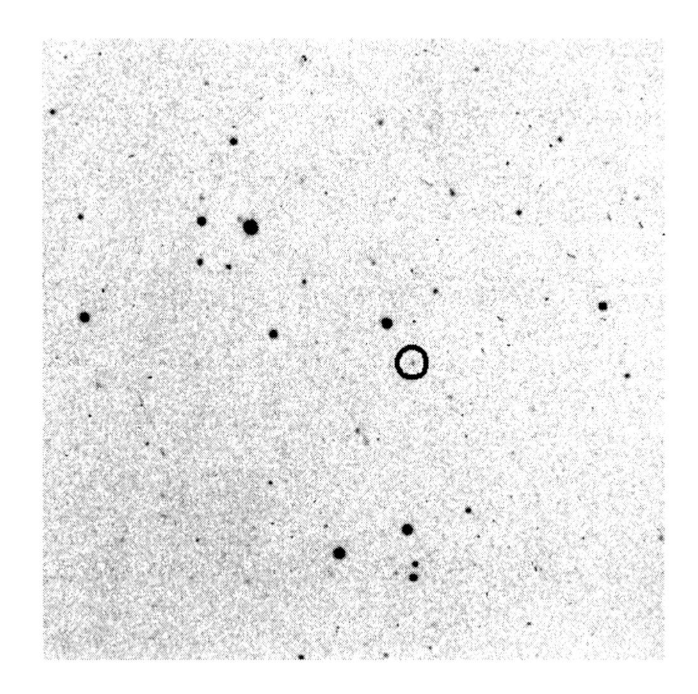

Figure 7.22 Negative image of afterglow of the GRB 010222, imaged by amateur astronomer Gary Billings, Calgary, on 22 February 2001 – one of the few GRBs to be imaged by an amateur astronomer. (From G. Billings.)

a model in which two orbiting neutron stars slowly spiral together and merge, forming a black hole and releasing a prodigious burst of energy.

Major advances in GRB research have continued with the launch of the *SWIFT* satellite by NASA in late 2004. This three-telescope, $250M collaboration between NASA and institutions in Italy and the UK is expected to discover 100–150 GRBs a year. A network is in place to ensure rapid ground-based follow-up. AAVSO is part of this network, and AAVSO observers are detecting a handful of afterglows, every year (figure 7.22). *SWIFT* has already been able to detect and study the afterglows, in the X-ray region of the spectrum, within a minute or two of the initial gamma-ray burst.

It has also discovered GRB050904 (in September 2005) which is 12.8 billion light years away, and therefore seen as it was when the universe was less than a billion years old. It is believed that massive stars, which may end as GRBs, were especially plentiful among the first generation of stars.

7.5 Active Galactic Nuclei (AGN)

AGNs are pointlike sources of light and other electromagnetic radiation which are located at the very centre or *nucleus* of many galaxies. In some cases

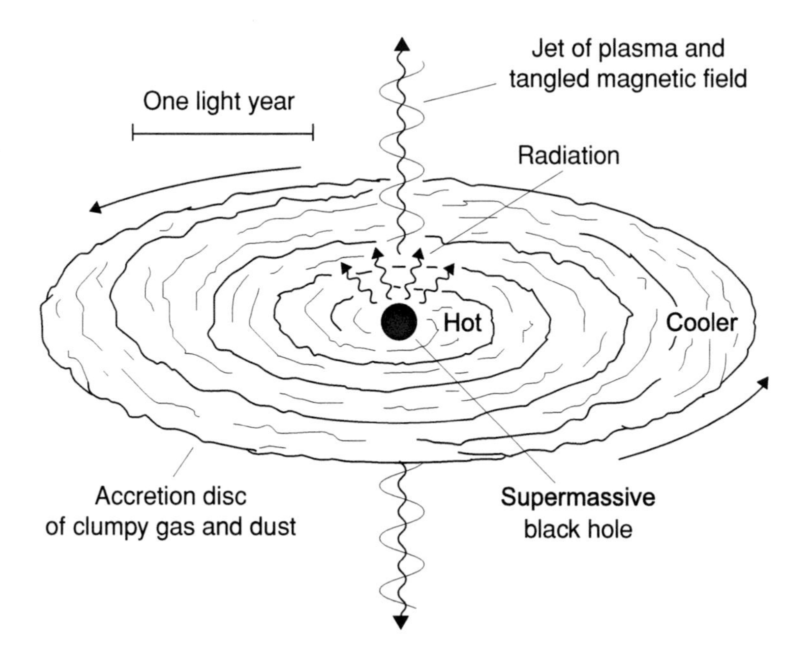

Figure 7.23 Model of an active galactic nucleus (AGN). Gas and dust in the accretion disc slowly approaches the supermassive black hole, becoming hotter and hotter as its gravitational potential energy is converted first into heat, and then into the radiation by which we see the AGN. Because the gas and dust is clumpy, the brightness of the AGN varies with time. Because the size scale of the AGN is about one light year, the typical timescale of variation is about a year. (Jeff Dixon Graphics.)

(the *quasars*), the AGN may outshine all of the hundreds of billions of normal stars in the galaxy. In most cases, however, its power is much less, but it may still be equivalent to that of millions or billions of stars.

AGNs are powered by super-massive black holes, up to millions of times more massive than our sun. Gas and dust spiral into the black hole, liberating gravitational energy (figure 7.23). The brightness of an AGN can vary on time scales of days to years so, if a black hole is a 'star', then an AGN is a variable star. Some AGNs have acquired variable star names, BL Lacertae being a prominent example. The variability of an AGN provides information about the process which powers it. A few amateur astronomers assist professionals in monitoring the changing brightness of AGNs, using CCD photometry, and hence contribute to understanding them. Since these objects are star-like, the techniques for monitoring them are similar to those for variable stars. AGNs are included in the GCVS4 as BLLAC (compact quasars showing almost continuous spectra), QSO (variable quasars, previously classified as variable stars), and GAL (optically variable quasi-stellar extragalactic objects).

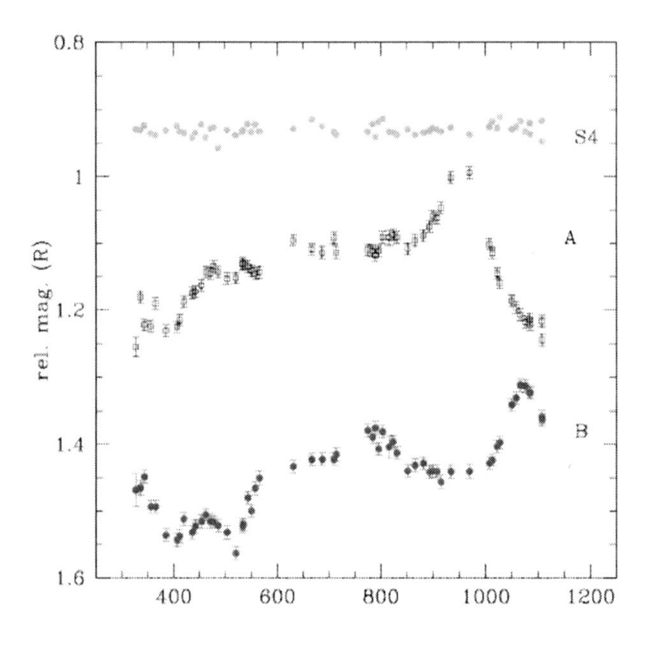

Figure 7.24 Light curves of the two images of the gravitationally lensed quasar SBS1520+530 A and B, and of a comparison star S4. A and B are each images of the quasar, and their variability is that of the quasar. But the light has travelled different paths, with different lengths and travel times, so the curves are offset by 130 days. (From Magain, 2005.)

An interesting application of variable star astronomy techniques to AGNs is *gravitational lenses*. If a massive galaxy, or cluster of galaxies, lies between us and a distant quasar, then the image of the quasar is 'lensed' into two or more separate images. Each is an image of the same quasar, but the light in the images has travelled along different paths, with different lengths. The variability of the different images will therefore be the same, but displaced in time by the difference in light travel time between the different images. The determination of this light travel time difference can be used to determine the distance to the quasar (figure 7.24).

The results of such monitoring show that AGNs vary by up to several tenths of a magnitude, on a time scale of years. This provides some information on the size of the region which emits the light: if the light varies on a time scale of X years, then the emitting region is no larger than X light years. Some AGNs flicker slightly, on a much shorter time scale. This may be due to a *hot spot* on the accretion disc surrounding the black hole.

AGNs make possible an interesting technique called *reverberation mapping* or *echo mapping*. As the variable brightness of the AGN scatters off discrete clouds of gas and dust near the nucleus, these clouds vary in brightness or spectrum, but with a time delay (relative to the AGN) which is their light-travel-time distance from the nucleus.

It may be stretching the definition to call an AGN a variable star. But they are variable; the variability can be studied using standard variable star techniques; and, in a sense, a black hole is a star!

Box 7.3 Star sample – V838 Monocerotis

V838 Monocerotis is a good example of the fact that some variable stars defy simple explanation. It could well be included under 'miscellaneous'. Considering that it was only discovered in 2002, however, perhaps that is not unreasonable. It was discovered by Australian amateur Nicholas Brown, who was carrying out a photographic nova search. On 6 January, it was about magnitude 10, then faded slightly. In February, it brightened to magnitude 6.5 – naked-eye visibility – shining blue, within a day or two! Again, it faded but, in March, it brightened rapidly from magnitude 9 to

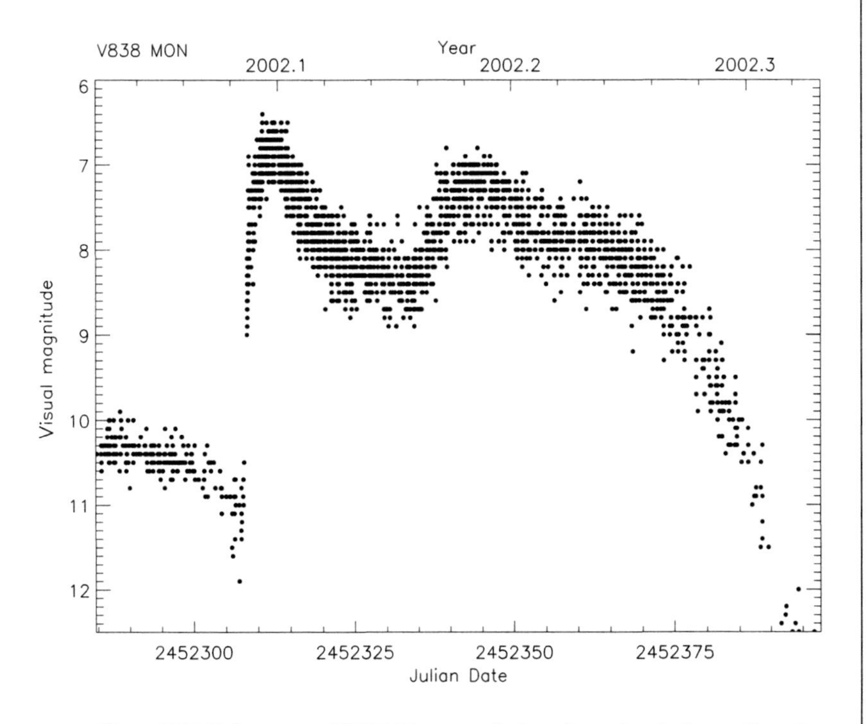

Figure 7.25 Light curve of V838 Monocerotis, based on visual observations from the AAVSO International Database. The variable was discovered by Australian amateur Nicholas Brown. It brightened almost to naked-eye visibility in February and again in March 2002, and has since faded to magnitude 15.5. (From AAVSO.)

Figure 7.26 Hubble Space Telescope image of V838 Monocerotis. The pulse of light, emitted when the star brightened in March 2002, moved outward through the nebula of gas and dust, ejected during the star's lifetime, illuminating successive layers of the nebula, producing a sort of 'CT scan' of the nebula. (HST/NASA.)

7.5, shining red. Since then, it has faded to magnitude 15. Archival data show that its previous magnitude had been about 16. Latest estimates are that the star is 18 000 light years away. Figure 7.25 shows the light curve of V838 Monocerotis, based on observations from the AAVSO International Database.

V838 Monocerotis's next 'claim to fame' occurred in late March. An image of the star showed a *light echo* (figure 7.26). The sphere of light emitted in February had travelled outward, and illuminated layers of dusty material around the star. A spectacular image of the light echo by HST was seen by millions of people around the world. Light echos were also observed around a nova in 1901, and around Supernova 1987A. In those cases, the illuminated material had been emitted from the star, earlier in its evolution. It is not clear whether the same is true for V838 Monocerotis, but it is highly suspected that these shells were ejected a million or more years ago.

The star continues to puzzle astronomers. Although it is listed on SIMBAD as a nova, it does not appear to be a classical thermonuclear-runaway nova, nor does it appear to be some kind of dwarf nova. Given the apparent shells of material around it, it is tempting to assume that it was an AGB star, and has perhaps undergone a final helium flash. That explanation does not work perfectly either, but, right now, it is the leading contender. The merger of two main sequence stars has been suggested, but that does not quite work either, because the spectrum appears to be that of a B3 main sequence star with a red supergiant companion visible during the outburst. This may be a less extreme star which has swelled up for some reason.

Only two other stars have behaved in a similar way to V838 Monocerotis. There are some very rare, bizarre species in the astrophysical zoo!

8

Pre-main-sequence variable stars

In section 2.19, we outlined what was presently known about star formation, and the early evolution of stars. There are several types of variability which are *specifically* or *predominantly* found among young stars. Most or all of these would be found in a given star, such as the sun, at various times in its pre-main-sequence (PMS) lifetime. We must always remember that the universe we see is a snapshot of millions of stars, seen at various random stages of their evolution. The PMS stage lasts only a few million years, so stars spend only a small fraction of their lifetime in it. But every star passes through this stage once.

PMS variables are often called *nebular variables*, because, being young, they are usually found in or near the nebulae from which stars are born. Or they may be called *Orion population* because the Orion region is a nearby, active site of star formation. One of the challenges in studying these stars is the fact that they are usually found within clouds of gas and dust, which will obscure or hide them at visible wavelengths. Radio observations have therefore been useful, and new sub-millimeter and mid-IR facilities such as the *Atacama Large Millimeter Array* and the *James Webb Space Telescope*, respectively, will be ideally suited for studying star formation.

A PMS star could also be an eclipsing or rotating variable, if it had a close companion, or a spotted surface. It could even be a pulsating variable, if it was located in an instability strip. We shall concentrate here on types of objects, and of variability, which are unique to PMS stars.

The study of PMS variables began in the 1920s, with the recognition that there were large numbers of irregular variables associated with the Orion Nebula. Studies of the spectra of these objects began about 1945, and are especially linked with the names of Guillermo Haro, George Herbig, Alfred Joy, and Otto Struve.

The Crimean Observatory and the Sonneberg Observatory (notably Paul Ahnert) were others associated with the study of these stars, along with astronomers such as P.N. Kholopov and P.P. Parenago (both also associated with the development of the GCVS), Cuno Hoffmeister who wrote a classic textbook on variable stars, and Viktor Ambartsumian. An important observation, by Kholopov and others in the 1950s, was that these variables were found in loose *associations*, which were usually not gravitationally bound, and which therefore remained intact for only a few million years. Kholopov coined the term *T associations* for these; they were a clue to the youth of these stars. It gradually became apparent that the nature of these variables was connected with their PMS status, and with processes such as accretion.

8.1 T Tauri stars

The largest group of pre-main-sequence stars are the T Tauri stars. In 1972, 323 of these had been catalogued, and many more are known today. They are found in regions of gas and dust along the Milky Way where stars are being formed. Often, they are found in loose groups called associations, parts of which may be gravitationally bound. The most famous of these is the Orion association, but there are other associations along the Milky Way. Most of these associations are referred to as *OB associations*, because they contain these hot, luminous stars with very short lifetimes. There are also T Tauri stars in young clusters such as NGC 2264.

Although the T Tauri stars derive their name from a variable star, they are *defined* by the appearance of their spectrum (figure 8.1). It shows:

- the Balmer and Ca II H and K lines in emission;
- Fe λ4063 and 4132 in emission;
- forbidden S II λ4068 and 4076 lines, usually in emission;
- Li λ6707 strong.

Some of these features are associated with the stars' youth, others with their accretion and activity. Incidentally, forbidden lines are ones that can only arise in low-density gases. They arise from transitions, within the atom, which require about 1 second, rather than the usual 10^{-8} seconds. In dense gases, collisions are too frequent for them to be able to occur.

The T Tauri stars are slowly contracting to the main sequence, so they lie just above it in the H–R diagram (figure 8.2). The variability and the spectral peculiarities appear to be connected with vigorous stellar activity on their surfaces. This may be due to their rapid rotation (which, in turn, is connected with their youth) and, in some cases, the effects of continued accretion of matter. It should

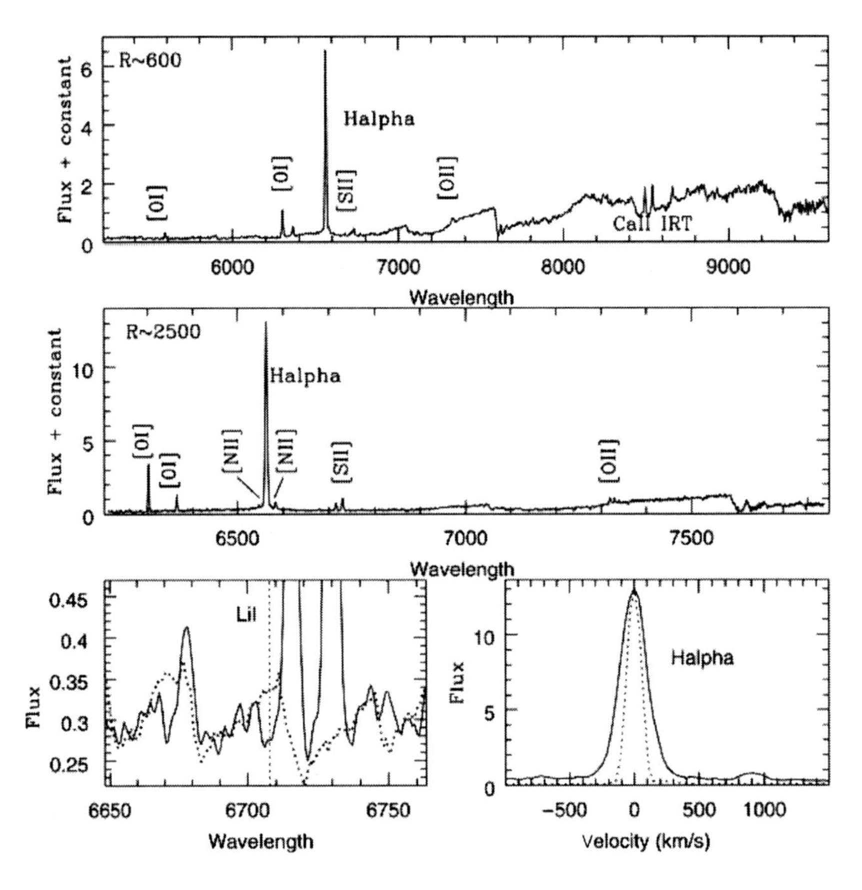

Figure 8.1 Spectra of T Tauri stars. The top panel shows the entire spectrum; the other panels show successively smaller regions of the spectrum. The panel in the bottom right shows a detailed profile of the Hα profile, broadened by turbulent accretion. (From R. Jayawardhana and A. Scholz, private communication.)

be realized that their rotation is rapid, compared with that of the sun, but not rapid compared with the O, B, and A type stars. The presence of Li lines in the spectrum of a star, indicating an abundance at least 100 times that in the sun, is a characteristic of youth. This element absorbs hydrogen in a fusion reaction, and forms helium, at a temperature of only about a million K. Circulation currents in a star normally mix the Li to deeper layers, where it is destroyed, within a few million years. So lithium is only abundant on the surface of stars which have not had time to mix the lithium to deeper layers.

In the GCVS, T Tauri stars are classified according to whether the star is associated with nebulosity (n), whether it has rapid, irregular variations (s), whether it has a typical T Tauri spectrum (T), or whether it has an early spectral class (a), or one later than A (b). This scheme has little or no physical relevance today.

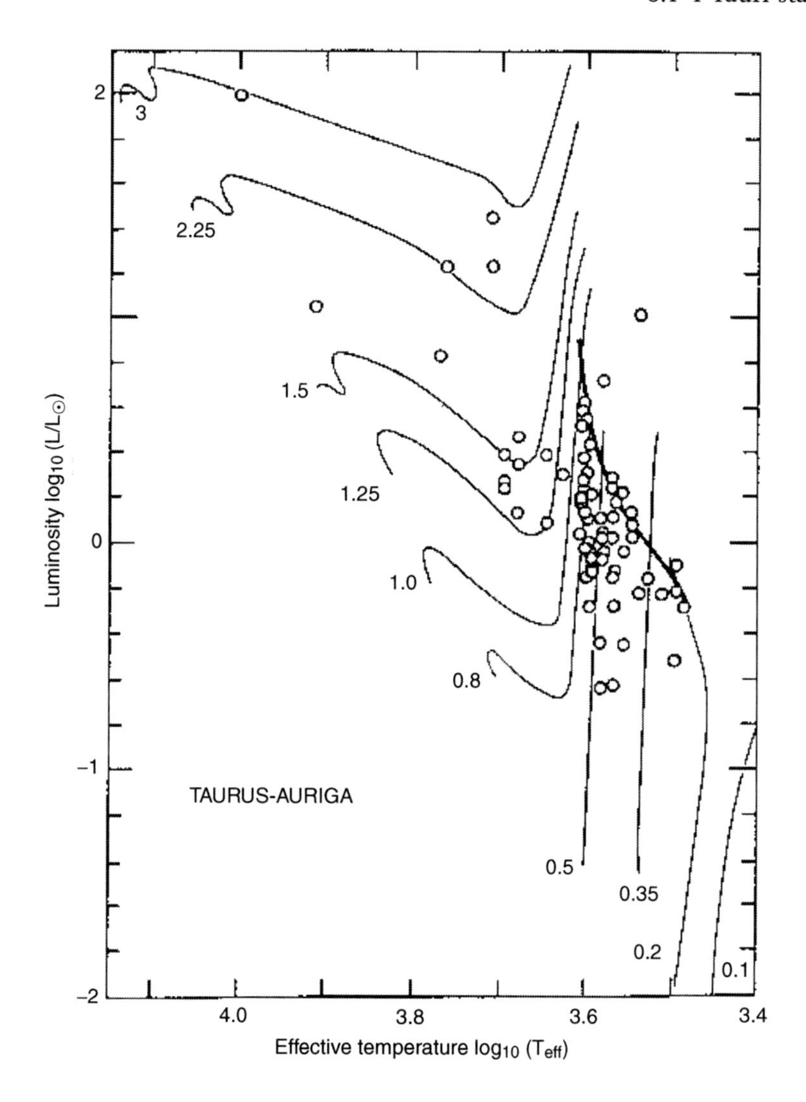

Figure 8.2 The position, in the H–R diagram, of a sample of T Tauri stars in the Taurus-Auriga molecular cloud complex. These are compared with theoretical evolution tracks, with the appropriate masses (in solar units) marked. (From Stahler, 1988.)

There are several T Tauri subtypes, or relatives, recognized today: classical T Tauri stars (CTTS) with evidence of an accretion disc; weak-lined (WTTS) or 'naked' T Tauri stars, which have little or no spectroscopically visible accretion disc (though there may be a cool, outer 'debris disc' still present); Herbig Ae/Be (HAEBE) stars, which are hotter, higher-mass analogues of T Tauri stars; and FU Orionis stars, which are T Tauri stars that exhibit significant brightenings, followed by slow declines (section 8.2). Most references to T Tauri stars are actually

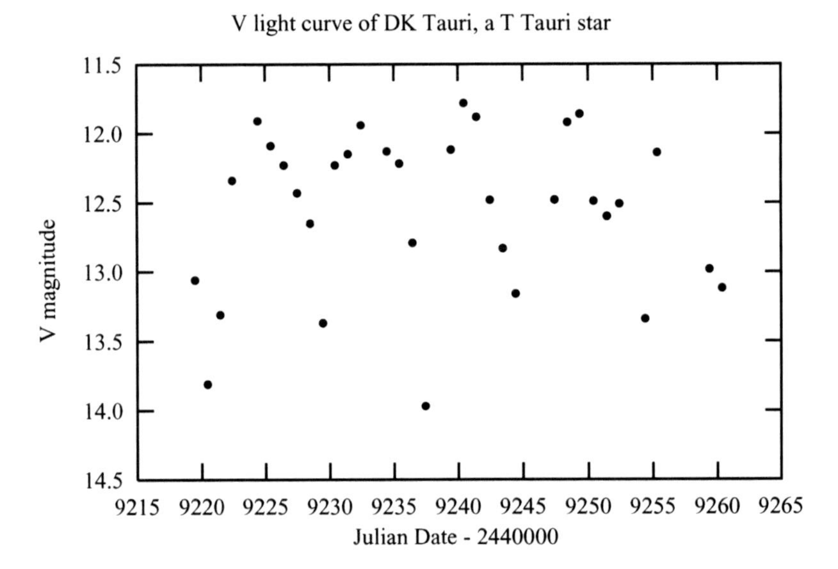

Figure 8.3 Light curve of the T Tauri star DK Tauri. The star shows very large periodic variations as a result of the rotation of the star. The surface of the star is highly non-uniform, as a result of magnetically channelled accretion from the disc around it. The period is 8.0 days. (Data from W. Herbst.)

to *classical* T Tauri stars. These evolve into WTTS, which then evolve smoothly into sunlike rotating variables, after their accretion disc is gone.

8.1.1 *Photometric variability*

Most WTTS have simple periodic variations, with occasional flares. V410 Tau is a well-studied example; it has a period of 1.871970 days, and an amplitude that varies in a 5.4-year cycle, which might be analogous to the solar cycle. It also flares in both light and X-rays (Stelzer *et al.*, 2003).

The brightness variations in classical T Tauri stars can be complex and irregular, with amplitudes ranging from 0.01 to many magnitudes, and taking place on time scales from days, through hours, to minutes. It is interesting that these three time scales resemble the natural time scales of rotation of the star (or orbit of circumstellar material), free fall, and flaring of the star, respectively. They are indicative of the broad range and complexity of the mechanisms in these stars. There may also be variations on time scales of weeks, months, or years. T Tauri stars that undergo FU Orionis outbursts vary on time scales of decades.

Figure 8.3 shows a typical light curve of a periodic T Tauri star. It has an unusually large amplitude, which is probably due to the rotation of a hot spot produced by accretion.

RW Aurigae is sometimes considered to be the photometric prototype star for this group (recall that T Tauri is actually based on spectroscopic criteria). This star is highly variable, both photometrically and spectroscopically, on time scales of hours to days.

Because the variations in T Tauri stars can be as large as several magnitudes, it is possible to observe them visually. The fact that they tend to be concentrated in small areas of the sky makes them efficient to observe, and some visual observers have made 10 000 or more observations of these stars in a year. In the 1960s, visual observers such as Albert Jones (New Zealand) made careful and significant measurements of some of these stars. The problem is that it is not always clear how such observations should best be analyzed. My student Rohan Pala has recently shown, using Fourier and self-correlation analysis, that it is possible to extract a period from visual T Tauri data archived by the AAVSO.

For decades, photography was the technique of choice for studying these stars. That was because they are usually concentrated in clusters or associations. Dozens or hundreds of stars can be recorded on a single photograph, for later study.

In 1952–53, there was a report of periodic variability in one T Tauri star, T Chamaeleontis. In the 1950s, periods were found in a few more stars. Important pioneering spectroscopic and photometric studies of the rotation of T Tauri stars were made by J. Bouvier (Bouvier *et al.*, 1986). Hundreds of pre-main-sequence stars – including T Tauri stars – have been photometrically monitored, especially by William Herbst and his collaborators (Herbst *et al.*, 1994; Herbst 2001), and this has resulted in the discovery of periods for many of them. His data and periodograms are available on-line. Periods would not be unexpected if the star had a close companion, or especially if its surface was patchy; in these cases, it would be an eclipsing and/or a rotating variable. T Tauri itself has a faint, distant, infrared companion; see the 'bio' of T Tauri, later in this chapter.

But almost all of the periodic variations are due to rotation. A few longer periods may be due to eclipses, or to orbiting inhomogeneities in the inner part of the accretion disc, but this variability is usually – but not always – small.

Previously, T Tauri stars were subclassified photometrically as: (i) more frequently bright than faint (RY Lupi stars); (ii) more frequently at mean brightness (T Chamaeleontis stars); (iii) more frequently faint than bright (RU Lupi stars); or (iv) no preference. It is not clear whether there is any physical significance to this subclassification, because not much attention was devoted to monitoring T Tauri stars, and to determining the causes of the photometric variability, until relatively recently. It appears that some of the short-term irregular variations can be explained in terms of the superposition of flares, or the type of flickering which occurs when an accretion stream impacts the stellar photosphere.

Herbst has developed a more refined classification system, based on his long-term photoelectric monitoring:

- Type I variables: cyclic variations with periods of 0.5 to 18 days or more, with amplitudes of a few tenths of a magnitude, seen mostly in WTTS, and due to rotational modulation by cool spots;
- Type II variables: generally irregular variations on time scales of hours or more, with amplitudes of typically a magnitude (occasionally larger), seen almost entirely in CTTS, and due to variations in mass accretion rate producing 'hot spots' and some rotational variability;
- Type IIp variables: like Type II, but quasi-periodic;
- Type III variables: generally irregular variations on time scales of days or weeks, with amplitudes of typically a magnitude (occasionally larger), seen in extreme T Tauri stars, and probably due to variable circumstellar obscuration.

Periodicity is an important criterion for classifying and understanding T Tauri stars, and the processes which occur within them. But to quote Herbst (2001):

> It soon became clear that most of the bright CTTS and HAEBE stars were not going to yield periodicities easily, if at all. Unlike the WTTS, where dogged monitoring is usually rewarded with a period ... the CTTS and HAEBE stars only rarely show significant periods. Claims to the contrary ... were based on too-optimistic interpretation of noise peaks in periodograms.

My students and I have recently been using self-correlation analysis in conjunction with Fourier analysis to re-analyze T Tauri variability. This combination is very useful in stars in which periodicity is irregular or sporadic, and veiled by other types of variability. It provides a check on the periods determined by Fourier analysis, and also produces a 'profile' of the variability, including the characteristic time scale of the irregular variability. This can range from days to weeks or more.

Multicolour photometry can provide additional information about the *cause* of the variability, such as whether the rotational variability is due to hot spots, cool spots, or both.

8.1.2 *Spectroscopic variability*

Until recently, it was difficult to study the spectroscopic variability of these stars, because of their faintness. Sensitive detectors on large telescopes have alleviated this problem. They can provide detailed information about rotation

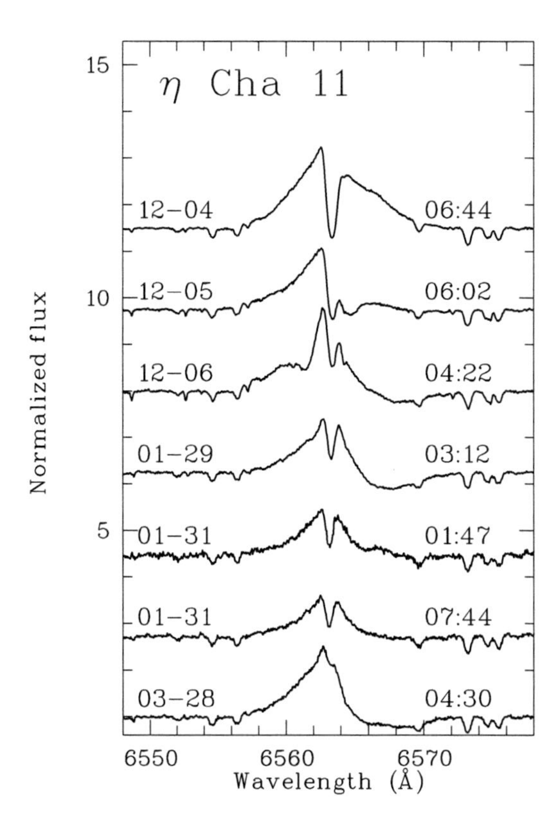

Figure 8.4 Spectroscopic variations in the T Tauri star η Chamaeleotis 11. The dates are shown on the left, the times on the right. (From R. Jayawardhana and A. Scholz, private communication.)

and accretion on the star, as well as about the motion and state of the gas in the disc.

Figure 8.4 shows the spectra of a T Tauri star at different times. The strengths and shapes of the emission lines may change with time, indicating that the flow of gas is highly variable. The shape of the lines may indicate outflow, inflow, both, or neither! The spectral type of the star does not change appreciably, which suggests that the variability of these stars is not particularly deep-seated. But the dominant physical process is accretion, and spectroscopic monitoring is very useful for detecting and studying the variable accretion rates in young T Tauri stars.

8.1.3 *Physical properties*

T Tauri stars were initially defined as F, G, and K stars, lying above the main sequence. More recently, it has been possible to identify T Tauri properties in M stars. Even the cooler L and T type stars are being investigated, though the results are inconclusive, so far. The warmer HAEBE stars are of special interest.

It is possible to estimate the stars' physical properties by using the continuous portion of the spectrum (between the emission lines) to estimate the effective temperature and reddening, and the total flux (including the ultraviolet and infrared) to estimate their luminosity. Since most are nearby and/or members of clusters, it is also possible to estimate the distance. The stars can then be plotted on the H–R diagram, and their positions can be compared with theoretical pre-main sequence evolution tracks to determine their evolutionary state, mass, and age. When this is done (figure 8.1), the following range of results is obtained for the FGK stars: $L = 4 - 50$ $L_{\odot}$,R$= 2 - 6R_{\odot}$, M $= 1 - 3M_{\odot}$, age $= 1 - 5 \times 10^6$ years. The M stars have lower masses and luminosities. These properties place the stars on the convective portion of the pre-main-sequence evolutionary tracks. (In cool stars, energy is transported by convection in the outer parts of the star, hence the term 'convective portion'.)

There are many uncertainties in this procedure. On the observational side: the spectrum and luminosity may still be affected by interstellar and circumstellar reddening and absorption, and by veiling by light emitted due to accretion; the energy distributions are complex and highly variable. The theoretical models do not allow for rotation or magnetic fields; this may seriously limit their applicability to these active young stars.

Further properties of these stars can be deduced from the details of their spectra. The absorption lines are broad, which is indicative of rapid rotation (compared with the sun), and/or possibly turbulence. The lithium abundance is high, as it is in other young stars; otherwise, the chemical composition appears to be normal. The emission line profiles were classically interpreted in terms of *mass outflow* (or in the case of the YY Orionis stars, mass inflow). The mass inflow rates can be as high as $10^{-7}M_{\odot}$ per year, but they are more typically 10^{-8}; the mass outflow rates are typically much smaller. The emission lines, which are the defining characteristic of T Tauri stars, indicate the presence of energized circumstellar gas in the chromosphere. Emission lines are also seen in the ultraviolet portion of the spectrum, as observed with satellites such as IUE. These come from even hotter regions in the upper chromosphere and corona.

Further evidence for circumstellar matter comes from observations in the visual, UV, IR, and radio portions of the spectrum. There is an 'UV excess' (in the visible or near UV) which is correlated with the emission in the Balmer lines, and which is presumably due to Balmer continuous emission. There is also excess IR emission, sometimes amounting to up to two-thirds of the observed total flux from the star. This appears to come from warm dust in a disc around the star. Note that the inner part of the disc, near the star, will be warmer than the outer part. There is also radio emission from a few T Tauri stars, apparently from an extended shell of ionized gas around the star. In the most recent, most sensitive surveys, almost all T Tauri stars - especially those with apparently weaker activity in other portions of the spectrum - are also X-ray sources. Recall that rapidly rotating solar-type stars have hot coronas, energized by stellar activity - the flares produced by violent reconnections of the magnetic fields. It is not clear whether this is the sole mechanism for generating activity in these very young stars.

8.1.4 *Binary and multiple T Tauri stars*

Given that most normal stars are in binary or multiple systems, it would be expected that many stars would be born in such systems. It is now possible, with modern technology, to detect and study multiple PMS stars and their discs. For about 20 of these, masses have been determined for the stars — a key piece of information for our understanding of these stars.

One of the most remarkable T Tauri stars is KH 15D. Kearns and Herbst (1998) discovered that it was an eclipsing binary with unusually deep and long eclipses ($\sim$3.5 magnitudes in the near-IR, lasting $\sim$16 days in 1999/2000). The binary orbit is highly eccentric, and has a period of 48.37 days. The eclipse light curve changes from year to year, probably because there is a disc of dust and gas around the system, and the disc slowly shifts due to precession. During the deep eclipses, it appears that the brighter star is completely eclipsed; the remaining light is scattered light (like the light that we see from the sun, just after it sets below the horizon).

8.1.5 *Interpretation*

Observations in different portions of the spectrum have revealed different aspects of T Tauri stars - different parts of the T Tauri model. IR variability may arise in cool material in an accretion disc; UV variability may arise in hot spots, as matter accretes on to the star at a variable rate. Visible and IR variability may arise from both of these, as well as from the rotation of cool and hot spots on the star. But how do these parts go together? What produces the T Tauri phenomenon? And how is this related to the evolutionary state of these stars?

Historically, there were two models for T Tauri stars: (i) a star is surrounded by outflowing hot gas which gradually slows down and cools, forming a shell around the star, and allowing dust to form therein; (ii) a star has an active, expanding chromosphere, where the temperature rises sharply to 100 000K or more. Above the chromosphere is an even hotter corona, which emits the observed X-rays. Embedded in the chromosphere and corona are loops of

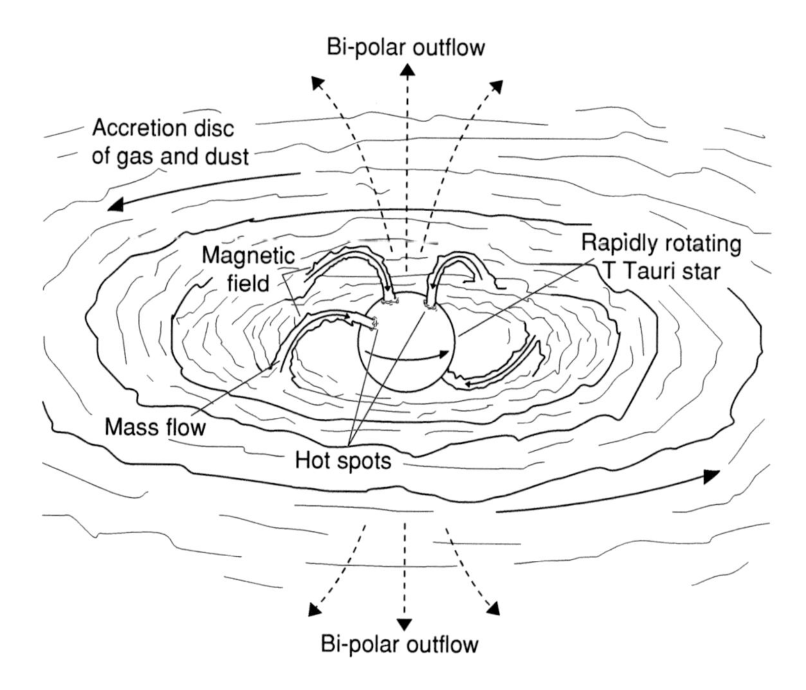

Figure 8.5 Schematic model of a T Tauri star. Gas and dust slowly approaches the star via the accretion disc. It may be channelled on to the star by magnetic field lines, producing hot spots on the star. As a result, the star is a rotating variable; the typical period is 1–10 days. After the T Tauri stage, the star may have cool spots, like the sun, and continue to be a rotating variable. (Jeff Dixon Graphics.)

magnetic field, along which the gas flows inward and outward. Above the corona is a cooler region where dust forms.

The important feature is the mechanism which drives the model. There was much to be said for the chromospheric model, with rotation as the ultimate driving mechanism, and this model was eventually incorporated into the present one. T Tauri stars were observed to be relatively rapidly rotating (or turbulent). Rotation and activity seemed to decline with age in T Tauri stars, as they do in older stars. Models of T Tauri chromospheres were able to explain many of the observed properties of the chromospheric spectrum (e.g. Ca II), but they did not explain the hydrogen lines.

We now know that the essential feature of a classical T Tauri star (figure 8.5) is the accretion disc (or the remains of the accretion disc) from which the star formed. (WTTS have little or no accretion disc, but they may have a 'debris disc' of material left over from the planet-forming process.) This realization developed in the 1980s, in part because accretion discs were being encountered and studied in a wide variety of other contexts, from cataclysmic

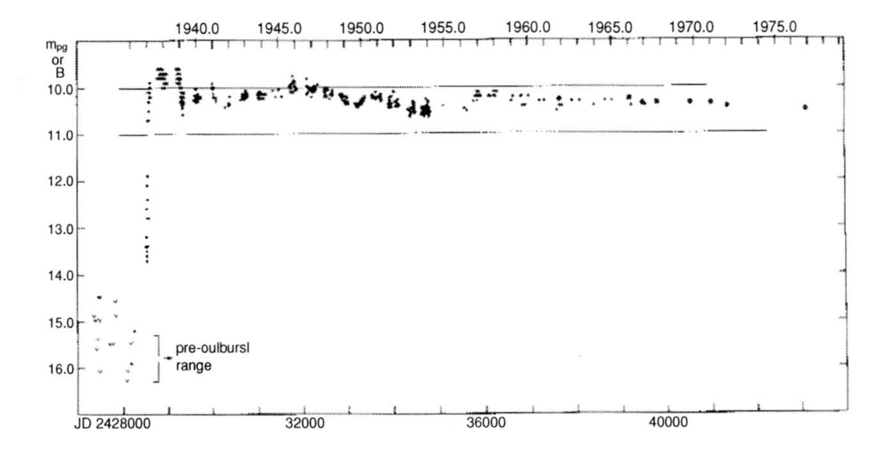

Figure 8.6 The photographic (small points) or B (large points) light curve of FU Orionis through 1976. After brightening in the 1930s, the star has faded slowly. (From Herbig, 1977.)

variables to active galactic nuclei. T Tauri discs are found among the lowest-mass PMS stars, and even around brown dwarfs. The key properties of a T Tauri star are therefore a result of its recent formation. The infall or accretion of material, from the disc, powers most of the features of these stars, and their variability.

8.2 FU Orionis stars

In the late 1930s, the star now known as FU Orionis brightened from m_{pg} +16 to m_{pg} +10 – a factor of more than a hundred. Since 1940, the star has faded somewhat, as shown in the light curve in figure 8.6. It is an F2 I–II star with an absolute magnitude of about −2. Thus, it appears, from its spectrum, to lie several magnitudes above the main sequence (though the true brightness of the star itself may still be that of a main sequence star). The nature of FU Orionis before the brightening, and the cause of the brightening itself were not known, but it was tempting to think that it represented some unique event in the star's evolution – perhaps even the 'birth' of the star. This was a common (mis)conception in the 1960s.

The situation was changed somewhat by the discovery of two more rather similar objects: V1057 Cygni and V1515 Cygni. V1057 Cygni was a T Tauri star which, in 1969, rose from m_{pg} +16 to +10, then slowly faded. By mid 1982, it was fainter than m_{pg} +13; it faded by one magnitude in five years, whereas FU Orionis faded by only 0.5 magnitude in 40 years. FU Orionis can no longer be considered unique. Strangely enough, V1057 Cygni is not included under

FUors in the GCVS4, but the stars V710 Casslopelae and V1735 Cygni are. It is interesting to speculate on what effects FUOr outbursts would have on a proto-planetary system (if any) around such a star!

Each of these three stars brightened by several magnitudes over a period of several years. FU Orionis faded slightly in 40 years. V1057 Cygni faded by two magnitudes in only five years.

Prior to brightening, each of the FU Orionis stars had an absolute magnitude of about +3 to +4. In the case of V1057 Cygni, the spectral type is known; it was a T Tauri star, with emission lines visible, but with no visible absorption lines. After brightening, each of the stars had an absolute magnitude of −1 to −2, and a spectral type of F-G I-III – apparently above the main sequence. But it may be that the accretion disc has brightened, not the star. The spectra show strong Li lines, as in other young stars, broad absorption lines due to rotation or possibly turbulence – also characteristic of young stars – and line profiles indicative of outflowing matter.

The T Tauri nature of V1057 Cygni, before outburst, demonstrates that the FU Orionis phenomenon is 'something that T Tauri stars do'. The declining brightness of FU Orionis and V1057 Cygni after maximum shows that the maximum phase is not a permanent one. Furthermore, the appearance of three FU Orionis stars in our neighbourhood of the Milky Way in 50 years would be highly improbable if each T Tauri star became an FU Orionis star only once, or not at all. It is much more likely, as George Herbig suggested, that the average T Tauri star became an FU Orionis star approximately every 10^4 years during its 10^6 year lifetime, remaining in the FU Orionis state for 10–100 years. Now, with over a decade of monitoring of thousands of T Tauri stars, we can study and question this suggestion. If each T Tauri star erupted as frequently as Herbig suggested, then we should be observing more FU Orionis brightenings. And we are not.

But what is the nature of the brightenings? They bear some resemblance to smaller brightenings in some other T Tauri stars (EX Lupi, UZ Tauri E, VY Tauri), but the cause of these brightenings is still not known.

One hypothesis is that outbursts occur in the rapidly rotating accretion disc when the rate of mass accretion is particularly high – typically 10^{-4} $M_{\odot}$ per year. Shantanu Basu's group at the University of Western Ontario has carried out simulations which support this hypothesis: blobs in the accreted material produce a 'burst mode' that can explain the FU Orionis phenomenon. In this sense, FU Orionis stars would be like dwarf novae; the outburst is connected with the flow of matter through the disc. Another hypothesis, based on the observation that FU Orionis and V1057 Cygni may be especially rapid rotators, is that the FU Orionis phenomenon occurs when a star reaches the limit of rotational stability (Herbig *et al.*, 2003).

8.3 Herbig–Haro objects

Herbig–Haro (HH) objects are not stars, but are wisps and knots of nebulosity with H and [O I] and other species in emission, and some continuum. As the name suggests, they were first identified and studied by two of the pioneers of the field - George Herbig and Guillermo Haro. HH objects have also been detected in other regions of the spectrum: IR and UV molecular hydrogen emission, and even X-rays. They can be subdivided into high and low-excitation objects, depending on the levels of excitation observed. These objects are found in the same heavily obscured regions as T Tauri stars, and their spectra bear some resemblance to the emission spectra of T Tauri stars. But they are actually excited wisps of gas.

Many HH objects are variable in brightness by up to several magnitudes on time scales of 10–20 years – which is justification for including them in this book. But what are they? What causes them to shine? And why are they variable?

An early hypothesis was that they are irradiated by hidden T Tauri stars. T Tauri and HL Tauri are examples of visible T Tauri stars which illuminate nearby nebulosity, but, in these cases, the T Tauri stars are not hidden. The HH 'stars' would therefore be very young T Tauri stars, visible by reflection in the nebulosity. This would provide an excellent opportunity to see even earlier phases of stellar evolution than the T Tauri stars. The variability of the HH objects would be a 'reflection' of the variability of the HH stars, which vary like T Tauri stars or perhaps even like FU Orionis stars.

The reflection hypothesis, however, has been replaced by a widely accepted hypothesis that HH objects are nebulosities which are excited by shock waves passing through the interstellar material from nearby forming stars. Pre-main-sequence stars are now known to be surrounded by accretion discs, and to produce bi-polar winds at right angles to the disc (figure 8.5). The HH wisps may be wind material and/or gas from the environment, accelerated to speeds of 20 to 200 km s^{-1} - far beyond the speed of sound in the gas. The greater the wind speed, the greater the excitation of the wisp.

8.4 Herbig Ae and Be stars

Be stars in general are discussed in the next chapter: they are B stars which have shown hydrogen emission at some time in their lives. They appear to be mature stars, ones which have above-average rotation. Herbig Ae and Be stars are defined as A and B stars with emission lines, lying in obscured regions, and associated with reflection nebulae. Some may be 'classical' Be stars, but the Herbig Ae and Be stars are specifically defined as pre-main-sequence stars, and

Table 8.1. *Bright and/or interesting pre-main-sequence stars*

Star	GCVS class	Herbst class	Range	Spectrum	Comments
RW Aur	INT	CTTS	9.6-13.6	G5Ve(T)	large-amplitude, active
AB Aur	INA	HAEBE	6.9-8.4	B9neqIV-V	bright
AE Aur	INA	–	5.78-6.08	O9.5e	bright
RU Lup	INT	CTTS	9.6-13.4	G5ep(T)	interesting name!
V582 Mon	–	–	16.1–	K7V	KH15D, eclipsing, see text
UX Ori	ISA(YY)	HAEBE	8.7-12.8	A2ea	prototype of UXors
YY Ori	INST(YY)	CTTS	13.2-15.7	K2IV-K5e(T)	prototype YY Ori star
EX Ori	LB	–	10.0-12.3	M7III	prototype of EXors
FU Ori	FU	FUOR	9.6-16.5	F2I-IIpeaq	prototype
T Tau	INT	GTTS	9.3-13.5	F8-K1IV-Ve(T)	prototype
V410 Tau	INSB	WTTS	11.3-12.4	K3-7Ve	largest-amplitude periodic T Tau
V892 Tau	INA	–	5.55-6.07	A6e	bright

appear to be massive, luminous counterparts of the T Tauri stars. The emission lines are associated with their youth, though (like the classical Be stars) rapid rotation may play a role. Herbig Ae/Be stars are rare, because A and B stars contract to the main sequence quickly, but they are also bright and easy to see and study.

Classical Be stars are variable on many time scales. The dominant variability is associated with the variable rate at which matter is being ejected from them. Herbig Ae and Be stars were not originally thought to be variable, but careful monitoring has shown that they exhibit some of the same forms of variability as classical T Tauri stars. No Herbig Ae/Be star is known to be a rotational variable. This is not surprising, since A and B type stars do not have solar-type magnetic fields because they do not have an outer convective zone. At least two Herbig Ae/Be stars – V628 Cassiopeiae and T Orionis – may be eclipsing variables.

8.5 Putting it all together

The various forms of pre-main-sequence variability are a consequence of the process of star formation. This occurs in regions such as the nearby Taurus Molecular Cloud, and the Orion Nebula, where we find a dazzling assortment of young objects, of all masses and ages. The newly formed stars are still surrounded by an accretion disc of material, slowly spiralling inward with a revolution period which is a function of the distance from the star. There is no reason why the disc should be smooth and uniform, so the rate of accretion may not be constant. At the same time, the star is rotating rapidly, as a result of the initial rotation of the cloud from which it formed, and of conservation of angular momentum as the cloud contracted due to gravity. The star's rapid rotation generates a magnetic field which in turn generates stellar activity – spots, flares, prominences, corona. The magnetic field may also be strong enough so that it guides the accretion of matter on to the star, producing 'hot spots'. So there may be periodic variability due to the rotation of cool or hot spots on the star. There may be irregular variability as the rate of accretion on to the star and the release of gravitational energy varies. There may be rapid variability due to flares. And there may be variable obscuration by the surrounding dust and gas; part of this variability may be periodic if there are blobs of dust orbiting in the accretion disc, or if the star has a companion.

Finally, the active star may produce a wind which, because of the presence of the accretion disc, is confined to the polar regions; it is a *bi-polar wind*. This wind can energize the Herbig–Haro objects in the nebulous regions around the star.

This is a simple picture of a complex process. Neither rotation nor magnetic fields are easy to understand. And the star's neighbourhood is a crowded mixture of stars and interstellar clouds. And over half of stars are binaries. How does that affect the process of star formation, and the many and complex forms of variability which occur?

And this picture applies to low-mass stars like the sun. High-mass forming stars, with their strong and energetic radiation, are surrounded by small *HII regions* of ionized hydrogen gas. As soon as the hot young star is exposed, its radiation and wind may quickly destroy the discs around low-mass stars in its vicinity.

How long does the formation process last? IR observations suggest that the dust disc is gone within a few million years. During this time, planets will form – if they can. A bit later in the star's history, it may have a *debris disc* of dust, resulting from collisions between left-over chunks of material which did not form into planets or satellites. Vega and β Pictoris have famous discs of this kind.

There is still much to learn about the earliest stages of stars' lives. These stages are hidden within the clouds of gas and dust that surround them. Future telescopes such as the Atacama Large Millimeter Array will be able to probe these mysteries.

Eventually, the rotation and activity decrease. The rotation is braked by interaction with the disc, and also by loss of angular momentum through the wind.

The spectacular phenomena which make up 'solar activity' are pale remnants of a once more-active star.

Box 8.1 Star sample - T Tauri

T Tauri [HD 284419, F8V-K1IV-Ve, V = 9.3-13.5], in Taurus near ϵ Tauri, was discovered as a variable star in 1852 by John Russell Hind. It is situated close to a nebula (now known as Hind's Nebula, or NGC 1555) which shines by reflection of T Tauri's light – so it is a variable nebula! Within ten arc sec of T Tauri is a smaller nebula. And less than an arc sec away is an infrared companion star. A second IR companion has recently been discovered; T Tauri is a triple system. This all goes to show that star formation takes place in interesting neighbourhoods. It also reminds us that over half of all stars are members of binary or multiple systems, so we must somehow explain their origin.

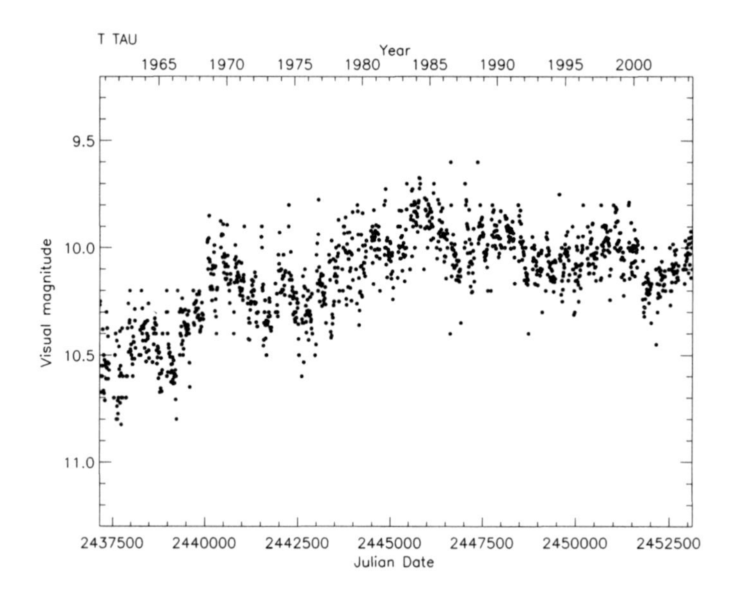

Figure 8.7 Light curve of T Tauri over 15 000 days, based on visual measurements from the AAVSO International Database; there are also very small variations on a time scale of 2.8 days, which are not visible on this light curve. (From AAVSO.)

T Tauri has varied between 9.3 and 13.5 in V, with a deep minimum between 1888 and 1891, probably due to obscuration by dust. Currently it varies between about 9.3 and 10.7 (figure 8.7). T Tauri is observed visually by amateur astronomers; you can construct a light curve with the Light

Curve Generator on the AAVSO web site. Its spectrum has varied from F8Ve to K1IV-Ve, so it is a spectrum variable as well! Its mass is probably similar to the sun's, but its age is only a few tens of millions of years. It shows small-amplitude variations with a period of 2.8 days, presumably due to the star's rotation, as well as larger, slower, more irregular variations.

9

Miscellaneous variable stars

All of the types of stars in this chapter could probably be fitted in, arbitrarily, elsewhere in this book. But they are not easily categorized, so we discuss them here – and admit that 'pigeon-holing' variable stars is not simple.

9.1 Be stars – Gamma Cassiopeiae variables

Be stars are defined as non-supergiant stars with temperatures between about 10 000K and 30 000K which have shown *emission lines* in their spectra on at least one occasion. As the definition suggests, the spectra of Be stars can vary with time (figure 9.1), as can their brightness. In the GCVS4, variable Be stars are known in most cases as *Gamma Cassiopeiae (GCAS) variables*, after the prototype, but others are arbitrarily classified as BE. About 20 per cent of B stars are Be stars and, because these stars are very luminous, there are about 200 of them among the 10 000 naked-eye stars. This makes them one of the most conspicuous classes of variables. See Percy (2001) for a brief review of Be stars, and Smith *et al.* (2000) for a recent conference proceedings. For current research information, the *Be Star Newsletter*, published by a working group of the IAU, is very useful:
http://www.astro.virginia.edu/~dam3ma/benews/

The emission lines in the spectra of Be stars arise in a disc of gas around the star. One thing that distinguishes Be stars from normal B stars is their rapid rotation – typically up to 500 km s^{-1} at their equator. This reduces the effective gravity at the equator, and makes it easier for other processes, such as radiation pressure and possibly pulsation, to produce a slowly expanding, outwardly spiralling disc of gas above the equator. So the disc is an excretion

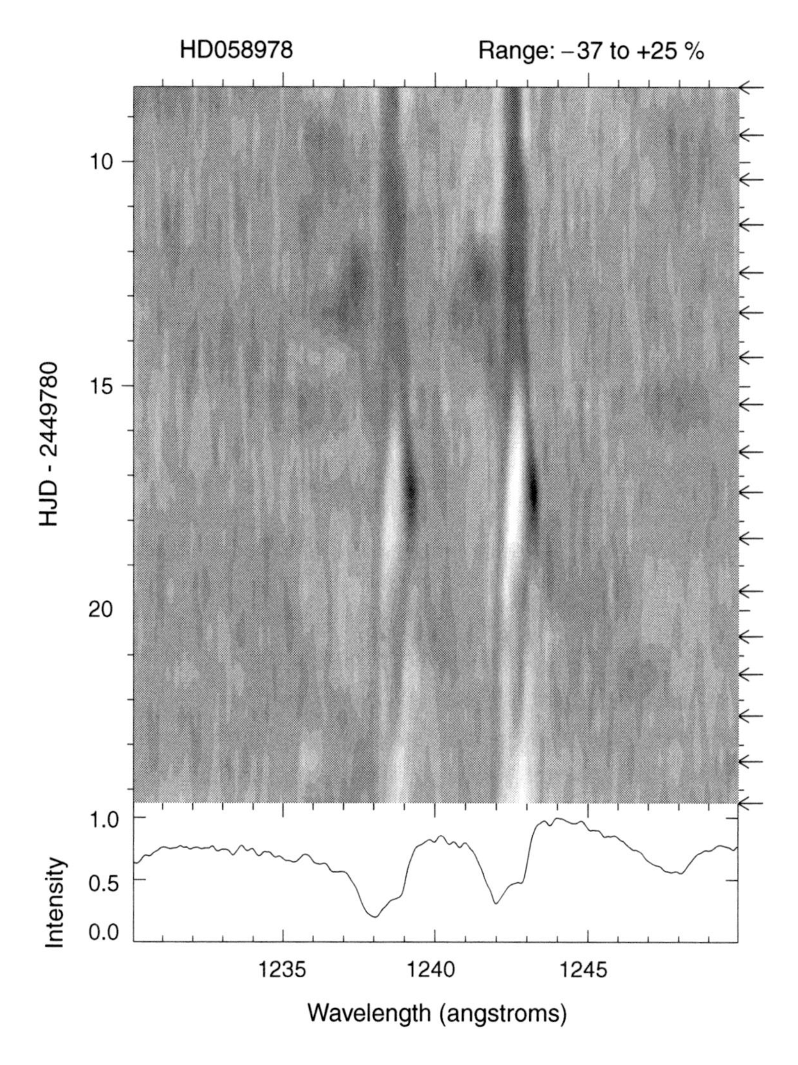

Figure 9.1 Spectrum variations in the UV NV lines of HD 58978 (FY Canis Majoris), a Be star. On this gray-scale plot, time increases downward (see scale on left), spanning about 15 days; light is emission, and dark is absorption. The average spectrum is shown at the bottom. (From G. J. Peters.)

disc, not an accretion disc. Note that the motion of the gas in the disc is primarily rotational; the relative outflow velocity is small.

Related to the Be stars are the *shell stars* – rapidly rotating B stars with deep, narrow absorption lines in their spectra, superimposed on the normal broad absorption lines of hydrogen and helium which are normally seen in the spectra of hot stars. The shell lines are due to scattering or absorption by atoms in the disc, in Be stars which are seen almost equator-on. Stars can actually change

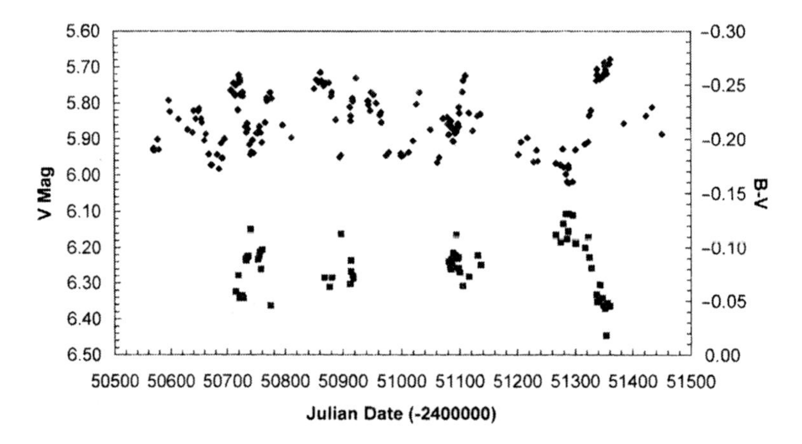

Figure 9.2 Long-term light curve of CX Draconis, a Be star. The V magnitude scale is on the left, and the $(B - V)$ colour scale is on the right. This light curve spans 1000 days, during which time there were several brightenings of about 0.2 in V. (From J. R. Percy.)

from B to Be and back again, or from B to Be to Be (shell) if their disc is seen edge-on, as the disc grows and decays.

Be stars vary in brightness, and spectrum, on several timescales. There are variations on timescales of weeks to decades (figure 9.2), which are connected with the formation and dispersal of the disc. There are quasi-cyclic variations on timescales of months to years, which may be due to wave motions in the disc. There are sometimes variations on timescales of days to weeks, which are often connected with binary motion and interaction. (Many binary stars have discs around one component, due to mass transfer and accretion, but only a small fraction of Be stars are in binary systems.) There are variations on timescales of 0.3 to 2 days (figure 9.3), which are due to non-radial pulsation, and/or perhaps rotation. These short-term variable Be stars are sometimes called *Lambda Eridani variables.*

9.1.1 *History*

Between 1863 and 1867, before the introduction of photography, Father Angelo Secchi (1818–1878) carried out a remarkable *visual* study of the spectra of 4000 stars, using a spectroscope on the telescope of the forerunner of today's Vatican Observatory. In an 1867 publication, he noted that γ Cassiopeiae showed emission in the H Beta line in its spectrum. Thus began the mystery of the Be stars. Gamma Cassiopeiae varies in brightness by over a magnitude, so its photometric variability could be discovered through visual observations. Photoelectric observations of Be stars have revealed much more: long-term variations in almost

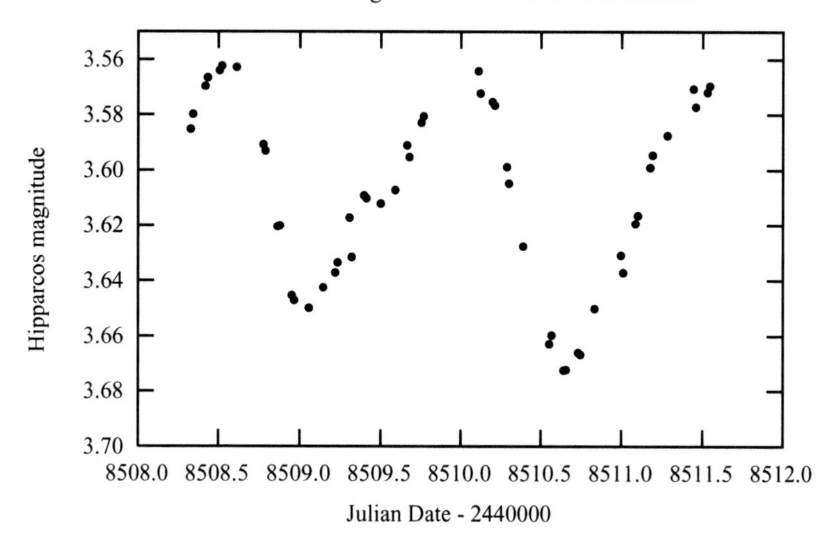

Figure 9.3 Short-term light curve of *o* Andromedae, a Be star, obtained by the *Hipparcos* satellite, in the *Hipparcos* photometric band. This was the first short-period Be star discovered; it has a stable period of about 1.6 days. (From the *Hipparcos* archive.)

all Be stars, small-amplitude medium- and short-term variations, the latter having periods of only a day or two. These periods are consistent with certain modes of non-radial pulsation, but they are also consistent with the rotation periods of the stars.

Since about 1980, high signal-to-noise spectroscopic observations have also been made in parallel with, and often together with, the photometric observations. In almost all of the hotter Be stars, absorption *line profile variation* (lpv) has been found (figure 9.4), and it is seen in a few of the cooler Be stars as well. Studies of these lpvs have been especially useful; they provide evidence for low-order non-radial g-mode pulsation in these stars. In a few stars, there is evidence for multiple periods. This is especially strong evidence for non-radial pulsation, since it would be very difficult to explain such multiple periods if they were due to rotation.

Polarimetric observations have confirmed the presence of the disc of gas around the stars, and optical interferometry has provided direct images of the disc. Still, many aspects of Be stars are still uncertain or controversial:

What causes the formation of a disc of gas at a specific time? Is it a combination of radiation pressure, 'centrifugal force' due to the rotation, and build-up of the amplitude of non-radial pulsation?

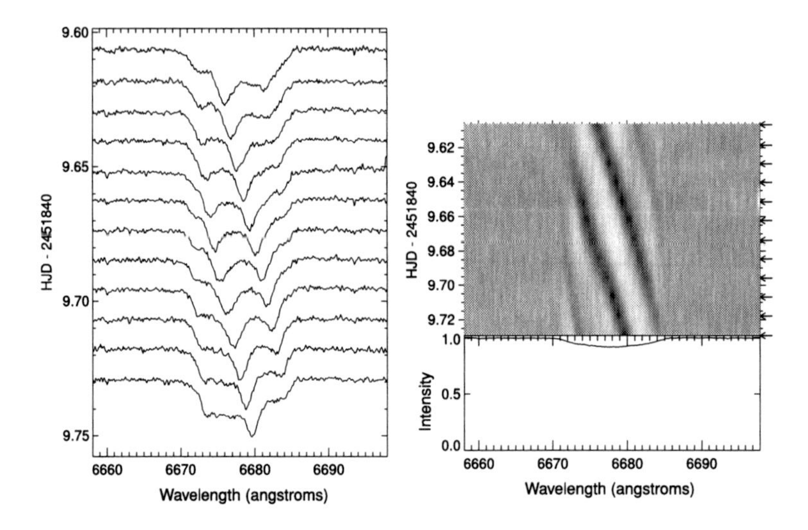

Figure 9.4 Short-term spectroscopic variations of π Aquarii, a Be star. Left: graphical plot; right: gray-scale plot. In both cases, time increases downward (scale is on the left), and spans one night. Note how the absorption features in the spectra (dark bands in the gray-scale plot) move across the line profile. The period is about two hours, and the cause is non-radial pulsation. (From G. J. Peters.)

What is the precise shape of the disc? Is it very flat, or only slightly flattened? Current evidence suggests that it is relatively thin.

How fast are the underlying stars rotating? Until recently, it was thought that the rotation rates were comfortably below the 'critical velocity' – the velocity at which matter on the equator would be flung off the star. But new observations of the Be star Achernar (α Eridani) suggest that its shape is more significantly distorted by rotation; the equatorial diameter is 1.6 times the polar diameter.

What causes the short-term variations – non-radial pulsation, or the rotation of a star with surface features, perhaps caused by magnetic features, as on the sun? Current evidence favours pulsation; there is a viable hypothesis that the build-up in pulsational amplitude when several modes add together is partly responsible for the formation of the disc. In an exciting new development, Walker *et al.* (2005), using the MOST satellite, observed the Be star HD 163868 for 37 days, and detected 60 significant periods which were clustered around seven and 14 hours, and eight days. The seven and 14 hour periods were successfully modelled as high-order *g* modes which are destabilized by the effect of the opacity of iron atoms, deep in the star; other researchers had previously determined that non-radial pulsations could be driven in this way. Nevertheless, the star *is* rotating, so the non-radial pulsation pattern will be carried around the star by that rotation.

Even if the variability is due to pulsation, do magnetic fields play *any* role in these stars? So far, magnetic fields have been difficult to detect, but there is evidence for a magnetic field in at least one Be star. And we cannot rule out the presence of *weak* magnetic fields in these stars.

What is the role of binarity in Be stars? In many close binary stars, there is an accretion disc of gas which forms around one of the components as a result of mass loss from the other. Is this a factor in some Be stars? CX Draconis (figure 9.2) is an example of a Be star in a binary system.

Perhaps the Be stars are even more complex than we think. Perhaps there are several mechanisms that can produce the Be phenomenon – even within the same star!

9.1.2 *'Bumpers'*

Cook *et al.* (1995) announced that, as a byproduct of the MACHO project to search for gravitational microlenses, they had identified a population of blue variable stars in the Large Magellanic Cloud. These stars brighten and redden over the course of days to weeks, then fade – usually more slowly. The brightenings were typically 0.2–0.4 magnitude, the durations tens to hundreds of days. The authors proposed the informal name *bumpers* for these. Spectroscopic observations showed that almost all of these stars showed hydrogen-line emission; they were Be stars. But are they different from the bright, nearby Be stars that have been studied so long and so well? They certainly seem closely related.

9.2 Wolf–Rayet stars

Wolf–Rayet (WR) stars are Population I stars, with luminosities up to 10^6 $L_\odot$, and effective temperatures of 30 000 K and higher, lying just above the upper main sequence in the H–R diagram. They are notable for their peculiar spectra which combine the absorption lines of an OB star with strong, broad, high-ionization C, N, O, and He emission. This emission is so strong that the first WR stars were discovered (by Georges Rayet and Charles Wolf in 1866) by *visual* spectroscopy. They have since been divided into two or three spectroscopic groups: the WC stars, subclassified WC4-WC9, with strong C and O emission, and the WN stars, subclassified WN2-WN9, with strong N and He emission (figure 9.5); there are a few which are classified as WO – strong O emission. Although they are very rare, they play an important role in the chemical evolution of galaxies because of their very high mass loss – up to 10^{-4} $M_\odot$ a year. Because of their high luminosities and bright emission lines, WR stars are

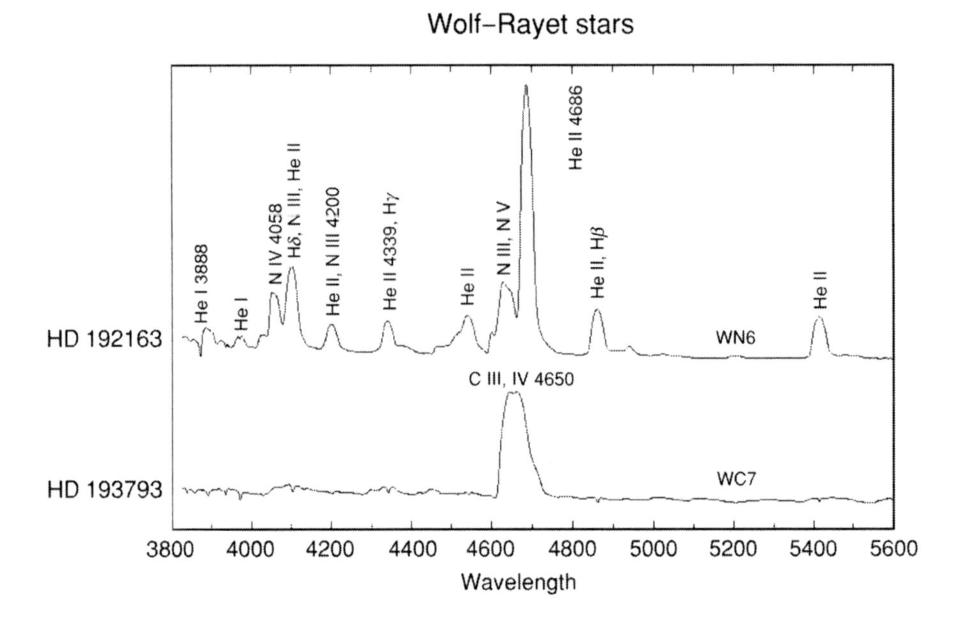

Figure 9.5 Spectra of two WR stars, HD 192163 which is of WN type, and HD 193793 which is of WC type. Note the dominance of the emission lines in each case. (From Richard O. Gray.)

easily identified, even in relatively distant galaxies, so they are a useful tool for studying stellar and chemical evolution in a variety of galactic environments.

Box 9.1 Star sample – Gamma Cassiopeiae

Between 1863 and 1867, before the introduction of photography, Father Angelo Secchi (1818–1878) carried out a remarkable study of the spectra of some 4000 stars, using a *visual* spectroscope on the telescope of the forerunner of today's Vatican Observatory. Most of the stars showed an absorption-line spectrum but, in an 1867 paper, Secchi noted that γ Cassiopeiae showed *emission* in the hydrogen-beta absorption line.

Like most Be stars, γ Cassiopeiae is variable in both brightness and spectrum. From 1866 to 1932, the star was relatively quiescent, with moderate to strong emission lines undergoing small variations. This was followed by an episode of spectacular variations from 1932 to 1942, when the star became a rather normal B star! By 1981, it regained its irregularly varying emission lines.

This star is a member of a visual double-star system, and it was recently found to be a member of a spectroscopic binary system with an orbital period of 203.59 days. The companion could be a normal star, but it could also be a white dwarf or even a neutron star. The system is also an X-ray source, and the X-ray emission could be due to matter accreting on to a compact star. Self-correlation analysis reveals a possible 0.4-day period, which may be due to non-radial pulsation, or possibly rotation.

Strangely enough, this prototype Be star is not well monitored. It is observed visually by the AAVSO, but its photometric variability is small. It is rather bright for photoelectric photometry, but some photoelectric observations are now being made. As for spectroscopy, very few professional astronomers monitor the changing spectra of Be stars. But amateur spectroscopists are beginning to fill the gap.

WR stars appear to be the helium-burning cores of stars with initial masses of 30–40 $M_\odot$. The outer layers of the star – up to 3/4 of the initial mass – have been stripped away by wind-driven mass loss. The typical rate of mass loss is 10^{-5} $M_\odot$ per year. These hot, dense stellar winds reach terminal velocities ranging from 750 to 5000 km s^{-1}. The sequence O–WN–WC–WO would then represent a sequence of the peeling off of successive layers. WR stars can show IR and radio emission from the gas and dust around the star.

About half of WR stars are noticeably variable, with the WN8 stars being considered to have the highest level of variability. There are 32 WR stars listed in the GCVS4. The brightest is γ^2 Velorum (WC8 + O9I) at V = 1.81. Two have periods listed: 3.763 days for EZ Canis Majoris, and 2.0 days for V919 Scorpii. See Moffat and Shara (1986) for light curves of a complete sample of northern WR stars.

The photometric variability of WR stars is generally small and extremely complex. They are hot and luminous and, on the H–R diagram, they are close to the region where theory suggests that they may be unstable to pulsation (figure 9.6). Like other massive stars, many are binaries, so they may be eclipsing or ellipsoidal variables. They may have small variations on time scales of minutes to hours, because of numerous random small-scale density enhancements ('puffs') embedded in their massive winds. If the emission lines vary in strength, then the brightness and colour will appear to vary also. One or two WR stars are known to erupt, notably HD 5980 in the Small Magellanic Cloud. This is an eclipsing binary, so the eruption may have occurred within an accretion disc around the companion. V444 Cygni (V range 7.92–8.22), with a period of 4.212424 days, is another well-known WR variable. Yet another complex example is DI Crucis (Oliveira *et al.*, 2004), which has a 0.3319-day period which may be an orbital period, and additional periods which could be due to non-radial

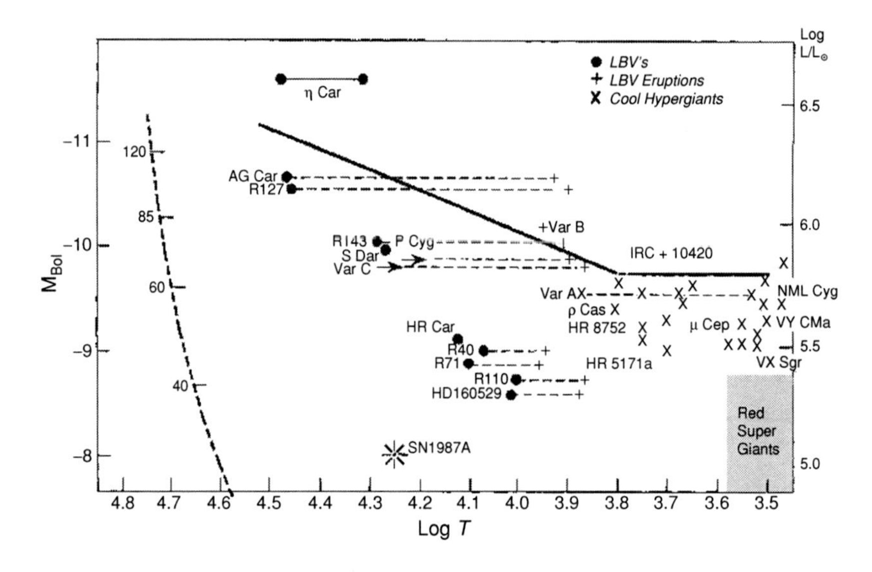

Figure 9.6 Location of hypergiants and WR stars on the H–R diagram. On the left is the upper main sequence, with the masses of the stars in solar units marked. The W-R stars lie in this region. Other blue, yellow, and red hypergiants are marked, including P Cygni, η Carinae, ρ Cassiopeiae, and the progenitor of SN1987A. The dashed lines represent the changes in the stars as they undergo major eruptions. The solid line is the empirical Humphreys–Davidson limit. (From R. M. Humphries.)

pulsation. But these authors warn that 'its true binary nature ... has not yet been demonstrated'.

The question of pulsation in WR stars is still not settled, though any pulsation must be of low amplitude. WR stars are among the targets for ultra-precise photometry by the MOST satellite (chapter 3). Figure 9.7 shows the MOST light curve of WR123 [HD177230]. It is a WN8 star; these tend to have the highest level of intrinsic variability. There is little or no variability on very short timescales; most of the variability is on timescales greater than a day, but is not periodic. There is a relatively stable 9.8 h period which may be due to pulsation, but this period does not agree with that of any expected radial or non-radial mode (Lefevre *et al.*, 2005).

See van der Hucht (2001) for a *Catalogue of Galactic Wolf-Rayet Stars.*

9.3 Hypergiant variable stars

Astronomers have known for many decades that the most luminous stars tend to be variable in brightness and spectrum (figure 9.8). Alpha Cygni (Deneb) was monitored extensively for radial velocity variations by Paddock and

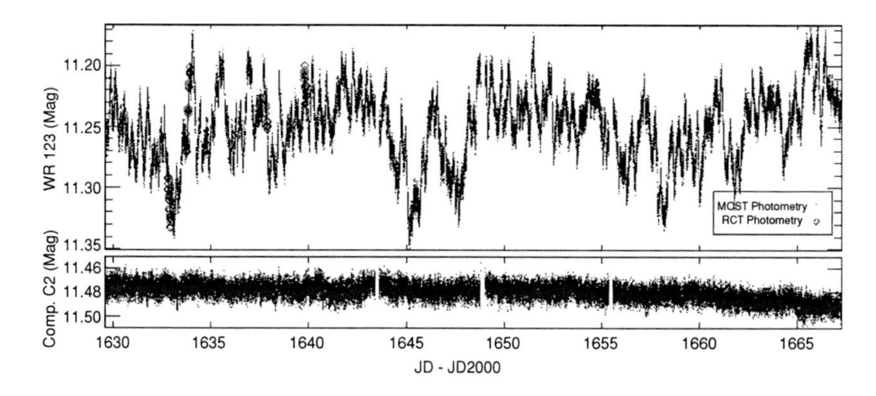

Figure 9.7 Light curve of WR123 (HD177230), a WR star, obtained by the MOST satellite. This is the most precise and detailed light curve yet obtained for a WR star (see text). (From S. Lefevre. T. Moffat, and the MOST Science Team.)

others, and found to vary on time scales of weeks. Blue hypergiant variables are sometimes called *Alpha Cygni variables* (they are classified ACYG in the GCVS4, and are described as non-radial multiperiodic), but I prefer the descriptive term *luminous blue variables.*

Burki (1978) carried out a comprehensive survey of photometric variability in supergiants. He found that, the more luminous the star, the larger the amplitude of variability. Furthermore, the characteristic time scale of the variability (which was in some cases periodic) was consistent with pulsation.

The most luminous supergiants, however, can vary in an even more dramatic way. That is true of blue, yellow, and red hypergiants. This is not surprising, since the luminosity of a hypergiant can be so great that the star is unstable against the pressure of its own radiation. This is sometimes described in terms of the *Eddington limit* – the upper limit to the ratio of the luminosity to the mass of a stable star. Arthur Eddington showed that this limit is about 40 000 in solar units. In these extreme conditions, the usual principles of stellar stability do not apply. Pulsational motions become chaotic, especially if they are non-radial, or if convection cells are important. Pulsation may drive off shells of material. Variable absorption, by this ejected gas and dust, is another source of variability in these stars.

9.3.1 *Luminous blue variables*

Here is another class of variable stars in which nomenclature is a problem. These are extremely luminous variables which vary on time scales from days to decades. They are sometimes called *S Doradus variables*, since S Doradus is one conspicuous member; at $M_v = -10$, it is the most luminous star in the Magellanic Clouds. Brighter members include P Cygni (figure 9.9) and η Carinae,

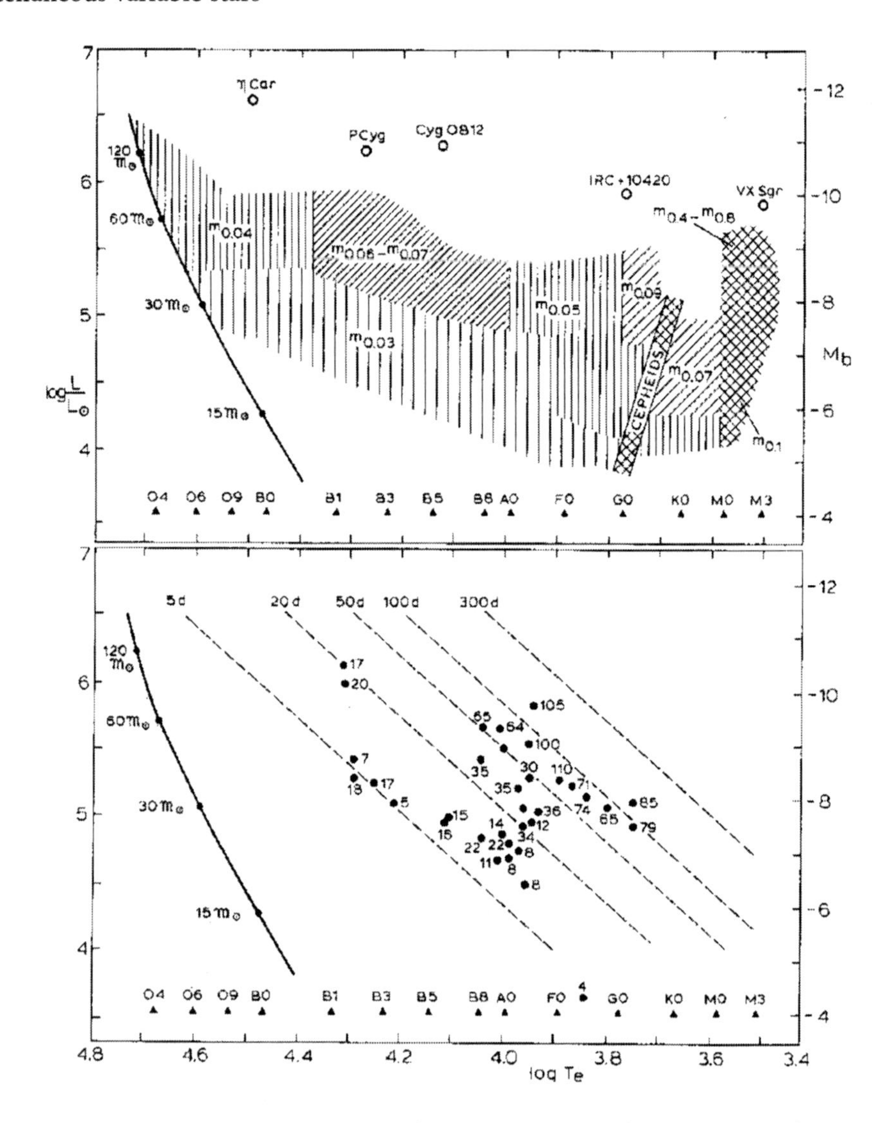

Figure 9.8 Top: the average full range of brightness variation for supergiants; a few extreme 'hypergiants' are marked. The amplitude of variability increases with increasing luminosity. Bottom: the characteristic time scales of the variability (numbers in days beside the dots); the dashed lines show the trends. The time scales are consistent with pulsation. (From Maeder, 1980.)

discussed below, and AG Carinae and HR Carinae, with photographic ranges of 7.1–9.0 and 8.2–9.6 respectively.

LBVs are rare but, because they are so luminous, they can be seen at great distances, unless they are hidden behind the obscuring dust in the disc of our galaxy. The prototype S Doradus is actually in another galaxy!

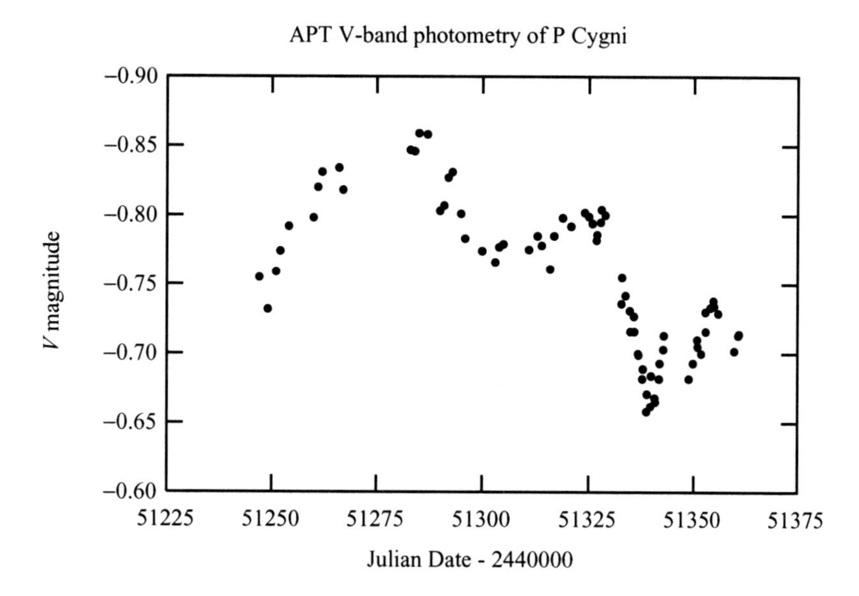

Figure 9.9 The photometric *V*-magnitude variations of P Cygni, obtained with a robotic telescope. (From Percy *et al.*, 2001b.)

P Cygni, a B1 hypergiant, is the prototype of 'stars with expanding atmospheres', as determined from their spectra. In 2000, the 400th anniversary of its discovery was celebrated (de Groot and Sterken, 2001). The strong lines in its spectrum consist of sharp emission lines, accompanied by strong, violet-displaced absorption lines which indicate that matter is moving away from the star. By modelling these line profiles, astronomers can conclude that P Cygni is losing mass at the rate of 5×10^{-6} $M_\odot$ a year, a rate which has been independently confirmed by radio astronomical techniques.

The expansion velocity increases with increasing distance from the star, being only 20–30 km s^{-1} in high-excitation lines formed near the photosphere, but 100–200 km s^{-1} in lower-excitation lines formed further out. This indicates that the atmosphere is being accelerated outward, due to a combination of radiation pressure and dynamic instability (figure 9.10).

The photometric history of P Cygni indicates that it underwent major outbursts in brightness in the 1600s. During these outbursts, its M_v approached −10; it is now two magnitudes fainter. There are also photometric and spectroscopic variations on time scales of 15–100 days (figure 9.9); these may be pulsational in nature. Some of the most comprehensive studies of LBV variability have been carried out by Christiaan Sterken and his collaborators, as part of the Long-Term Photometry of Variable Stars project. See Sterken and Jaschek (1996) for an excellent summary of the light curves of LBVs.

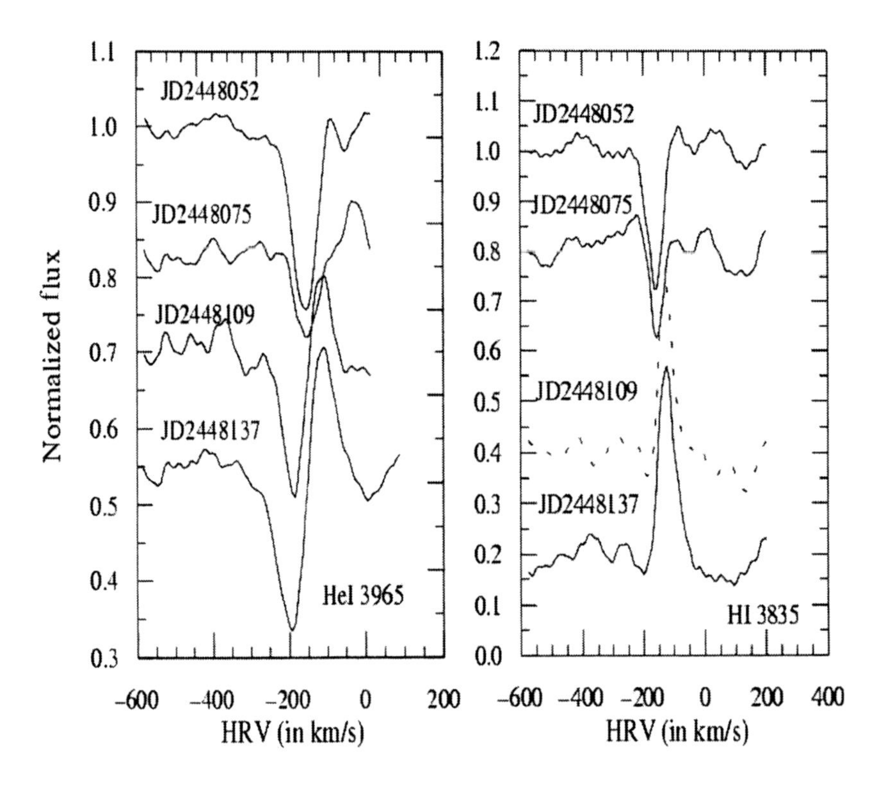

Figure 9.10 The HeI λ3965 and HI λ3835 lines in the spectrum of P Cygni, and their variation with time (the Julian Dates are marked). Note the absorption features displaced to negative radial velocities ('violet-shifted'). These are called *P Cygni profiles*, and are characteristic of stars losing mass through a stellar wind. (From Markova, 2000.)

Eta Carinae is the most spectacular of the S Doradus variables. Its photometric history is shown in figure 9.11. At one time, it was the second-brightest star in the night sky. It subsequently faded by 10 magnitudes, as dust formed around it, but it has gradually brightened in the last century. This may be due, in part, to the gradual dispersion of the dust around the star. The star also varies in visual and near-IR brightness with a period of about five years, probably due to binarity. There are also variations on a time scale of weeks, the cause of which is uncertain. R81 is another LBV which is a binary – an eclipsing binary with a period of 74.59 days, and a primary eclipse depth of 0.5 magnitude.

The nebula around η Carinae is even more spectacular than its light curve. Some of the material may have been ejected in the 1850 outburst, and the star appears to be losing mass at the prodigious rate of 10^{-2} to 10^{-1} $M_{\odot}$ a year. The spectrum shows P Cygni line profiles, with evidence for a succession of ejected shells. The star is a strong source of 3–10 μ IR emission – the strongest such source outside the solar system. This emission presumably arises from dust which has been warmed by stellar radiation.

LBVs in other galaxies are often called *Hubble–Sandage variables*, after Edwin Hubble and Allan Sandage who identified them in M31, M33, and a few other galaxies, starting in 1953.

LBVs are massive stars in an advanced stage of evolution. Because of their high luminosity, the balance between inward gravity and outward radiation pressure is tenuous at best. Although the underlying stars appear to be B–F type hypergiants, their spectra and colours are not normal, because of the large amounts of ejected gas and dust around them.

The major mass-losing eruptions, which LBVs undergo, consist of sudden and dramatic decreases in apparent temperature, due to the formation of a thick stellar wind (Humphries *et al.*, 2006). These occur at approximately constant luminosity, close to 10^6 $L_\odot$. The example described by Humphries *et al.* (2006) is Variable *A* in M33, one of the original Hubble–Sandage variables. In the 1950s, its V magnitude was close to +15, corresponding to M_v of about −9.5. It subsequently faded to V = +18, and also became much redder; its $B - V$ colour changed from +0.5 to +1.5. Since 1990, the colour has gradually become less red, and is approaching its former value, as its cool, dense wind disperses after a 50-year eruption!

What causes these eruptions? One possibility is that the star's position in the H–R diagram – luminosity near the Eddington limit, and possibly pulsational instability – makes it susceptible to dynamical instabilities in which steady mass loss becomes a stable state – at least for a finite period of time.

Despite their rarity, LBVs play an important role in the chemical evolution of galaxies, and in the turbulent dynamics of their interstellar material.

Box 9.2 Star sample – Eta Carinae

Eta Carinae (HD 93308, V = −0.8 to 7.9) is a most remarkable object – one of the two most conspicuous examples of a *luminous blue variable* (see Morse *et al.* (1998) for a comprehensive review). Prior to the 1800s, it had a magnitude of 3±1 but, in 'the great eruption of 1843', it rose to magnitude −0.8, and became the second-brightest star in the night sky (figure 9.11). At one time, it was considered to be some kind of supernova! It has left behind a bipolar 'Homunculus (Little Man) Nebula', which has become one of the most famous HST images. This nebula is the result of a fast stellar wind, which has ejected at least 5 $M_\odot$ of material, probably due to the radiation pressure from this massive (100 to 150 $M_\odot$), ultra-luminous (up to a 10^6 $L_\odot$) star. The star is close to or above the *Eddington limit* – the luminosity at which the star would be unstable against the outward pressure of its own radiation.

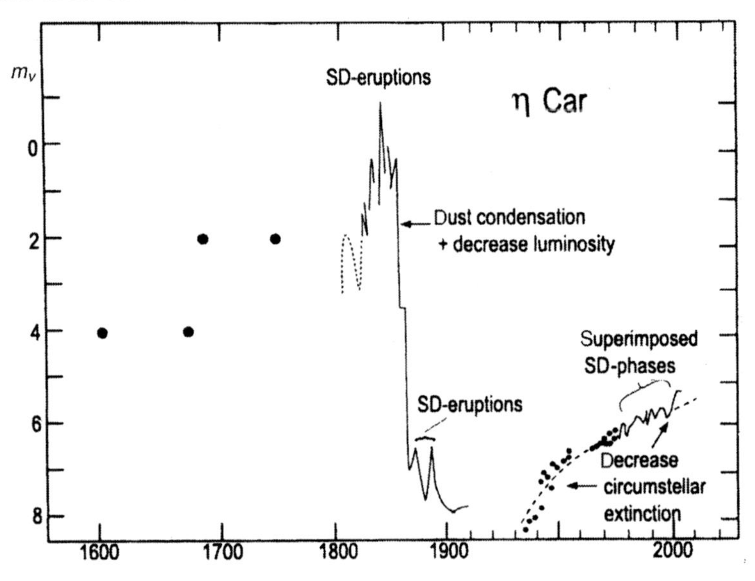

Figure 9.11 Long-term light curve of η Carinae, showing the great eruption in the nineteenth century, the subsequent fading, and the more recent recovery, as the obscuring material disperses. (From Van Genderen *et al.*, 2001.)

Eta Carinae continues to perform, and is being observed across the electromagnetic spectrum, from the ground and from space. After fading to magnitude 7.9 a century ago, it has gradually brightened, and is presently (2006) visible to the unaided eye, and brightening at an accelerated rate of 0.3 magnitude per decade. Astronomers believe that the brightening is due to the dispersal or destruction of obscuring dust. If the star continues to brighten at the present rate (which is possible but unlikely), it could dominate the night sky once again. There is a problem measuring this star, however: ground-based observations of the 'star' actually include bright portions of the nebula; HST observations, which can resolve the star, show that it is brightening even faster than the nebula. Another problem is that the spectrum of this source is dominated by emission lines, so the brightness depends on whether the wavelength range being used by the observer includes these emission lines.

Eta Carinae, being in the southern sky, is accessible to the most prolific visual observer of all time, Albert Jones – 'the man with the photoelectric eyes'. Jones' observations provide a unique, multi-decade picture of this slowly varying star.

Eta Carinae varies spectroscopically with a period of 5.5 years, probably due to binarity; the companion may possibly be almost as massive as the primary star. There is both visible and X-ray variability which occurs on this same time scale. Most likely, the orbit is highly elliptical, and seen almost edge-on. Binarity could explain why η Carinae's wind is bi-polar. It could also result in the collision between two winds, which obscure our view of the stars, and could also explain the strong X-ray emission from the star.

But what triggered the Great Eruption, when the star (or possibly even the companion star) 'went super-Eddington'? Was it triggered when the two stars were at their closest, in their elongated orbit? Will it happen again? Keep watching!

9.3.2 *Rho Cassiopeiae and the yellow hypergiants*

Rho Cassiopeiae is a yellow hypergiant which is one of the most luminous stars in our galaxy. It is also one of the largest: if it replaced the sun, its outer atmosphere would stretch well beyond Mars. It lies on the upward extension of the Cepheid instability strip in the H–R diagram, so it is not surprising that it pulsates – somewhat like a Cepheid, but on a much longer time scale, and with much less regularity. The most prominent time scale for variability ranges from about 300 to about 820 days. The amplitude is typically 0.3 mag in *V*. Every decade or so, however, there are larger variations. These are accompanied by spectroscopic 'episodes' in which the mass loss from the star increases significantly, throwing off a shell of material. These episodes are manifested, in the light curve, by cycles with larger than average amplitudes, and longer than average time scales. The star may become unusually red, as the photosphere expands and cools; the spectrum of ρ Cassiopeiae has been classified at times as cool as type M. The star has even been classified as an R Coronae Borealis, because the large-amplitude cycles can appear like small fadings.

Humphries *et al.* (2006) call attention to the similarities between the eruptions of LBVs, and those of ρ Cassiopeiae. The thick stellar wind produces a cool 'pseudo-photosphere' which mimics that of a much cooler star. And ρ Cassiopeiae and LBVs are both hypergiants which show evidence of pulsational instability.

In late 2000, ρ Cassiopeiae varied by over a magnitude in *V*, the photosphere expanded and cooled from 7250 to cooler than 3750 K, and the mass loss reached 0.05 $M_{\odot}$ per year, or 10 000 earth masses in only 200 days, during the outburst (Lobel *et al.*, 2004). This was deduced from systematic, careful observations of the star's spectrum, which showed both pulsational expansion, and evidence of an expanding shell of material (figure 9.12). These spectroscopic observations were

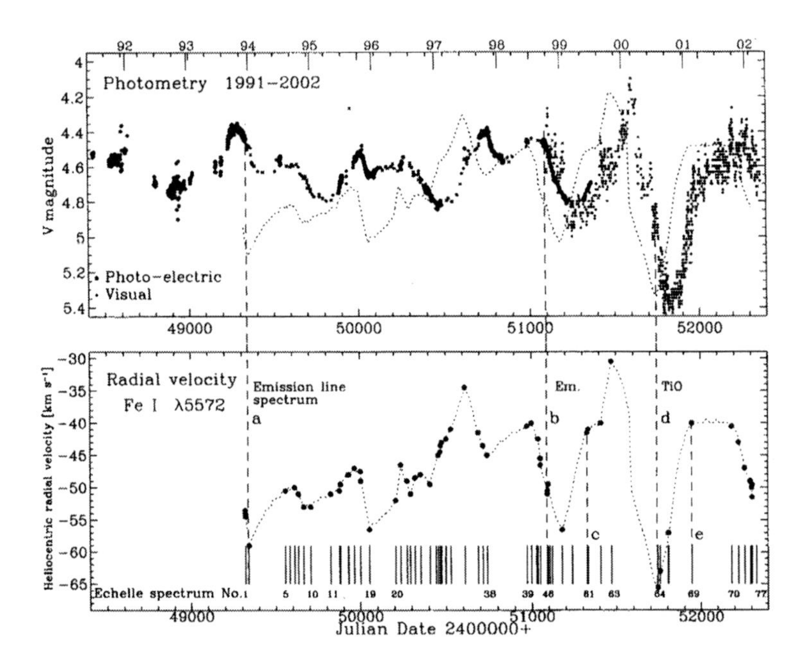

Figure 9.12 The light curve of ρ Cassiopeiae (filled symbols, top) compared with the radial velocity curve (dotted line, bottom). The radial velocity curve for Fe λ5572 (connected filled circles) shows a strong increase of the photospheric pulsation amplitude before the 'outburst' of fall 2000 when TiO bands develop (marked 'TiO'). The light curve is a mixture of photoelectric (Percy *et al.*, 2000,) and AAVSO visual observations. (From Lobel *et al.*, 2004).

combined with visual and photoelectric photometry from the AAVSO International Database. Stars like ρ Cassiopeiae are major players in the recycling of matter in our galaxy.

Lobel concludes, tantalizingly: 'could the "millennium eruption" have been only the precursor of a stronger eruption yet to blow'?

Rho Cassiopeiae has a spectroscopic 'twin' V509 Cassiopeiae (HR 8752) which has very similar physical and variability properties (as well as being next door to each other in the sky). There are a dozen or more other yellow hypergiants, but none as well studied as ρ Cassiopeiae.

9.4 R Coronae Borealis (RCB) stars

These variables are unlike any others, and it is only for convenience (or out of frustration) that we include them in this chapter. They are rare low-mass yellow supergiant variables that have bizarre chemical compositions. They spend years or decades at normal (maximum) brightness. Suddenly and unpredictably,

Table 9.1. *Notable R Coronae Borealis stars*

Name	GCVS Type	Spectral Type	Range	Most Recent Fading; Other Comments
S Aps	RCB	C(R3)	9.6–15.2	2451700
XX Cam	RCB	G1I(C0–2,0)	8.09–9.8	no recent fadings
V854 Cen	RCB	–	7.13–14.1	2453380
R CrB	RCB	C0,0(F8pep)	5.71–14.8	2452680, prototype, bright
W Men	RCB	F8:Ip	13.4–18.3	2451870, in LMC
RY Sgr	RCB	G0Iaep(C1,0)	5.8–14.0	2453270, bright

they *decrease* in brightness by up to ten magnitudes, then slowly return to normal. So their light curves look like novae in reverse, though there is absolutely no similarity in the mechanism involved.

R Coronae Borealis itself was discovered to be variable by Pigott, in 1795, and it has been observed regularly ever since. Because the fadings are large and unpredictable, amateur astronomers can make an important contribution to the study of these stars. They should monitor them regularly and, if they observe them to be significantly fainter than normal, they should notify organizations such as the AAVSO. These organizations can then notify professional astronomers, who can study the progress of the fadings with more specialized equipment.

Peter L. Cottrell and his group at the University of Canterbury, New Zealand, have carried out important long-term studies of RCB stars, and other kinds of slow variable stars, using the photometric and spectroscopic facilities of the Mount John University Observatory, which is available to them on a continuous and long-term basis. This illustrates the advantage of having access to a high-quality 'local' observatory with both photometric and spectroscopic instruments – and, in the case of the MJUO, excellent resident telescope operators and observers. Skuljan and Cottrell (2004) summarize their recent results, and provide a brief discussion of current issues in the understanding of these stars.

The sudden decreases in brightness in these stars are usually accompanied by changes in colour. The relation between the absorption ΔV and the reddening $\Delta(B - V)$ and $\Delta(U - B)$ usually follows the normal relation for dimming by interstellar-type dust. Occasionally, however, 'neutral extinction' (with no change in colour) is observed. There are several possible interpretations: the dust particles causing the dimming may have sizes different from normal, or the dimming may be due to a geometrical eclipse of part of the star's surface by a dense cloud of obscuring material.

IR studies of these stars are extremely important, because the dusty material around the star absorbs starlight, and re-emits the energy as IR radiation.

Unfortunately there are relatively few IR telescopes (and observers) able to monitor specific stars for months, or years. IR astronomy is one field which is becoming possible and desirable for amateur astronomers.

When the visual brightness of the star begins to fade, the 3.5 μ IR emission is initially unaffected; it does vary, on a time scale of years, but this variability is not correlated with the variations in visual brightness; it is probably more related to the average rate at which the star forms dust, over many years. One or two of these stars are known to have large fossil dust shells around them – a consequence of dust formation over many millennia.

Spectroscopic changes also occur. In its normal state, an RCB star is a (usually) F or G type supergiant, with a very large abundance of carbon (ten times solar) and a very small abundance of hydrogen (1/100 000 solar). A few days after a fading begins, the normal absorption-line spectrum is replaced by a narrow-line emission spectrum characteristic of a stellar chromosphere; the sodium D and calcium II H and K lines are prominent. As the fading progresses, the emission lines come from atomic levels of progressively lower excitation. At the bottom of the fading, these lines may show outflow velocities of up to 200 km s^{-1}. Later, broad (200 km s^{-1}) emission lines of sodium D, and calcium H and K may occur. By this time, the lines have become very weak, and the star has become very faint.

Some and perhaps all RCB stars, including R Coronae Borealis and RY Sagittarii, the two brightest, show another kind of variability – pulsation – which may or may not be connected with the fadings. RY Sagittarii has a period of 38.6 days and a range of 0.5 in *V*. R Coronae Borealis has a period of about 40 days and a range of 0.1, but its variability is much less regular than that of RY Sagittarii. The pulsation could, in principle, be used to derive the mass and luminosity of the star. It is *possible* that the pulsation is responsible, in some way, for triggering the ejection of the obscuring material from the star; in RY Sagittarii and V854 Centauri, the fadings seem to be correlated with the pulsation phase (figure 9.13).

There are only about 35 RCB stars listed in the GCVS4, and 37 spectroscopically confirmed variables in Clayton's (1996) review. They are a very heterogeneous group, ranging from spectral type O to the coolest spectral type R (the carbon-rich equivalent of type M). Most, however, are F or G supergiants; many may be misclassified. Or all of them may share a common 'RCB process'; this would include the four hot RCB stars.

Their spectral types are all qualified as 'peculiar' because of the over-abundance of carbon and under-abundance of hydrogen. They seem to be old Population I stars, with masses about equal to that of the sun. None is known

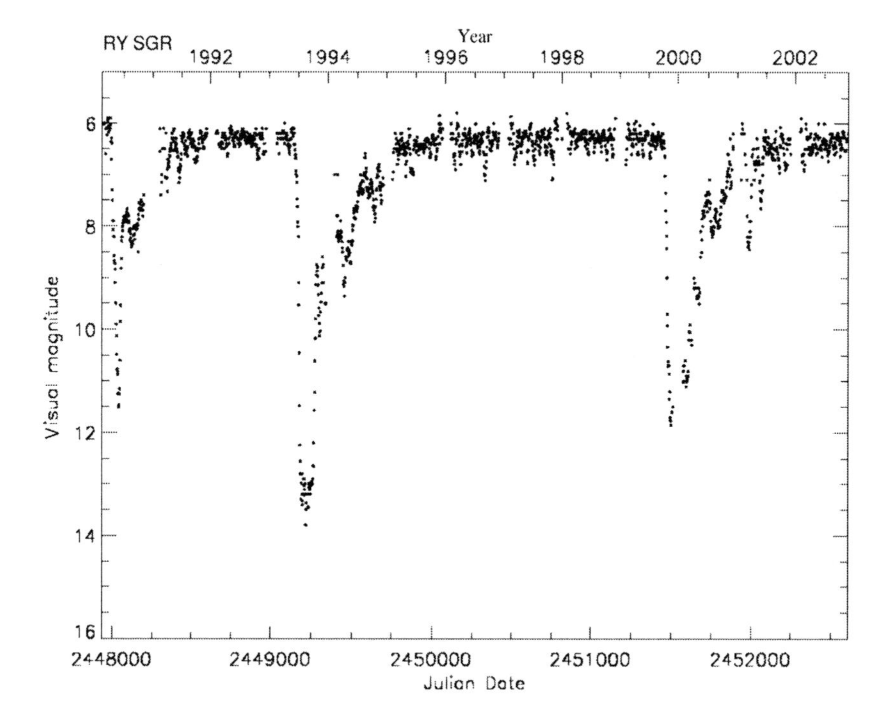

Figure 9.13 The light curve of RY Sagittarii from 1991 to 2003. Three declines are visible; their depths, durations, shapes, and separations are apparent. There is also a 39-day pulsational variability which is marginally visible at the maximum of this light curve, even though its amplitude is small. (From AAVSO.)

to be a binary star. Surveys of RCB stars in the Large Magellanic Cloud suggest that there may be many undiscovered variables of this type in our own galaxy. Although the characteristic time between fadings is 1100 days, there is a wide variation from star to star. Some individual variables can go for decades without fading, and may therefore not yet have been discovered.

The most interesting and important property of RCB stars is their chemical composition. The ratio of carbon to hydrogen is 25; in normal stars, hydrogen is vastly more abundant. The ratio of carbon to iron is 35 times normal. How did they acquire their abnormal composition? Was carbon synthesized in the interior, and then mixed to the surface by convection currents? Or were the surface layers removed by mass loss, revealing the carbon-rich layers underneath? The masses and compositions of these stars, along with theoretical models, suggest the latter. These stars are thus the cores of 'double-shell-source' stars which have lost their hydrogen-rich envelopes. This envelope might appear as a planetary nebula, except that the star was probably

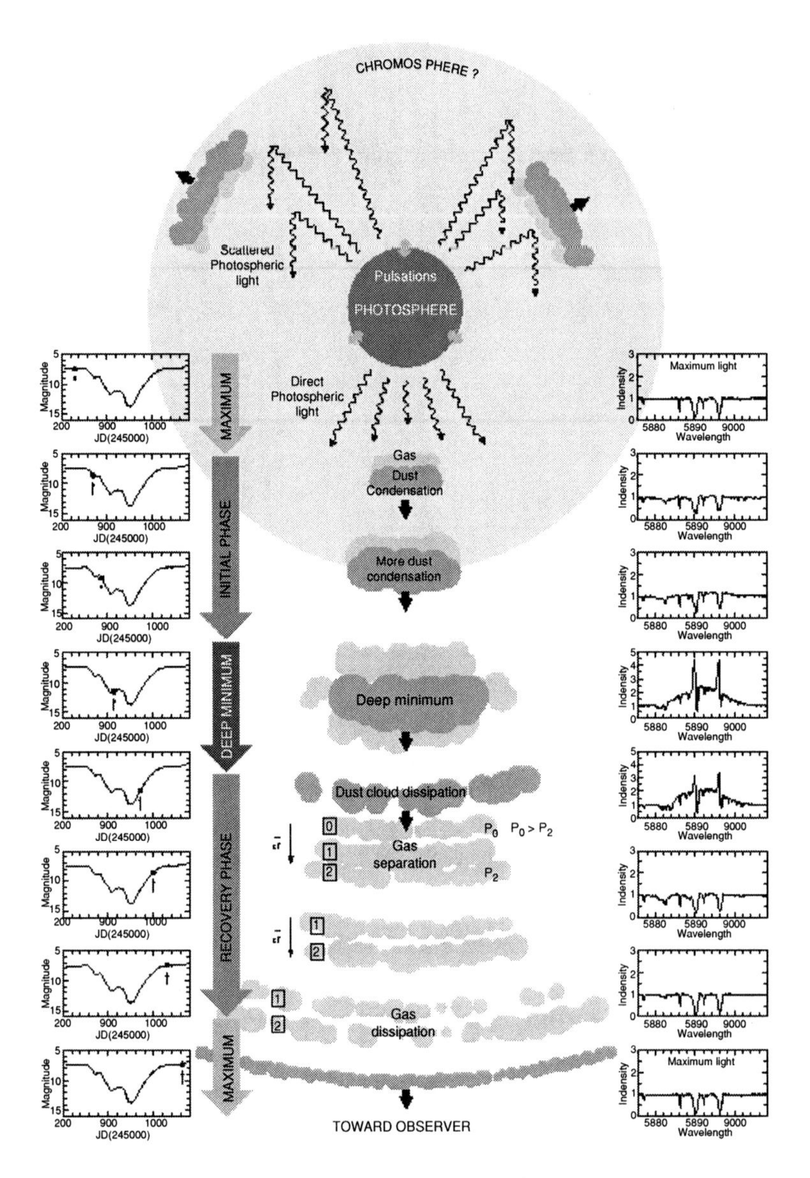

Figure 9.14 Spectrum and model for an R Coronae Borealis star. The star is a carbon-rich, hydrogen-deficient yellow supergiant. A small, discrete cloud of dust forms in the photosphere, perhaps as a result of pulsation, and gradually moves outward at up to 200 km s^{-1}. If the observer lies in the right direction, they will see the cloud eclipse the photosphere, then the chromosphere. Eventually, it becomes optically thin, and joins the multitude of other outward-moving cloud remnants. In this complex diagram, the shape of the Na D line is shown on the right; the corresponding phase of the light curve is shown on the left; the central panel shows the successive phases in the ejection of the dust cloud. (From P. L. Cottrell and L. Skuljan.)

never hot enough to light it up. There is some evidence, from pulsational period changes, that RY Sagittarii is evolving from red (cool) to blue (hot) in the H–R diagram, which suggests that, eventually, it will become a hot *helium star*.

There are at least two possible models for the evolutionary history of RCB stars. In one, a binary system of two white dwarfs - a helium white dwarf and a carbon-oxygen white dwarf - merges, and one of the stars is inflated to supergiant size as a result. Given the frequency of binary systems, the duration of the merger process, and the rarity of RCB stars, this hypothesis is not unreasonable. In the other model, a single star on its way from the AGB to the white dwarf stage undergoes one final helium shell flash within its interior, also inflating it to supergiant size. At least four stars show evidence for being in a final helium shell flash: V4334 Sagittarii (Sakurai's Object), FG Sagittae, V605 Aquila, and V838 Monocerotis. All four are located within planetary nebulae, or other shells of gas and dust. There is no reason why both mechanisms could not be viable (Clayton, 2002).

Next, we might ask what is the cause of the fadings, the defining characteristic of an RCB star? An *orbiting* dust cloud is unlikely, because it would produce *periodic* fadings. A cool binary companion was once proposed as the explanation for the IR variations but, again, there is no evidence for periodicity. As early as the 1930s, E. Loreta and John A. O'Keefe independently suggested that the carbon-rich nature of these stars might be a clue: clouds of amorphous carbon (soot) might form above the photosphere of the star, obscuring the light until the clouds were blown away by the pressure of the star's radiation. This might explain the spectroscopic variations: after the photosphere disappears, the higher chromosphere remains visible, but then gradually loses its source of excitation.

This model does *not* explain why the IR radiation remains constant when the visual light fades. If the soot forms a shell around the star, then the light absorbed by the soot should be re-emitted as IR radiation.

This suggests an alternative model, in which the dust forms in discrete random clouds above the photosphere. Occasionally, a cloud will eclipse the photosphere, leaving the higher chromosphere visible. As the cloud expands, and cools, it blocks the chromosphere as well. The dust grains absorb starlight, which helps to push them (and the gas in the atmosphere) outward, forming a large shell around the star. It is from this expanding shell of gas and dust that the high-velocity spectroscopic features are formed.

But there was a problem with this model: carbon would normally not condense until it was at least 20 stellar radii away from the star. Several lines of evidence – including the possible connection between the fadings and the phase of the star's pulsation – suggest that the clouds must form much closer to the star. Perhaps the pulsation produces shock waves which facilitate the formation of the dust, as it does in Mira stars. Final confirmation of this model awaits a better understanding of the chemical and dynamical processes in these pulsating stars' atmospheres.

10

Epilogue

10.1 Variable stars and astronomical research

Throughout this book, we have described the basics of variable star astronomy, the many ways in which variable stars have increased our understanding of the universe, the frontier areas of this field, and the questions that remain. What part will variable stars play in astronomical research in the next generation? Here are a few possibilities:

- The sun is the nearest and most important star, because its energy output and its activity both affect the earth, and life on it. It may be 'average' and relatively benign, but the nature and cause of its activity are still poorly understood. Study of other sun-like stars will continue to help.
- Sky surveys will continue to expand in terms of depth and continuity of coverage, turning up rare stars which are 'Rosetta stones' for understanding the processes which occur in stars.
- Continued studies of pulsars – especially binary pulsars – will provide unique information about the laws of physics which apply in the most extreme environments in the universe.
- Optical interferometry and the *GAIA* astrometric satellite will usher in a new era in 'precision astrophysics' by measuring the physical properties of stars to better than 1 per cent.
- Asteroseismological satellites such as *MOST* and *COROT* will provide stringent tests of theories of the structure and evolution of stars.
- Faint, distant supernovae will lead to a better understanding of the nature of dark energy. But astronomers will need to know their peak

magnitude and intrinsic colours with better accuracy, and with fewer possible systematic errors.

- A better understanding of gamma-ray bursts will come from more and better observations of these rare, bizarre events, which are the most energetic phenomena in the universe since the Big Bang.
- The mechanism of supernova explosions is still poorly understood, since the stars themselves are complex, and the explosive process is fast and violent. The increasing power of computers will be necessary to solve this and related problems.
- New infrared and sub-millimetre facilities such as the Atacama Large Millimeter Array (ALMA) and the James Webb Space Telescope will probe the earliest phases of star formation and evolution.

But these will be the 'front page' discoveries. At the same time, there are vast numbers of loose ends to be tied up, unspectacular but important questions to be answered, improvements in knowledge to be made through more and better studies of both large samples and individual objects. And, until we have surveyed every star, on every time scale and in every region of the spectrum, we must realize that there may be questions which we have not yet thought to ask.

Some of the front-page discoveries will be made by 'big science' facilities such as HST. But others will be made by creative astronomers using more modest facilities and techniques. The technique for discovering exoplanets was honed on a 2m-class telescope. Adaptive optics techniques are enabling ground-based telescopes to achieve resolution comparable to that of HST.

10.2 Variable stars and amateur astronomy

Astronomy is unique, among the sciences, in the extent to which *amateurs* can contribute to it, and nowhere is this more true than in the study of variable stars. Amateurs can discover variable stars, and observe them. Many variable stars are irregular or unpredictable. Many have time scales which are too long to fit into the typical observing program of the professional. There are so many variable stars that professionals cannot monitor them all. Throughout this book, I have highlighted areas in which amateurs can and have made contributions.

What is an amateur astronomer? By one definition, it is someone who loves astronomy, and cultivates it as a pastime or hobby. Historian Tom Williams has proposed a more stringent definition: an amateur astronomer is someone who does astronomy with a high degree of skill, but not for pay. Amateur astronomers usually do not have extensive education in astronomy, though some do. Many are professionals in other fields, and bring their own unique and valuable skills to astronomy.

In North America, there are hundreds of thousands of people who are interested in astronomy, mostly as 'recreational sky observers' or 'armchair astronomers'. These people play an important role in astronomy, by providing grass-roots interest and support. A few, however, may have an interest in contributing to research, and may wish to climb the ladder from novice, to journeyman, to master. Perhaps this book will inspire them to begin. Visual observing of variable stars is easy, once you start!

The distinction between amateur and professional was not always so clear. A century or two ago, there were few professional scientists. It was possible for a broadly educated person to work at the frontier of almost any branch of knowledge. Herschel, Piggott, and Goodricke were all amateurs. Nowadays, science is more specialized, but there is still a continuous spectrum of astronomical workers, from 'pure amateur' to 'pure professional'. And there is more to astronomy than pure research. There are variety of important non-research careers, not all of them requiring a Ph.D., which are essential for the support and advancement of astronomy: engineers, technicians, computer scientists, managers, librarians, archivists, teachers, writers, other communicators, and many more. Amateurs can contribute in all these areas.

At the same time, astronomy has become a 'big science', with facilities costing a billion dollars or more. (This is not necessarily a bad thing for amateur astronomy, as we shall note later.) But there are still many observatories which fall between the extremes of the backyard spyglass and the world's great observatories. Many colleges and some schools have small telescopes, often used mainly for teaching by astronomers who are not strongly research oriented. Indeed, many astronomy instructors in both schools and small colleges (at least in North America) consider themselves to be amateur astronomers. In the US, there is NASA and NSF funding to support student research with these facilities. There are programs which provide summer research opportunities for undergraduates; these were pioneered by the Maria Mitchell Observatory, on Nantucket Island, using variable star projects. In Europe and Japan especially, there are 'public observatories', devoted mainly to public and school education, but used also for research by the staff or by skilled local amateurs.

Amateurs may work alone, or in groups. The groups may be local clubs, branches of national organizations such as the Royal Astronomical Society of Canada, variable star 'sections' of larger organizations such as the British Astronomical Association (BAA), and the Royal Astronomical Society of New Zealand (RASNZ), or specialized organizations devoted almost entirely to variable star observing. The groups offer a number of advantages: shared skills and knowledge, feedback and motivation, coordination, and – not least – camaraderie. The largest and best-known group devoted to variable star observing is the American Association of Variable Star Observers (AAVSO). I should stress, however, that the AAVSO is not the only group, and its philosophy is not the only philosophy. The BAA, for instance, has gathered observations of stars for the specific purpose of analyzing and interpreting them, and publishes the results regularly in the *Journal of the BAA*. The Belgian Vereniging Voor Sterrenkunde has had an active variable star observing program for many decades, and has published three decades of results in Broens *et al.* (2000).

The AAVSO was founded in 1911, primarily through the efforts of William Tyler Olcott, a lawyer and amateur astronomer, and Edward C. Pickering, director of the Harvard College Observatory. Its purpose is to coordinate variable star observations — made primarily by skilled amateur astronomers — evaluate the accuracy of the observations, compile, process, and publish them, and make them available to researchers and educators. AAVSO data and services are sought by astronomers to: (a) receive real-time, up-to-date information about unusual stellar behaviour; (b) assist in scheduling and executing variable-star observing programs, using ground-based or space telescopes; (c) request simultaneous optical monitoring of variable stars being observed by other techniques; (d) correlate optical data on variable stars with other photometric, spectroscopic, or polarimetric data; (e) carry out analysis of the (especially) long-term behaviour of variable stars; and (f) enhance astronomy education programs. Today, the AAVSO has over 1000 members, and annually receives 500 000 observations from 550 observers worldwide. The measurements are entered into the AAVSO International Database, which contains over ten million measurements of several thousand stars. Many of these measurements are made by other variable star observing groups, and sent to the AAVSO for archiving. The measurements are primarily visual, but the demand for these visual measurements has actually *increased* by a factor of 25 since the 1970s – partly as a result of major collaborations in space astronomy. (This is one way in which the 'big-science' aspects of astronomy have had a positive effect on amateur astronomy.) The AAVSO also has photoelectric and CCD observating programs, nova and supernova search programs, and a solar program. More recently, the AAVSO has coordinated a network of observers to study the short-lived optical transients of gamma-ray bursters (GRBs) – the most powerful explosions in the universe. The AAVSO works with observers, mentoring them and helping them to develop their observing programs. A lot of useful information is on their website (www.aavso.org). Observations, and their interpretation, are published in the AAVSO *Journal* and *Monographs*. One of the most popular and useful features of the AAVSO website is the Light Curve Generator. The user can plot a light curve for any star in the database, for whatever time period they choose. They can highlight their own observations, and see how they compare with others.

For me, one of the great advances in variable star astronomy came when it became possible to plot the entire database of AAVSO visual observations for any star, going back many decades. Suddenly, the *long-term* behaviour of variable stars was easy to see!

Some variable star observing groups have made their observations immediately available to professionals and others using the Internet. Other groups, such as the AAVSO, have been more reluctant to do so, partly for reasons of quality control. The AAVSO has recently completed a project to validate (quality control) all of its measurements, going back for a century, and make them freely available on the Internet. So both past and present observations are freely available, and quality controlled.

This highlights another possible contribution of amateur astronomers to variable star research: data analysis. The AAVSO International Database is only one of many photometric databases now available on-line. Software for time-series analysis is also freely available, some of it developed and disseminated by the AAVSO. Many amateur astronomers have the expertise in computing and statistics to master the 'black art' of time-series analysis. There is lots to do.

In recent years, amateur astronomy has been revolutionized by several technical developments: good-quality, reasonably priced off-the-shelf computerized small telescopes; affordable CCD cameras; powerful desktop computers; and e-mail and the Internet as tools for communication and sources of information. For $US 100 00, the price of a good home theatre or used car, an individual or club can acquire such an equipment package. This revolution, and related developments are outlined in the proceedings of a conference on *Amateur–Professional Partnerships in Astronomy* (Percy and Wilson, 2000), especially the paper by Leif Robinson. A sparkling example is the *Center for Backyard Astrophysics* network coordinated by Professor Joe Patterson, a pro-am network which has published numerous papers on cataclysmic variables. Many professionals encourage and support pro-am collaboration; the American Astronomical Society, for instance, has a Working Group on Pro-Am Collaboration, in which I have participated for many years (see http://www.aas.org/wgpac).

But there is another trend which is not so healthy. In most countries, amateur astronomy is facing – in my opinion – a severe challenge: most amateur astronomers are middle-aged white males. Let us hope that the lure of variable stars, and the rest of the sky's treasures, can attract a more diverse following, especially young people, and women! One possible way of attracting more young people is by incorporating variable stars into science and math education.

10.3 Variable stars and science education

Science education and literacy are essential in today's world, not just for the health of our economy, but for the health of our environment, our bodies, and even our culture. Citizens must be scientifically literate in order to make rational decisions in a technology-based and information-based society. Surveys show that, while most people in the industrialized countries are interested in science, their scientific literacy is low. They have a variety of deeply rooted misconceptions about scientific subjects, and deeply rooted beliefs in pseudo-science and superstition.

Education is also important for attracting and training the next generation of scientists. In the US, for instance, there is concern about the low number of students who study science – especially physical science – in high school and university. These low numbers result, in part, from the popular image of science, and scientists, and from the way that science is taught in schools. In my own university department, we always hope that there are a goodly number of students who are potentially interested in becoming astronomers, and they have the necessary knowledge, skills, and motivation to do so.

Educators know a lot about effective teaching and learning of science. The problem is in applying this information. Rote memorization of textbooks and lectures is insufficient. Hands-on activities are a start, but students' minds must be motivated and engaged for effective learning.

Astronomy has many general applications to education. It provides an example of an alternative to the usual scientific method (experiment and theory), one involving observation and simulation as well. It can be used to illustrate topics such as gravitation, light, and spectra. It enables students to think more abstractly about scales of time, distance, and size. If properly taught, it can promote rational thinking, and an understanding of the nature of science, through examples from the history of astronomy, and from present issues such as widespread belief in the pseudo-sciences. It is the ultimate interdisciplinary subject, and cross-curricular connections are highly valued in modern school curriculum development. It can attract young people to science and technology, and hence to careers in these fields. It is an enjoyable hobby for millions of people. But the specific applications of variable stars to education, and to astronomy as a hobby, go beyond these general applications, so variable stars could actually help to improve science and math education! To follow the latest trends and issues in astronomy education, see the *Astronomy Education Review* (http://aer.noao.edu).

The best way to learn science (and learn about science) is to do it. For over 30 years, I have been supervising student research projects on variable stars.

Variable star observation and analysis are inherently simple, but the actual techniques of analysis and interpretation involve a wide range of scientific and mathematical skills – some of which would be understood and appreciated by a 12-year-old, and some of which would challenge an expert in the field. Thus, the use of variable star data in science, math, and computing courses and projects can develop and integrate a variety of skills: literature research and background reading; research planning and strategy; problem-solving and judgement; calibration and measurement; identifying and understanding random and systematic errors; computer programming for data management, analysis, and display; construction, analysis, and interpretation of graphs; concepts of regularity and prediction; significance, curve-fitting, and other statistical procedures; preparation of oral and written reports.

There are also motivational factors. Students can use real research data, with real errors, either from an archive, or from their own observations. They can potentially make new discoveries, and publish research papers. Activities and projects can be organized in such a way that each student has 'their own' star for study, in parallel with other students. Thus, each student plays an individual role but, at the same time, is part of a team.

All in all, these projects are real science; the process is everything. They integrate math and science, and many other disciplines. Technology is introduced in appropriate ways. There is nothing artificial or contrived; there is no predetermined 'right' or 'wrong' answer. Often when I ask students or teachers to make measurements of a variable star, on a print or slide or in the real sky, they ask 'what is the answer supposed to be?'. I ask them to think about the question that they asked. Then they realize the true nature of science!

Most of my students are undergraduates, in research-project courses, work-study positions, or summer research assistantships. In terms of data, we use photoelectric data from a small telescope on the roof of my university building, and from a robotic telescope in Arizona. We have used photoelectric and visual observations from the American Association of Variable Star Observers and other amateur organizations. We have used epoch photometry from the European Space Agency's *Hipparcos* satellite and, more recently, archival data from the OGLE microlensing survey. In the 1980s, I began supervising younger students, through the University of Toronto's Mentorship Program, which enables outstanding senior secondary school students to work on research projects at the university. Dr Janet Mattei, Director of the AAVSO, and I independently realized that variable star astronomy could be used to enhance science and math education at the secondary school level. My interest was partly based on my frequent visits to high schools and small colleges, where instructors were looking for genuine laboratory activities for introductory astronomy.

We therefore developed *Hands-On Astrophysics*, with partial support from the US National Science Foundation. HOA (hoa.aavso.org) includes 35-mm constellation slides and prints for classroom practice in variable star measurement, charts for observing from slides or in the real sky, data and software for analysis, instructional videos, and a 600-page teachers' and students' manual. This has been used in a variety of contexts – even as an introduction to research for students in astronomically developing countries. HOA is currently being upgraded and web-based, and, with the recent formation of an AAVSO Education Committee, we hope that the AAVSO's formal education activities will increase.

Ideally, students should do at least some observing. A successful model was done as part of the testing of *Hands-On Astrophysics*. Two dozen teachers from across the US were asked to observe δ Cephei as often as possible. Some were able to observe it several times, some only a few. But the observations were pooled, and it was easy to see the pattern of the star's periodic variation. And if the observations were phased with that period, the time of maximum could be estimated accurately enough to detect the star's evolutionary period change, by the $(O - C)$ method! If you have less-than-perfect skies, and can not see third-magnitude stars, then you can observe α Ori. You can pool the class observations, and combine them with other visual observations from the AAVSO website. Students love to compare their measurements with those of others!

Another interesting approach, which was tried successfully by one of my high school research students, was to photograph variable stars with a 35-mm camera, on as many nights as possible. This turned out to be quite feasible, even in my climatically underprivileged city. The slides were then used in the classroom for variable star measurement by the whole class. The measured time of maximum of the classical Cepheid RT Aur agreed to within a fraction of a day with the predicted value.

In 2004, I tried yet another approach. I distributed archival photoelectric observations of about 30 sun-like stars to a group of a dozen senior high school physics students, along with self-correlation analysis software. The measurements had already been analyzed by Fourier techniques, and the results had been published. The students re-analyzed the measurements with self-correlation, and were able to obtain new insights into a few of the stars.

Observing variable stars would be highly desirable, both to develop students' skills, and to make them aware of the origin of the archival data which they are using for analysis. But observing requires clear, dark skies, and there is the problem that 'the stars come out at night; the students don't'. But it is increasingly possible for students to access remote/robotic telescopes. There are now many of these, and they are crying out for real science projects to do. Variable stars can fill that niche! My eventual goal is to create a web-based system in which students can learn about variable stars and, using on-line software, analyze and interpret the on-line observations of thousands of variable stars which have heretofore not been adequately studied. The challenge is to put judgement, mentoring, and enthusiasm on-line!

There is another aspect to variable stars and education: the education of the amateur astronomers who make such an important contribution to the field. We professionals – especially those of us who use amateur astronomers' data – have an obligation to provide the observers with feedback on how their observations contribute to science. We do that by giving talks at their meetings, by writing articles for their publications, and by communicating with them one-on-one. There is also a need to mentor new observers, and organizations like the AAVSO take this need very seriously.

10.4 Variable stars and the general public

Astronomers are concerned about public awareness, appreciation, and understanding of their science – partly because that may affect the support of astronomy by taxpayers, and partly because astronomers want to know that their work is appreciated. Fortunately, interest in astronomy tends to be high, especially among young people.

What use is the study of variable stars? An astronomer would answer that variable stars provide important information about astrophysical processes, about the nature and evolution of stars, and even about the size, age, and evolution of the universe. But is this a sufficient answer for the taxpayers who support most astronomy research? Or is the study of variable stars, and astronomy in general, more broadly and deeply useful?

Astronomy is deeply rooted in culture, religion, and philosophy. Amongst the scientific revolutions of history, astronomy stands out. In the recent lists of 'the most influential people of the millennium', a handful of astronomers – including variable star astronomer Edwin Hubble – were always included. The practical applications of astronomy are so obvious that we sometimes forget them. Nowadays, astronomy advances physics, by providing environments unattainable in the laboratory. Astronomy has led to technological advances, ranging from radio receivers, to imaging and image-processing techniques. In the past, it spurred the development of trigonometry, logarithms, and calculus. Now, it drives high-performance computing.

Astronomy also has aesthetic and emotional dimensions. It deals with our cosmic roots, and our place in time and space, and with the origins of the universe, galaxies, stars, planets, the atoms and molecules of life, perhaps of life itself. It reveals a universe which is vast, varied, and beautiful. It has inspired artists and poets. It harnesses curiosity, imagination, and a sense of shared exploration and discovery. It promotes environmental awareness, through spacecraft images of our fragile planet, and through the realization that we *may* be alone in the universe. And there are the many benefits to education which we have already discussed.

Variable stars, of course, are only a small facet of astronomy, but they have connections to many of the more general 'uses' of astronomy. Here are some that occur to me:

- People (especially young boys) seem to love things which explode. And we would rather see things explode in far-off space than here on earth. Gamma-ray bursts are the biggest blasts in the universe since the Big Bang.
- Supernovae are responsible for us. Most of the chemical elements except hydrogen and helium, including the elements of life, were synthesized in the cores of stars, and ejected into space to become part of the next generation of stars, planets, and life.
- Although, contrary to popular belief, the sun will not explode and turn into a black hole, it *will* do spectacular things at the end of its life. It will swell up into a red giant, engulfing the inner solar system. Mira-star pulsation will drive off its outer layers, contributing to the chemical enrichment of our galaxy (and perhaps to the origin of life elsewhere), and baring the white-dwarf core. The gases around it will light up, thanks to the ultraviolet light from the core, and the sun will become a beautiful planetary nebula.
- Some of the most aesthetically pleasing images in astronomy, such as the Crab Nebula, planetary nebulae, and the nebula around V838 Monocerotis, are consequences of the fact that the original star was a variable star.
- Historically, Cepheid pulsating variable stars have been the most important cosmic yardstick for measuring the size scale and age of the universe. More recently, supernovae have joined them as even more powerful yardsticks, and have led to the discovery of dark energy.
- RR Lyrae stars have played a particularly important role in determining the distance and age of globular clusters, and the nature and evolution of stars within them. Globular clusters are the oldest structures in our galaxy, and their ages provide an important constraint on the age of the universe. They are fossils of the origin of our galaxy.
- Even a garden-variety pulsating star can boggle the imagination. BW Vulpeculae, a Beta Cephei star, is expanding and contracting at over 150 km/s – over ten times faster than the Space Shuttle in orbit. And

some of the peculiar A stars have magnetic fields, thousands of times stronger than the earth's, covering their entire surfaces.

- Supernovae give rise to neutron stars in which a cubic centimetre of matter would weigh megatons! And these stellar corpses may be spinning a hundred times a second. How bizarre!
- By studying pre-main-sequence stars such as T Tauri stars, we are seeing what our sun was like 4.5 billion years ago. The study of sun-like stars in general has shown that rotation and stellar 'activity' are highest in young stars, and decline with age. Thus, our sun was much more active in its youth. This may have had some implications for the origin and evolution of life on earth.
- The techiques for studying variable stars have led to the discovery of exoplanets by the transit method. This method *might* lead to the discovery of the first earth-sized planet beyond our solar system.
- At least one biological extinction on earth, the cretaceous-tertiary event, was most likely due to the impact of an asteroid or comet. But it is possible that the others could be connected with long-term processes on the sun, or even perhaps the explosion of a supernova or gamma-ray burster nearby.
- Many of the public, especially (but not exclusively) older people, are interested in star lore, and the appearance of the night sky. For them, variable stars are 'action in the sky'. In particular, people are interested to learn that the best-known star in the northern sky, Polaris, is actually a Cepheid variable star – and a very unusual one.
- Astronomy overflows with human-interest stories, and some of them have been hinted at in this book: John Goodricke, who overcame adversity, only to die at an early age; William Herschel and his astronomer sister Caroline; 'Pickering's women' at Harvard College Observatory, who made such notable contributions to astronomy at substandard salaries.

This brings us to the wider significance of the study of variable stars. They are among the many fascinating phenomena which are found in our world, and in our universe. Curiosity about nature and the origin of these and other wonders of nature is what distinguishes us from other species on earth. Science is a process for exploring questions about nature, from the mundane processes which govern our daily existence, to the deepest philosophical questions about our origins, our nature, and our fate. Science is as important to the common citizen as it is to scientists. The study of variable stars can be shared by all. To paraphrase the song: 'the variable stars belong to everyone'.

Appendix: Acronyms

AAVSO: American Association of Variable Star Observers
AGB: asymptotic giant branch (in the Hertzsprung–Russell diagram)
AGN: active galactic nucleus
ASAS: All-sky Automated Survey
AU: astronomical unit
BAA: British Astronomical Association
BC: bolometric correction
BV: Bamberg variable
CCD: charge-coupled device
CGRO: *Compton Gamma-Ray Observatory* (satellite)
CDS: *Centre de Données astronomiques de Strasbourg*
COROT: *COnvection, ROtation, and planetary Transits* (satellite)
CSV: *Catalog of Suspected Variable Stars*
CTTS: classical T Tauri stars
CV: cataclysmic variable
EUVE: *Extreme Ultraviolet Explorer* (satellite)
GAIA: *Global Astrometric Interferometer for Astrophysics* (satellite)
GALEX: *GALaxy evolution EXplorer* (satellite)
GCVS: *General Catalogue of Variable Stars*
GPB: *Gravity Probe B* (satellite)
GRB: gamma-ray burst
HC: heliocentric correction
HD: *Henry Draper (catalogue)*; HDE: *HD Extension*
H–R: Hertzsprung–Russell; H–R diagram;
HST: *Hubble Space Telescope* (satellite)
IAU: International Astronomical Union
IBVS: *IAU Information Bulletin on Variable Stars*
IR: infrared
IUE: *International Ultraviolet Explorer* (satellite)
JD: Julian Date

LMC: Large Magellanic Cloud (galaxy)

$L_\odot$: solar luminosity – 3.85×10^{26} Watts

MACHO: MAssive Compact Halo Object (also project of this name)

MOST: *Microvariability and Oscillations of STars* (satellite)

$M_\odot$: solar mass – 1.9891×10^{30} kg

NASA: National Aeronautics and Space Administration (US)

NSF: National Science Foundation (US)

NSV: *New Catalog of Suspected Variable Stars*

$O - C$: observed – computed (method)

OGLE: Optical Gravitational Lensing Experiment

PL: period–luminosity (relation)

PMS: pre-main sequence

RA: right ascension; Dec.: declination

RASNZ: Royal Astronomical Society of New Zealand

ROSAT: *Röntgen Satellite*

ROTSE: Robotic Optical Transient Search Experiment

$R_\odot$: solar radius – 696265 km

SIMBAD: Set of Identifications, Measurements, and Bibliography for Astronomical Data

SMC: Small Magellanic Cloud (galaxy)

TASS: The Amateur Sky Survey

UBV(RIJHKL): Johnson photometric system

UHURU: (satellite)

UV: ultra-violet

VLA: *Very Large Array* (radio telescope)

VLT: *Very Large Telescope*

WIRE: *Wide-field InfraRed Explorer* (satellite)

WTTS: weak-lined T Tauri stars

WWW: World Wide Web

Bibliography

Alard, C. and Lupton, R.H. (1998), 'A method for optimal image subtraction', *ApJ*, **503**, 325–33.

Alcock, C. *et al.* (1998),'The MACHO project LMC variable star inventory: VII. The discovery of RV Tauri stars and new Population II Cepheids', *AJ*, **115**, 1921–33.

Andersen, J. (1991), 'Accurate masses and radii of normal stars', *ARAA*, **3**, 91–126.

Andersen, J., Clausen, J.V., Gustafsson, B., Nordstrom, B. and Vandenberg, D. (1988), 'Absolute dimensions of eclipsing binaries: XIII. AI Phoenicis: a case study in stellar evolution', *A&A*, **196**, 128–40

Arellano Ferro, A. (1983), 'Period and amplitude variations of Polaris', *ApJ*, **274**, 755–62.

Baker, N. (1966), 'Simplified models of Cepheid pulsation', in *Stellar Evolution*, ed. R.F. Stein and A.G.W. Cameron, New York: Plenum Press, 333.

Barnes, T.G. and Evans, D.S. (1976), Stellar angular diameters and visual surface brightness. II. Early and intermediate spectral type', *MNRAS*, **174**, 503–12.

Bedding, T.R., Zijlstra, A., Jones, A. and Foster, G. (1998), 'Mode-switching in the nearby Mira-like variable R Doradus', *MNRAS*, **301**, 1073–82.

Belserene, E.P. (1986), '$(O - C)$ by computer', *JAAVSO*, **15**, 243–8.

Belserene, E.P. (1989), 'Precision of parabolic elements', *JAAVSO*, **18**, 55–62.

Berdyugina, S.V., Berdyugin, A.V., Ilyn, I., and Tuominen, I. (2000). The long-period RS CVn binary IM Pegasi. II. First surface image', A & A, **360**, 272.

Bode, M.F. and Evans, A. (1989), *Classical Novae*, Chichester, New York.

Böhm-Vitense, E. (1992), *Introduction to Stellar Astrophysics*, volume 3, Cambridge: Cambridge University Press.

Bono, G. *et al.* (1997), 'Non-linear investigation of the pulsation properties of RR Lyrae variables', *A&AS*, **121**, 327–42.

Bowen, G. (1988), 'Dynamical modelling of long-period variable star atmospheres', *ApJ*, **329**, 299–317.

Bouvier, J., Bertout, C., Benz, W. and Mayor (1986), *A & A*, **165**, 110.

Bowen, G.H. and Willson, L.A. (1991), 'From wind to super-wind – the evolution of mass-loss rates for Mira models', *ApJ*, **375**, L53–L56.

Breger, M. and Montgomery, M.H. (eds) (1999), *Delta Scuti and Related Stars*, San Francisco: ASP Conference Series, **210**.

Breger, M. *et al.* (2004), 'The δ Scuti star FG Vir: V. The 2002 photometric multisite campaign', *A&A*, **419**, 695–701.

Breger, M. *et al.* (2005), 'Detection of 75+ pulsation frequencies in the δ Scuti star FG Virginis', *A&A*, **435**, 955–65.

Broens, E., Diepvens, A. and Van Der Looy, J. (2000), *Variable Stars: Visual Light Curves*, Brugge, Belgium: Vereniging Voor Sterrenkunde.

Buchler, J.R. and Kovacs, G. (1987), 'Period-doubling bifurcations and chaos in W Virginis models', *ApJ*, **320**, L57–L62.

Buchler, J.R., Kollath, Z. and Cadmus, R.R. Jr. (2004), 'Evidence for low-dimensional chaos in semi-regular variable stars', *ApJ*, **613**, 532–47.

Burki, G. (1978), 'The semi-period-luminosity relation for supergiant variables', *A&A*, **65**, 357–62.

Buzasi, D.L. *et al.* (2005), 'Altair: the brightest δ Scuti star', *ApJ*, **619**, 1072–6.

Cannizzo, J.K. and Kaitchuck, R.H. (1992), 'Accretion discs in interacting binary stars', *Scientific American*, January, 92–9.

Cannizzo, J.K. and Mattei, J.A. (1998), 'A study of the outbursts in SS Cygni', *ApJ*, **505**, 344–51.

Castor, J.I. (1971), 'On the calculation of linear, non-adiabatic pulsation of stellar models', *ApJ*, **166**, 109–29.

Charbonneau, D., Brown, T.M., Latham, D.W. and Mayor, M. (2000), 'Detection of a planet transit across a sun-like star', *ApJ*, **529**, L45–L48.

Charpinet, S., Fontaine, G. and Brassard, P. (2001), 'A theoretical explanation of the pulsational stability of subdwarf B stars', *PASP*, **113**, 775–88.

Chen, W.P., Lemme, C. and Paczyński, B. (eds) (2001), *Small-Telescope Astronomy on Global Scales*, San Francisco: ASP Conference Series, **246**.

Christensen-Dalsgaard, J. and Petersen, J.O. (1995), 'Pulsation mdels of the double-mode Cepheids in the Large Magellanic Cloud', *A&A*, **299**, L17–L20.

Christensen-Dalsgaard, J. and Dziembowski, W. (2000), 'Basic aspects of stellar structure and pulsation', in *Variable Stars as Essential Astrophysical Tools*, ed. C. Ibanoglu, Dordrecht: Kluwer Academic Publishers, 1.

Christy, R.F. (1964), 'The calculation of stellar pulsation', *Rev. Mod. Phys.*, **36**, 555–71.

Christy, R.F. (1966), 'A study of pulsation in RR Lyrae star models', *ApJ*, **144**, 108–79.

Clark, D.H. and Stephenson, F.R. (1977), *The Historical Supernovae*, Oxford: Pergamon Press.

Clausen, J.V. *et al.* (1991), 'Four-colour photometry of eclipsing binaries: XXXIII. Light curves of TX Fornacis', *A&AS*, **88**, 535–44.

Clayton, G.C. (1996), 'The R Coronae Borealis stars', *PASP*, **108**, 225–41.

Clayton, G.C. (2002), 'The R Coronae Borealis connection', *Ap. Sp. Sci.*, **279**, 167–70.

Cook, K.H. *et al.* (1995), 'Variable stars in the MACHO collaboration database', in *Astrophysical Applications of Stellar Pulsation*, ed. R.S. Stobie and P.A. Whitelock, *ASP Conference Series* **83**, 221.

Corradi, R.L.M., Mikolajewska, J. and Mahoney, T.J. (2003), *Symbiotic Stars Probing Stellar Evolution*, San Francisco: ASP Conference Series, **303**.

Cox, J.P. (1980), *Theory of Stellar Pulsation*, Princeton: Princeton University Press.

De Ruyter, S., Van Winckel, H., Dominik, C., Waters, L. B. F. M. and Dejonghe, H. (2005), 'Strong dust processing in circumstellar discs around 6 RV Tauri stars. Are dusty RV Tauri stars all binaries?', *A&A*, **435**, 161–5.

Dworetsky, M. (1983), 'A period-finding method for sparse, randomly spaced observations, or "How long is a piece of string?"', *MNRAS*, **203**, 917–24.

Eaton, J.A. (1993), 'On the chromospheric structure of Zeta Aurigae', *ApJ*, **404**, 305–15.

Eddington, A.S. and Plakidis, L. (1929), 'Irregularities of period in long-period variable stars', *MNRAS*, **90**, 65–71.

Evans, N.R. (1992),'A magnitude-limited survey of Cepheid companions in the ultraviolet', *ApJ*, **384**, 220–33.

Evans, R., van den Bergh, S. and McClure, R.D. (1989), 'Revised supernova rates in Shapley-Ames galaxies', *ApJ*, **345**, 752–8.

Feast, M. (1981), in *ESO Workshop on the most Massive Stars*, ed. S. D'odorico, D. Baade and K. Kjaer, Garching: European Southern Observatory.

Feast, M. (2004), in *Variable Stars in the Local Group*, ed. D.W. Kurtz and K.R. Pollard, San Francisco: ASP Conference Series, **310**, 304.

Fokin, A.B. 1994, 'Non-linear pulsations of the RV Tauri stars', *A&A*, **292**, 133–51.

Fontaine, G., Brassard, P. and Charpinet, S. (2003), 'Outstanding issues in post-main sequence evolution', *ApSS*, **284**, 257–68.

Foster, G. (1995), 'The cleanest Fourier spectrum', *AJ*, **109**, 1889–902.

Foster, G. (1996), 'Wavelets for period analysis of unevenly-sampled time series', *AJ*, **112**, 1709–29.

Freedman, W.L. *et al.* (2001), 'Final results from the Hubble Space Telescope key project to measure the Hubble constant', *ApJ*, **553**, 47–72.

Gaidos, E.J., Henry, G.W. and Henry, S.M. (2000), 'Spectroscopy and photometry of young solar analogs', *AJ*, **120**, 1003–13.

Garrison, R.F. (ed.) (1983), *The MK Process*, Toronto: The David Dunlap Observatory.

Gies, D.R. and Kullavanijaya, A. (1988), 'The line profile variations of Epsilon Persei: I. Evidence for multi-mode non-radial pulsation', *ApJ*, **326**, 813–31.

Gilliland, R. and Dupree, A.K. (1996), 'First image of the surface of a star with the Hubble Space Telescope', *ApJL*, **463**, L29–L32.

Goldberg, B.A., Walker, G.A.H., and Odgers, G.J. (1976), 'The Variations of BW Vulpeculae', *AJ*, **81**, 433–44.

Goldberg, L. (1984), 'The variability of Alpha Orionis', *PASP*, **96**, 366–71.

Golub, L. and Pasachoff, J.M. (2002), *Nearest Star: The Surprising Science of our Sun*, Cambridge, MA: Harvard University Press.

de Groot, M. and Sterken, C. (eds) (2001), *P Cygni 2000: 400 Years of Progress*, San Francisco: ASP Conference Series, **233**.

Hall, D.S. (1976), 'The RS CVn binaries, and binaries with similar properties', in *Multiple Periodic Variable Stars*, ed. W.S. Fitch, Dordrecht: D. Reidel Publishing, 287.

Hall, D.S. and Genet, R.M. (1988), *Photoelectric Photometry of Variable Stars* (2nd edition), Willmann-Bell Inc.

Handler, G. (1999), 'The domain of the γ Doradus stars in the Hertzsprung–Russell diagram', *MNRAS*, **309**, L19–23.

Harmanec, P. (2002), 'The ever-challenging emission-line binary Beta Lyrae', *AN*, **323**, 87–98.

Harris, H.C. (1985), 'A catalogue of field Type II Cepheids', *AJ*, **90**, 756–60.

Hearnshaw, J.B. (1996), *Measurement of Starlight: Two Centuries of Astronomical Photometry*, Cambridge: Cambridge University Press.

Hearnshaw, J.B. and Scarfe, C.D. (1999), *Precise Stellar Radial Velocities*, San Francisco: ASP Conference Series, **195**.

Hellier, C. (2001), *Cataclysmic Variable Stars: How and Why They Vary*, Springer-Praxis.

Henden, A.A. and Kaitchuck, R.H. (1982), *Astronomical Photometry*, New York: Van Nostrand Reinhold.

Henden, A.A. and Kaitchuck, R.H. (2006), *CCD Photometry*, Willmann-Bell.

Henry, G.W., Marcy, G.W., Butler, R.P. and Vogt, S.S. (2000), 'A transiting "51 Peg-like" planet', *ApJ*, **529**, L41–L44.

Herbig, G.H. (1977), 'Eruptive phenomena in early stellar evolution', *ApJ*, **217**, 693–715.

Herbst, W., Herbst, D.K., Grossman, E.J., and Weinstein, D. (1994), 'Catalogue of UBVRI photometry of T Tauri stars and analysis of the causes of their variability', *AJ*, **108**, 1906.

Herbst, W. (2001), 'The rotation and variability of T Tauri stars: results of two decades of monitoring at Van Vleck Observatory', in *Small-Telescope Astronomy on Global Scales*, ed. B. Paczynski, W.-P. Chen and C. Lemme, San Francisco: ASP Conference Series, **246**, 177.

Herczeg, T. and Frieboes-Conde, H. (1968), 'A propos of Algol's changing period', *Sky and Telescope*, **35**, 288–9.

Hilditch, R.W. (2001), *An Introduction to Close Binary Stars*, Cambridge: Cambridge University Press.

Hoffleit, D. (1986), 'A history of variable star astronomy to 1900 and slightly beyond', *JAAVSO*, **15**, 77–106.

Hoffleit, D. (1987), 'History of variable star nomenclature', *JAAVSO*, **16**, 65–70.

Hogg, H.S. (1984), 'Variable stars', in *The General History of Astronomy*, vol. 4, ed. O. Gingerich, Cambridge: Cambridge University Press, 73.

Horne J.H. and Baliunas, S.L. (1986), 'A prescription for period analysis of unevenly-spaced time series', *ApJ*, **302**, 757–63.

Hoskin, M.A. (ed.), (1997), *The Cambridge Illustrated History of Astronomy*, Cambridge: Cambridge University Press.

Houk, N. (1963), 'V1280 Sgr and the other long-period variables with secondary periods', *AJ*, **68**, 253–6.

Howell, S. (2006), *Handbook of CCD Astronomy*, Second Edition, Cambridge: Cambridge University Press.

Humphries, R.M. *et al.* (2006), 'M33's Variable A – a hypergiant star more than 35 years in eruption', *AJ*, **131**, 2105–13.

Hutchings, J.B. and Hill, G. (1977), 'Copernicus OAO observations of Beta Cephei and Alpha Virginis', *ApJ*, **213**, 111–20.

Iglesias, C.A., Rogers, F.J. and Wilson, B.G. (1987), 'Re-examination of the metal contribution to astrophysical opacity', *ApJ*, **322**, L45–L48.

Johnson, H.L. and Morgan, W.W. (1953), 'Fundamental stellar photometry for standards of spectral type on the revised system of the Yerkes Spectral Atlas', *ApJ*, **117**, 313–52.

Jura, M. (1986), 'RV Tauri stars as post-asymptotic-branch objects', *ApJ*, **309**, 732–6.

Kaler, J. (1997), *Stars and their Spectra*, Cambridge: Cambridge University Press.

Kallrath, J. and Milone, E.F. (1999), *Eclipsing Binary Stars: Modeling and Analysis*, New York: Springer-Verlag.

Karttunen, H., Kröger, P. Oja, H., Poutanen, M. and Donner, K.J. (1996), *Fundamental Astronomy*, Berlin: Springer-Verlag.

Kawaler, S. and Bradley, P.A. (1994), 'Precise asteroseismology of pulsating PG 1159 stars', *ApJ*, **427**, 415–28.

Kaye, A.B., Handler, G., Krisciunas, K., Poretti, E. and Zerbi, F.M. (1999), 'Gamma Doradus stars: defining a new class of pulsating star', *PASP*, **111**, 840–44.

Kaye, A.B., Gray, R.O. and Griffin, R.F. (2004), 'On the spectroscopic nature of HD 221866', *PASP*, **116**, 558–64.

Kearns, K.E. and Herbst, W. (1998), 'Additional periodic variables in NGC 2264', *AJ*, **116**, 261–5.

Kemp, J.C. *et al.* (1986), 'Epsilon Aurigae: polarization, light curves, and geometry of the 1982-84 eclipse', *ApJ*, **300**, 11–.

Kenyon, S.J. and Webbink, R.F. (1984), 'The nature of symbiotic stars', *ApJ*, **279**, 252–83.

Kervella, P., Nardetto, N., Bersier, D., Mourard, D. and Coudé du Foresto, V. (2004), 'Cepheid distances from infrared long-baseline interferometry: I. VINCI/VLTI observations of seven galactic Cepheids', *A&A*, **416**, 941–53.

Kholopov, P.N. *et al.* (1998), *The Combined General Catalogue of Variable Stars*, 4th Edition, available on-line at SIMBAD.

Kilkenny, D., Koen, C., O'Donoghue, and Stobie, R.S. (1997), 'A new class of rapidly pulsating stars: I. EC 14026-2647, the class prototype', *MNRAS*, **285**, 640–4.

Kiss, L.L., Szatmary, K., Cadmus, R.R. and Mattei, J.A. (1999), 'Multiperiodicity in semi-regular variables: I. General properties', *A&A*, **346**, 542–55.

Kiss, L. L., Szabó, G. M. and Bedding, T. R. (2006), 'Variability in red supergiant stars: pulsations, long secondary periods, and convection noise', *MNRAS*, **327**, 1721–34.

Kitchin, C.R. (2003), *Astrophysical Techniques*, 4th edition, London: Institute of Physics.

Kochukhov, O., Piskunov, N., Ilyin, I., Ilyina, S., and Tuominen, I. (2002), 'Doppler imaging of stellar magnetic fields. III. Abundance distribution and magnetic field geometry of α^2 CVn', *A&A*, **389**, 420.

Kramer, M., Wex, N. and Wielebinski, R. (eds) (2000), *Pulsar Astronomy: 2000 and Beyond*, San Francisco: ASP Conference Series, **202**.

Kurtz, D.W. (1982), 'Rapidly-oscillating Ap stars', *MNRAS*, **200**, 807–59.

Kurtz, D.W. *et al.* (2003), 'High precision with the Whole Earth Telescope: lessons and some results from Xcov20 for the roAp star HR 1217', *Baltic Astron.*, **12**, 105–17.

Kurtz, D.W. and Pollard, K.R. (2004), *Variable Stars in the Local Group*, San Francisco: ASP Conference Series, **310**.

Latham, D.W., Nordstrom, B., Andersen, J., Torres, G., Stefanik, R.P., Thaller, M. and Bester, M.J. (1996), 'Accurate mass determinations for double-lined spectroscopic binaries by digital cross-correlation spectroscopy: DM Vir revisited', *A&A*, **314**, 864–70.

Lattimer, J.M. and Prakash, M. (2004), 'The physics of neutron stars', *Science*, **304**, 536-42 (and papers following, on pulsars and neutron stars).

Lee, Y.-W., Demarque, P. and Zinn, R. (1990), 'The horizontal branch stars in globular clusters', *ApJ*, **350**, 155–72.

Lefebre, L. *et al.* (2005), 'Oscillations in the massive Wolf-Rayet star WR123 with the MOST satellite', *ApJ*, **634**, L109–112.

Lesh, J.R. and Aizenmann, M.L. (1978), 'The observational status of the β Cephei stars', *ARAA*, **16**, 215–40.

Lebzelter, T., Hinkle, K.H., Joyce, R.R. and Fekel, F.C. (2005), 'A study of bright southern long-period variables', *A&A*, **431**, 623–34.

Libbrecht, K.G. and Woodard, M.F. (1990), 'Solar cycle effects on solar oscillation frequencies', *Nature*, **345**, 779–82.

Lobel, A. *et al.* (2003), 'High-resolution spectroscopy of the yellow hypergiant ρ Cassiopeiae from 1993 through the outburst of 2000-2001', *ApJ*, **583**, 923–54.

Lu, W. and Rucinski, S.M. (1999), 'Radial velocity studies of close binaries. I', *AJ*, **118**, 515–526 (paper 1 in a series).

MacRobert, A. (1985), 'The puzzle of Epsilon Aurigae', *Sky and Telescope*, **70**, 527.

Madore, B.F. (ed.) (1985), *Cepheids: Theory and Observation*, Cambridge: Cambridge University Press.

Madore, B. F. and Freedman, W. L. (1991), 'The Cepheid distance scale', *PASP*, **103**, 933.

Maeder, A. (1980), 'Supergiant variability: amplitudes and pulsation constants in relation to mass loss and convection', *A&A*, **90**, 311–17.

Magain, P. (2005), in *The Light-Time Effect in Astrophysics*, ed. C. Sterken, San Francisco: ASP Conference Series, **335**, 207–214.

Manchester, R.N. (2004), 'Observational properties of pulsars', *Science*, **304**, 542–47.

Markham, T. (2004), 'Observing eclipsing variables: a beginner's guide', *J. Br. Astron. Assoc.*, **114**, 215–19.

Markova, N. (2000), 'New aspects of the line profile variability in P Cygni's optical spectrum', *A&AS*, **144**, 391–404.

Matthews, J.M., Wehlau, W.H., Walker, G.A.H., and Yang, S. (1988), 'Detection of radial velocity variations in the rapidly oscillating Ap star HR 1217', *ApJ*, **324**, 1099–105.

Messina, S. and Guinan, E.F. (2002), 'Magnetic activity of six young solar analogues. I: Starspot cycles from long-term photometry', *A&A*, **393**, 225–37.

Messina, S. and Guinan, E.F. (2003), 'Magnetic activity of six young solar analogues. II: Surface differential rotation from long-term photometry', *A&A*, **409**, 1017–30.

Meynet, G. and Maeder, A. (1997), 'Stellar evolution with rotation. I: The computational method, and the inhibiting effect of the μ gradient', *A&A*, **321**, 465–76.

Mikolajewska, J. (2003), 'Orbital and stellar parameters of symbiotic stars', in *Symbiotic stars Probins Stellar Evolution*, ed. R.L.H. Corradi, R. Mikolajewska, and T.J. Mahoney, San Francisco: ASP Conference Series, **303**, 9.

Milone, E.F. (ed.) (1993), *Light Curve Modelling of Eclipsing Binary Systems*, Springer-Verlag.

Morgan, W.W., Keenen, P.C., and Kellman, E. (1993), *An Atlas of Stella Spectra*, Chicago: University of Chicago Press.

Morse, J.A., Humphreys, R., and Damineli, A. (eds) (1999), *Eta Carinae at the Millennium*, San Francisco: ASP Conference Series, **179**.

Moskalik, P. and Dziembowski, W.A. (2005), 'Seismology of triple-mode classical cepheids of the large magellanic clouds', *A&A*, **434**, 1077–84.

Neill, J.D. and Shara, M.M. (2004), 'The Hα light curves and spatial distribution of novae in M81', *AJ*, **127**, 816–31.

Olech, A., Wozniak, P.R., Alard, C., Kaluzny, J. and Thompson, I.B. (1999), MNRAS, **310**, 759.

Oliveira, A.S., Steiner, J.E. and Diaz, M.P. (2004), 'The multiple spectroscopic and photometric periods of DI Crucis', *PASP*, **116**, 311–25.

Osaki, Y. (1974), 'An accretion model for the outbursts, of U Geminorum stars', *Publ. Astron. Soc. Japan*, **26**, 429.

Osaki, Y. (1996), 'Dwarf nova outbursts', *PASP*, **108**, 39–60.

Pamyatnykh, A. (1999), 'Pulsational instability domains in the upper main sequence', *Acta Astron*, **49**, 119.

Patterson, J. *et. al.* (2003), 'Superhumps in cataclysmic binaries. XXIV. Twenty more dwarf novae', *PASP*, **115**, 1308.

Percy, J.R. (1969), 'Light variations in ψ Orionis', *JRASC*, **63**, 233–7.

Percy, J.R. (1975), 'Pulsating stars', *Scientific American*, **232**, 66–75.

Percy, J.R. and Welch, D.L. (1982), 'Photometric observations of RS CVn stars', *JRASC*, **76**, 185–204.

Percy, J.R. (1986), 'Highlights of variable star astronomy: 1900–1986', *JAAVSO*, **15**, 126–32.

Percy, J.R., Ralli, J. and Sen, L.V. (1993), 'Analysis of AAVSO visual observations of ten small-amplitude red variables', *PASP*, **105**, 287–92.

Percy, J.R., Desjardins, A., Yu, L., and Landis, H.J. (1996), 'Small-amplitude red variables in the AAVSO photoelectric photometry program: light curves and periods', *PASP*, **108**, 139–45.

Percy, J.R., Bezuhly, M., Milanowski, M., and Zsoldos, E. (1997), 'The nature of the period changes in RV Tauri stars', *PASP*, **109**, 264–9.

Percy, J.R. and Hale, J. (1998), 'Period changes, evolution, and multi-periodicity in the peculiar Population II Cepheid RU Camelopardalis', *PASP*, **110**, 1428–30.

Percy, J. R. and Wilson, J. B. (2000), *Amateur-Professional Partnerships in Astronomy*, San Francisco: ASP Conference Series, **220**.

Percy, J. R., Wilson, J. B. and Henry, G. W. (2001), 'Long-term VRI photometry of small-amplitude red variables. I: Light curves and periods', *PASP*, **113**, 983–96.

Perlmutter, S. *et. al.* (1999), 'Measurements of Omega and Lambda from 42 high-redshift supernovae', *ApJ*, **517**, 565.

Petersen, J. O. and Christensen-Dalsgaard, J. (1999), 'Pulsation models of delta scuti variables. II: Delta scuti stars as precise distance indicators', *A&A*, **352**, 547–54.

Porter, J. M. and Rivinius, T. (2003), 'Classical Be stars', *PASP*, **115**, 1153–70.

Richards, M. T., Mochnacki, S. W. and Bolton, C. T. (1988), 'Non-simultaneous multi-colour light-curve analysis of Algol', *AJ*, **96**, 326–36.

Rivinius, Th., Baade, D. and Stefl, S. (2003), 'Non-radially pulsating Be stars', *A&A*, **411**, 229–47.

Rucinski, S. M. *et al.* (2005), 'Radial velocity studies of close binaries, X', *AJ*, **130**, 767–75.

Scargle J. D. (1982), 'Studies in astronomical time-series analysis, II: Statistical aspects of spectral analysis of unevenly-spaced data', *ApJ*, **263**, 835–53.

Schaller, G., Schaerer, D., Meynet, G. and Maeder, A. (1992), 'New grids of stellar models from 0.8 to 120 solar masses at Z = 0.02 and 0.001', *A&AS*, **96**, 269–331.

Schmidt, B. P. *et al.* (1998), 'The high-z supernova search; Measuring cosmic deceleration and global curvature of the universe using type Ia supernovae', *ApJ*, **507**, 46.

Schwarzschild, M. and Härm, R. (1970), 'On the evolutionary phase of Cepheids in globular clusters', *ApJ*, **160**, 341–4.

Shore, S. N., Starrfield, S. and Sonneborn, G. (1996), 'The ultraviolet and X-ray view of the demise of nova V1974 Cygni', *ApJ*, **463**, L21–L24.

Simon, N. R. and Schmidt (1976), 'Evidence favouring non-evolutionary Cepheid masses', *ApJ*, **205**, 162–4.

Skuljan, L. and Cottrell, P. L. (2004), in *Variable Stars in the Local Group*, ed. D. W. Kurtz and K. R. Pollard, San Francisco: ASP Conference Series, **310**, 511.

Smith, H. A. (1995), *RR Lyrae Stars*, Cambridge: Cambridge University Press.

Smith, M. A., Henrichs, H. and Fabregat, J. (eds) (2000), *The Be Phenomenon in Early-Type Stars*, San Francisco: ASP Conference Series, **214**.

Sparke, L. S. and Gallagher, J. S. (2000), *Galaxies in the Universe*, Cambridge: Cambridge University Press.

Stahler, S. W. (1988), 'Understanding young stars – a history', *PASP*, **100**, 1474.

Stairs, I. H. (2004), 'Pulsars in binary systems', *Science*, **304**, 547–52.

Starrfield, S. and Shore, S. N. (1995) (January), 'The birth and death of nova V1974 Cygni', *Scientific American*, **272**, 56–61.

Stellingwerf, R. F. (1978), 'Period determination using phase dispersion minimization', *ApJ*, **224**, 953–60.

Stelzer, B. *et al.* (2003), 'The weak-line T Tauri star V410 Tau: a multi-wavelength study of variability', A&A, **411**, 517.

Sterken, C. (1993), 'On the period history of the Beta Cephei star BW Vulpeculae', *A&A*, **270**, 259–64.

Sterken, C. and Jerzykiewicz, M. (1993), 'Beta Cephei stars from a photometric point of view', *Space Sci. Rev.*, **62**, 95–171.

Sterken, C. and Jaschek, C. (1996), *Light Curves of Variable Stars*, Cambridge: Cambridge University Press.

Sterken, C. (ed.) (2005), *The Light-Time Effect in Astrophysics*, San Francisco: ASP Conference Series, **335**.

Sterken, C. and Duerbeck, H.W. (eds) (2005), *Astronomical Heritages*, Brussels: Astronomical Institute, Vrije Universiteit Brussel.

Szabados, L. and Kurtz, D.W. (2000), *The Impact of Large-Scale Surveys on Pulsating Star Research*, San Francisco: ASP Conference Series, **203**.

Szeidl, B., Olah, K., Szabados, L., Barlai, K. and Patkos, L. (1992), 'Photometry of the peculiar Population II Cepheid RU Camelopardalis (1966–1982)', *Contr. Konkoly Obs.*, 97, 247.

Takeuti, M. and Petersen, J. O. (1983), 'The resonance hypothesis applied to RV Tauri stars', *A&A*, **117**, 352–56.

Templeton, M., Basu, S. and Demarque, P. (2002), 'High-amplitude Delta Scuti and SX Phoenicis stars: the effects of chemical composition on pulsation, and the period-luminosity relation', *ApJ*, **576**, 963–75.

Templeton, M. (2004), 'Time series analysis of variable star data', *JAAVSO*, **32**, 41–54.

Terrell, D. (2001), 'Eclipsing binary stars: past, present, and future', *JAAVSO*, **30**, 1–15.

Turner, D. G. (1998), 'Monitoring the evolution of Cepheid variables', *JAAVSO*, **26**, 101–111.

Turner, D. G. and Burke, J. F. (2002), 'The distance scale for classical Cepheid variables', *AJ*, **124**, 2931–42.

Turner, D. G. (2004), Paper presented at the 2004 meeting of the Canadian Astronomical Society, Winnipeg MB.

Van den Bergh, S. and Tammann, G. A. (1991), 'Galactic and extragalactic supernova rates', *Ann Rev. Astron Astrophys*, **29**, 363.

Van der Hucht (2001), 'The Seventh Catalogue of Galactic Wolf–Rayet stars', *New Astronomy Reviews*, **45**, 135–232 (available at SIMBAD: Catalogue III/215).

Van Genderen, A. M., de Groot, M. and Sterken, C. (2001), *P Cygni 2000*, ed. M. de Groot and C. Sterken, San Francisco: ASP Conference Series, **233**, 59.

Vassiliadis, E. and Wood, P. R. (1993), 'Evolution of low and intermediate mass stars to the end of the asymptotic giant branch with mass loss', *ApJ*, **413**, 641–57.

Walker, G. A. H. (1987), *Astronomical Observations*, Cambridge: Cambridge University Press.

Walker, G. A. H. *et al.* (2005), 'MOST detects g-modes in the Be star HD 163868', *ApJ*, **635**, L77–L80.

Wallerstein, G. and Cox, A. N. (1984), 'The Population II Cepheids', *PASP*, **96**, 677–91.

Wallerstein, G. (2002), 'The Cepheids of Population II and related stars', *PASP*, **114**, 689–99.

Warner, B. (1995), *Cataclysmic Variable Stars*, New York: Cambridge University Press.

Warner, B. (ed.) (2003), *Cataclysmic Variable Stars*, Cambridge University Press.

Warner, B. (2004), 'Rapid oscillations in cataclysmic variables', *PASP*, **116**, 115–132.

Webbink, R., Livio, M., Truran, J.W. and Orio, M. (1987), 'The nature of the recurrent novae', *ApJ*, **314**, 653–72.

Wehlau, A. and Bohlender, D. (1982), 'An investigation of period changes in cluster BL Herculis stars', *AJ*, **87**, 780–91.

Willson, L.A. (1986), 'Mira variables', *JAAVSO*, **15**, 228–35.

Willson, L.A. (1986), in *The Study of Variable Stars using Small Telescopes*, ed. J.R. Percy, Cambridge: Cambridge University Press, 219.

Wilson, R.E. (1994), 'Binary-star light curve models', *PASP*, **106**, 921–41.

Winget, D. *et al.* (1991), 'Asteroseismology of the DOV star PG 1159-035 with the Whole Earth Telescope', *ApJ*, **378**, 326–46.

Wolff, S.C. and Morrison, N.D. (1973), 'A search for Ap stars with very long periods', *PASP*, **85**, 141–9.

Wood, P.R. (2000), 'Variable red giants in the LMC: pulsating stars and binaries?', *PASA*, **17**, 18–21.

Wood, P.R., Olivier, E.A. and Kawaler, S.D. (2004), 'Long secondary periods in pulsating asymptotic giant branch stars: An investigation of their origin', *ApJ*, **604**, 800–16.

Yaron, O., Prialnik, D., Shara, M.M. and Kovetz, A. (2005), 'An extended grid of nova models. II: The parameter space of nova outbursts', *ApJ*, **623**, 398–410.

Zijlstra, A.A., Bedding, T.R. and Mattei, J.A. (2002), 'The evolution of the Mira variable R Hydrae', *MNRAS*, **334**, 498–510.

Zverko, J. (ed.) (2005), *The A-Star Puzzle*, Cambridge: Cambridge University Press.

Resources

American Association of Variable Star Observers, 49 Bay State Road, Cambridge MA 02138-1205, USA; www.aavso.org. Major source of visual and other data. Web site contains extensive information on individual variable stars, under 'Variable Star of the Season', and variable stars in general. Also an excellent manual on how to observe variable stars.

Arbeitsgemeinschaft für Veränderliche Sterne (BAV), Munsterdamm 90, 12169 Berlin, Germany; thola.de/bav.html

Association Francaise des Observateurs d'Etoiles Variables, c/o Observatoire Astronomique de Strasbourg, 11 rue de l'Université, 67000 Strasbourg, France; cdsweb.u-strasbg.fr/afoev

Astro-ph: an archiving system for astronomical preprints; widely used; arxiv.org/archive/astro-ph

Astrophysics Data System (ADS), an information archiving and retrieval service sponsored by NASA and based at the Harvard-Smithsonian Center for Astrophysics; most of the astronomical literature can be searched through this system; www.adsabs.harvard.edu

British Astronomical Association, Burlington House, Piccadilly, London W1J 0DU, UK; www.britastro.org. Variable Star Section: www.britastro.org/vss

Gaposchkin, S. and Payne-Gaposchkin, C., 1938, *Variable Stars*, Harvard College Observatory. Still a classic, full of archival treasures.

General Catalogue of Variable Stars: see Samus (2004); www.sai.msu.su/groups/cluster/gcvs/gcvs/

Hands-On Astrophysics, an education project developed by J.A. Mattei and J.R. Percy for the AAVSO, with funding from the US National Science Foundation, and much help from AAVSO staff and volunteers (hoa.aavso.org).

Hoffmeister, C., Richter, G. and Wenzel, W., 1985, *Variable Stars* (trans. S. Dunlop), Berlin: Springer-Verlag. Classic text on the subject.

IAU Information Bulletin on Variable Stars (IBVS); www.konkoly.hu/IBVS.html

International Astronomical Union (IAU); www.iau.org. International union of professional astronomers; co-ordinates astronomy; organizes conferences; its sub-sections include:

Division V (Variable Stars): www.konkoly.hu/IAUDV/
Commission 27 (Variable Stars): www.konkoly.hu/IAUC27/
Commission 42 (Close Binary Stars): www.konkoly.hu/IAUC42/

Kaler, J.B., 1989, *Stars and their Spectra*, Cambridge: Cambridge University Press. Excellent introduction to stars and their properties.

Kurtz, D.W. and Pollard, K.R., 2004, *Variable Stars in the Local Group*, Astronomical Society of the Pacific Conference Series, **310**. Proceedings of the most recent biennial conference on pulsating stars.

Levy, D.H., 2006, *David Levy's Guide to Variable Stars*, Cambridge: Cambridge University Press.

North, Gerald, 2004, *Observing Variable Stars, Novae, and Supernovae*, Cambridge: Cambridge University Press.

Royal Astronomical Society of New Zealand, P.O. Box 3181, Wellington NZ; www.rasnz.org.nz. The RASNZ has an active, distinguished Variable Star Section.

Samus, N.N. *et al.*, 2004, *Combined General Catalogue of Variable Stars* (GCVS4.2). Catalogue II/250 at SIMBAD.

SIMBAD: a stellar data service maintained by the Centre de Donneés astronomiques de Strasbourg: simbad.u-strasbg.fr/Simbad

Sterken, C. and Jaschek, C., 1996, *Light Curves of Variable Stars: A Pictorial Atlas*, Cambridge: Cambridge University Press.

VSNET. www.kusastro.kyoto-u.ac.pp/vsnet

Index

CPSIA information can be obtained at www.ICGtesting.com
Printed in the USA
LVOW050242151211

259447LV00003B/17/P